A TEXTBOOK OF
PRACTICAL BIOLOGY

A TEXTBOOK
OF PRACTICAL BIOLOGY

by

GEORGE USHER

B.Sc., Dip.Agric.Sci., D.T.A., M.I.Biol., F.L.S.
Senior Biology Master, Bedstone School

CONSTABLE   LONDON

Published by
Constable & Company Ltd
10 Orange Street, London, W.C.2

First Published 1970
Reprinted 1974

Printed in Great Britain by
Anchor Press, and bound by Wm. Brendon,
both of Tiptree, Essex
09 456370 5

Preface

This book is planned to be of assistance to the 'A' level candidate in the biological sciences, and the type specimens included are therefore suitable for the zoology and the botany student. It is essentially a practical book, for use in the laboratory, and theoretical discussion has been reduced to a minimum.

The experiments described have been chosen for their simplicity, so that they can be performed with the minimum of apparatus by an individual, or by a small group of students. Similarly, the diagrams have been made simple, to avoid, it is hoped, the temptation of copying them directly into practical note-books, without reference to actual specimens.

I hope that this book will be of help particularly to the student in the over-crowded class, where it is impossible for him to have the individual attention which is the ideal requirement.

As only a few of the experiments are original, this book may well be considered a compendium of the experiments I myself have used during several years of teaching. I would like to acknowledge the many sources which I must have used inadvertently but would show my appreciation especially for the help in teaching I have derived from 'Simple Experiments in Biology' by Cyril Bibby, 'Practical Biology' by C. J. Wallis, and 'Three Vertebrates' and 'The Rat' by T. A. G. Wells.

Finally I would like to thank my wife for her invaluable help in preparing the diagrams.

November 1968   GEORGE USHER

Contents

1 Physical Principles

Colloids

Experiment: To show the differences between a crystalloid solution and a colloidal solution.
Apparatus: Two test-tubes, Bunsen burner, glue, sodium chloride.
Method: Dissolve a little glue and salt in separate test-tubes, warming the former to dissolve the glue. Allow the tube to cool and observe any changes that take place.

The salt will form a clear crystalloid solution, but the glue solution, which is colloidal, will be turbid, and form a jelly (gel) on cooling.

Experiment: To examine the colloidal properties of egg albumen.
Apparatus: Egg white, three small beakers, two test-tubes, Bunsen burner, lead acetate, copper sulphate.
Method: Make concentrated solutions of lead acetate and copper sulphate in separate test-tubes. Divide the egg white into the three beakers, and add an equal volume of water to two of the beakers and mix well. Now add the copper sulphate solution to one of the beakers of egg white solution, and the lead acetate solution to the other. Warm the third beaker. Observe what happens in each of the beakers.

Egg albumen is practically pure protein. The protein mixed with salt solutions will form a colloidal solution. The heated albumen turns white and opaque. In the latter case, the protein has been denatured by the alteration of its molecular structure.

Experiment: To show the force exerted by the imbibition of water by colloids.

Apparatus: Dried peas, small tin box, (a throat pastille tin is ideal), beaker, weights (1 kg. upwards), plastic bowl.
Method: Punch a few holes in the bottom and sides of the tin, and pack it full of peas so that the lid shuts comfortably. Lower it into the beaker of water slowly, so that the water gets in through the holes. Leave overnight.

In the morning, the box will have been burst open due to the pea-colloids imbibing water. Notice the increase in the volume of the individual peas. A force must have been exerted to have pushed the lid open.

This experiment can be repeated, using a plastic bowl to hold the water, and placing weights on the lid of the tin. This will give some measure of the force exerted during imbibition. The weights may be pushed off the lid into the bowl, so do not use one which will break.

Experiment: To measure the increase in volume brought about by the imbibition of water by colloids.
Apparatus: A cube of gelatine (or ordinary table-jelly), beaker, ruler.
Method: Measure the volume of a piece of gelatine; place it in a beaker of water, and leave it for a few hours. Re-measure it, and allow it to dry out slowly. Then measure it again.

The jelly becomes softer, but increases in volume due to the imbibition of water. but on drying, it returns to its original volume.

Diffusion

Experiment: To demonstrate the diffusion of gases, and to compare their rates of diffusion.
Apparatus: Glass tube (not less than 18 in. long, nor less than $\frac{1}{2}$ in. internal diameter), stand and clamp, cotton-wool, ammonium hydroxide, concentrated hydrochloric acid, stop-watch, ruler.
Method: Fix the glass tube horizontally in the clamp. Soak one pice of cotton wool in ammonium hydroxide, and another in hydrochloric acid. Push these into opposite

ends of the glass tube simultaneously and measure the time it takes for a white solid to be formed on the inside of the tube. Measure the distances from the solid to the respective ends of the tube.

Hydrochloric acid and ammonia react to form ammonium chloride which is a white solid:

$$HCl + NH_4OH = NH_4Cl + H_2O.$$

Calculate the molecular weight of the NH_4^+ ion and the Cl^- ion. Is there any connection between the molecular weight of a gas and its rate (distance/time) of diffusion? Look up Graham's Law of Diffusion.

Experiment: To demonstrate the diffusion of a dissolved solid in a liquid.
Apparatus: Potassium permanganate or copper sulphate crystals, beaker.
Methods: Partly fill the beaker with water, and very gently drop one crystal of the coloured salt into it. Make your observations immediately, and then leave the experiment for a day.

The water becomes coloured, at first deeply around the crystal, but the colour gradually spreads uniformly throughout the water. Notice that the dissolved salt moves upwards, i.e. against the force of gravity, so that this is not simply a gravitational phenomenon, but some energy, inherent in the system, must be employed.

Experiment: To compare the rates of diffusion through a gel and a liquid.
Apparatus: Test-tubes, 10% sodium hydroxide solution, phenolphthalein solution, gelatin, stopwatch, ruler, burner.
Method: Make a solution of gelatin by boiling in water. Add a few drops of phenol-phthalin, shake well, and three-quarters fill one of the test-tubes with it. Allow the gelatin to set. Fill the second test-tube with water to the same height as the gelatin in the first tube. Add the same amount of phenolphthalein and shake. Gently place a few drops of the sodium hydroxide solution

on the top of both tubes at the same time, and time the colour-change down the tube containing water. Measure the distance the colour change has proceeded down the tube of gelatin in the same time, thus comparing the rates of diffusion through the two media.

Phenolphthalein is an indicator which is colourless in acid or neutral solution, but turns red-purple in alkaline solution. Thus the indicator shows the position of the sodium hydroxide in the two media.

Experiment: To demonstrate diffusion through a differentially permeable membrane.
Apparatus: Starch, sodium chloride, sodium sulphate, cellophane, "Sellotape", beaker, silver nitrate solution, barium chloride solution, iodine solution, burner.
Method: Make a tube of the cellophane by sealing the edges with 'Sellotape'. Seal off the bottom. Dissolve some starch in water by boiling. Allow the solution to cool and pour it into the cellophane bag. Dissolve some sodium chloride in water and pour this solution into the bag. Mix well. Dissolve the sodium sulphate in the beaker. Test each of the solutions for starch, chloride ion, and sulphate ion. Place the cellophane bag in the beaker of sodium sulphate solution and leave it for half an hour. Re-test the solutions as before. What conclusions can be drawn regarding the permeability of cellophane to starch, chloride ions, and sulphate ions?

Repeat the experiment placing the starch-salt solution in a cellophane bag and suspending it in running water. Leave it for 2–3 hours and test the solution for chloride ions and starch. This is called *dialysis* and is an important method of washing solutions, such as proteins, of any simple salts that may be present.

See Appendix 1 for the tests.

Experiment: To show the selective action of the membrane bounding living cells.
Apparatus: Microscope, test-tubes, Congo red solution, living yeast, Bunsen burner, dropper, slides and cover slips.

Method: Make a suspension of yeast cells in water, and divide it equally into two test-tubes. Boil one of the tubes for about half a minute to kill the cells, and allow to cool. Add a few drops of Congo red solution to each of the suspensions and observe a drop of each under the microscope. What differences can be seen between the cells on the two slides?

This experiment shows that only living cells are selectively permeable.

Osmosis

Experiment: To demonstrate osmosis through a semi-permeable membrane.
Apparatus: Sucrose, thistle funnel, piece of pig's bladder or parchment, beaker, pipette, cotton, clamp and stand.
Method: Tie the pig's bladder or parchment tightly over the bowl of the thistle funnel. Place the covered end in a beaker of water, and pipette a concentrated sugar solution down the stem of the funnel until the bulb and a little of the stem are full. Mark the position of the solution and leave for a few hours. Notice the new level of the solution.

Water enters the solution through the semi-permeable membrane by osmosis.

It is easier to soak the membrane before tying it over the funnel, this stretches it, and makes it easier to tie. Fill the funnel with the membrane under water; then the membrane is less likely to slip at the joint.

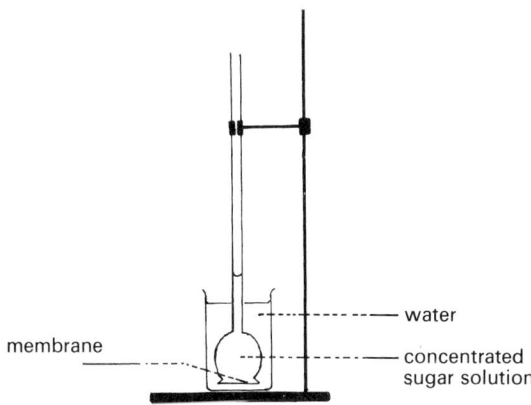

Fig. 1. To demonstrate osmosis through a semi-permeable membrane.

Experiment: To show that osmosis takes place through living tissue.
Apparatus: Three medium sized potatoes, three shallow dishes (or Petri dishes), sucrose, large beaker, 'Vaseline'.
Method: Kill one of the potatoes by placing it in boiling water for three-to-four minutes. Cut the top and bottom of each potato flat, and scoop out a hole to leave the sides about $\frac{1}{2}$ in. thick. Partially peel the potatoes so that when they are placed in the dishes, the peeled surfaces are covered with water. 'Vaseline' the upper cut surface and set up the experiment as shown in Fig. 2. (This should leave no cut surfaces exposed to the air, thus preventing the potatoes from drying-out and discolouring.) Place a little sugar in the dead potato and in one of the others. Leave for 24 hours.

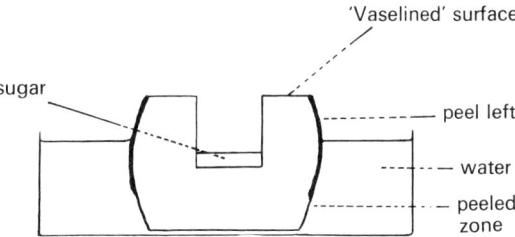

Fig. 2. To show that osmosis takes place through living tissue.

Water does not enter the potato without sugar, i.e. the living tissue resists the entry of water, but it fills and overflows the living potato with sugar, i.e. the cells act as a semipermeable membrane. Water enters the dead potato up to the level of that in the dish, i.e. it enters passively.

Experiment: To estimate the suction pressure of plant tissues.
Apparatus: Balance, potato, dandelion or daffodil scape, sucrose, beakers, filter paper.
Method: Make a molar solution of sucrose, and by successive dilutions make others of $\frac{1}{2}$, $\frac{1}{4}$, $\frac{1}{8}$, and $\frac{1}{16}$ molar. Cut the potato into cubes of about 1 cm. side. Dry the surfaces carefully with filter paper, and weigh the cubes separately. Place one in each beaker of

solution and leave for about two days. Dry and re-weigh. Calculate the *percentage* gain or loss in weight. Graph the results (as in Figure 3). At the point on the graph where there is no gain or loss in weight, the suction pressure of the cell is equal to the osmotic pressure of the solution.

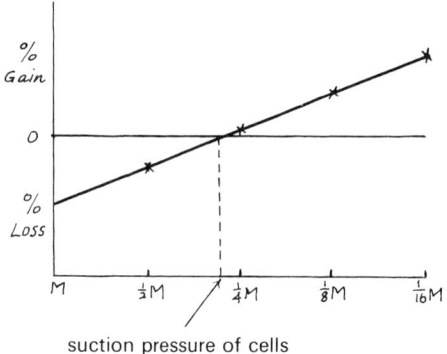

Fig. 3. Graph to show the estimation of the suction pressure of plant tissue.

Notice that the pieces that have gained weight are still turgid, and those that have lost weight are flaccid.

A solution of sucrose containing 1 gm. mol./litre (molar) exerts an osmotic pressure of one atmosphere.

Experiment: To show plasmolysis in living cells.
Apparatus: Spirogyra filaments, beetroot, blood, sucrose, sodium chloride, beaker, razor, microscope, cover slips and slides.
Method: Make a 15% sugar solution. Mount the *Spirogyra* on the slide in water, and examine it under the microscope. Irrigate the slide with sugar solution. Observe the protoplasm contracting away from the cell-wall by plasmolysis and the cells becoming smaller due to the contracting of the cell-wall.

Repeat the experiment with thin sections of beetroot that have been washed.

By repeating the experiment with different concentrations of sucrose, the osmotic pressure of the cell-sap can be estimated, by noting the concentration at which plasmolysis just occurs. Devise your own dilution series.

Repeat the experiment using blood and a 10% salt solution.

Turgidity

Experiment: To show the turgidity of plant tissues, and that it is due to the osmotic pressure of the cells.
Apparatus: Rhubarb petiole, dandelion scape, sodium chloride, beakers, ruler.
Method: Measure a piece of rhubarb (about 2 in.), peel off the outer skin and measure it again. Cut a cylinder from the rhubarb and after a few moments try to push it back into the hole. Split the dandelion scape longitudinally.

The pieces of rhubarb get bigger, and the dandelion scape pieces bend outwards, due to the removal of the restraint of the outer tissues from the turgid inner cells.

Now place the cut pieces into a concentrated salt solution and leave for an hour or so.

The pieces will have returned to their original size, or smaller due to the plasmolysis of the cells and loss of turgor.

Surface Properties

Experiment: To demonstrate adsorption on to surfaces.
Apparatus: Test-tubes, powdered charcoal, methylated spirit, methyl violet, solution.
Method: Half fill a test-tube with water and add a few drops of methyl violet. Add enough charcoal as would cover a sixpence and shake. What happens?

The dye is adsorbed on to the surface of the charcoal particles which, on settling, leave the solution colourless.

Make a similar solution of methyl violet and empty it out of the tube. Wash the tube thoroughly with water until the water is colourless. Now wash it with methylated spirit. What colour is the methylated spirit? Account for your findings.

Experiment: To demonstrate capillarity.
Apparatus: Glass tubing of various bores (about 9 in. each), water, coloured with red ink, beaker, black-board 'chalk'.

Method: Ensure that the tubing is clean, and stand the pieces vertically in the beaker of water. Observe the height to which the water rises in each tube.

The narrower the tube, the higher the water rises.

Stand a piece of 'chalk' vertically in the beaker. What happens, and why?

Experiment: To demonstrate surface tension.
Apparatus: Needle, filter-paper, large beaker.
Method: Make sure that the needle is clean and dry. Place it on the filter-paper and place the filter-paper on the surface of the water.

The paper sinks, because it becomes soaked, but the needle, which has a much higher density than water is left floating due to the surface tension of the water.

Brownian Movement

Experiment: To demonstrate Brownian movement.
Apparatus: Indian ink, slide, microscope, culture of bacteria.
Method: Dilute some Indian ink in a drop of water on the slide and observe it under the microscope.

The ink particles (fine carbon particles) move at random. This type of movement is called Brownian movement.

Repeat the experiment by mounting some bacteria in a drop of water. Observe what happens to the bacterial cells.

2 Chemistry of the Cell

Carbohydrates

Experiment: To test for starch.
Apparatus: Starch, sucrose, glucose, iodine solution, test-tubes, burner.
Method: Dissolve the starch in water by boiling, and allow to cool. Dissolve the sucrose and glucose in separate tubes. Add a few drops of iodine in potassium iodide solution to each tube. The iodine gives a blue-black colour in the starch, but not in the other tubes, showing the iodine is a specific test for starch.

Experiment: To test for reducing sugars.
Apparatus: Benedict's solution, Fehling's solution, glucose, fructose, sucrose, maltose, starch, test-tubes, burner.
Method: Dissolve the carbohydrates in separate test-tubes, heating when necessary, and allow to cool. Divide each into two separate portions. Add Benedict's solution to one of each of the tubes, and Fehling's solution to each of the others. Warm all the tubes. Tabulate your results. Which of the carbohydrates are reducing sugars?

Benedict's solution gives a red, green, or yellow precipitate with reducing sugars, and Fehling's solution gives a red precipitate.

Experiment: To test for sucrose.
Apparatus: Sucrose, Fehling's solution, Benedict's solution, sodium hydroxide solution, dilute hydrochloric acid, test-tubes, burner.
Method: Boil a solution of sucrose with a few drops of dilute hydrochloric acid. Add a similar quantity of sodium hydroxide solution to neutralize any excess acid. Test for reducing sugars with Fehling's solution and Benedict's solution.

In the previous experiment the test was negative, but it is now positive due to the hydrolysis of the sucrose into glucose and fructose by the acid.

Experiment: To test for cellulose.
Apparatus: Cotton wool (cellulose), iodine in potassium iodide solution, concentrated sulphuric acid, chlor-zinc-iodide solution, microscope, slides.
Method: Mount a few strands of cotton wool on a slide in the iodine solution, and irrigate with sulphuric acid. The strands swell and become stained bright blue.

Repeat the experiment, mounting the strands in chlor-zinc-iodide solution, when the strands swell and are stained violet.

Experiment: To show the presence of carbohydrates in a leaf.
Apparatus: Leaf, iodine in potassium iodide solution, Benedict's solution, chlor-zinc-iodide solution, slides and cover-slips, microscope, razor, watch-glass, burner.
Method: Cut sections of the leaf. (Sections of suitable thickness are obtainable by rolling the leaf and cutting across the roll.) Mount four groups of sections on separate slides. Leave one as a control, and stain the other three with the reagents, warming the Benedict's solution gently. Examine the reactions under the microscope, and by comparing with the control note the distribution of the various carbohydrates (starch, reducing sugars, and cellulose).

Experiment: To show that carbohydrates contain carbon, hydrogen, and oxygen.
Apparatus: Hard-glass test-tube, delivery tube, test-tube, lime water, anhydrous copper sulphate, starch (or other carbohydrate, clamp and stand.
Method: Set up the apparatus as shown in the diagram (Figure 4). Heat the carbohydrate strongly.

The lime water goes milky showing the presence of carbon dioxide, (hence carbon and oxygen in the carbohydrate). Test the drops of liquid in the delivery tube (after cooling) with anhydrous copper sulphate. This turns blue showing the presence of water (and hence hydrogen and oxygen in the carbohydrate.

At the end of the experiment remove the tube of lime water before removing the heat. If this is not done, the lime water will 'suck back' into the hard-glass tube and may crack it.

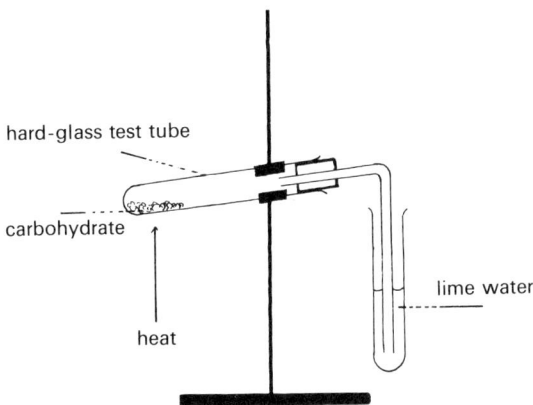

hard-glass test tube

carbohydrate

lime water

heat

Fig. 4. To show that carbohydrate contains carbon, hydrogen and oxygen.

Experiment: To test for fats.
Apparatus: Lard, (or olive oil), Filter paper, test-tubes, watch glass, Sudan III solution, osmic acid solution.
Method: Smear a little lard on the filter paper. It leaves a transluscent mark, typical of fats.

Place a few drops of osmic acid on the fat. This gives a blue-black coloration.

Shake the fat (olive oil is better) in a test-tube of water. Add a few drops of Sudan III solution and shake again. When the fat floats to the surface, it is pink, having been selectively stained by the Sudan III.

Protein
Experiment: To test for protein.
Apparatus: Egg albumen (protein), sucrose, glucose, sodium chloride, starch, olive oil, test-tubes, burner, concentrated nitric acid, ammonium hydroxide, Millon's reagent, 5% sodium hydroxide solution, 1% copper sulphate solution.
Method: Dissolve the carbohydrates, protein, and fat each in separate test-tubes, and divide each into three lots. Carry out the following tests on one of each of the groups of tubes. Tabulate the results. Which tests are specific for protein?

Add about 1 ml. of concentrated nitric acid and warm. Then, on cooling, add ammonium hydroxide. *(Xanthoproteic test)*.

Add a few drops of *Millon's reagent* and warm.

Add an equal volume of sodium hydroxide solution to the test solution and *one drop* of copper sulphate solution. *(Biuret test)*.

Protein gives a yellow colour with nitric acid, and an orange colour on addition of ammonium hydroxide in the xanthoproteic test; a red precipitate or colour with Millon's reagent; and a light violet colour in the biuret test.

Experiment: To show the presence of nitrogen in protein.
Apparatus: Hard-glass test-tube, soda-lime, red litmus paper, concentrated hydrochloric acid, any protein (e.g. cheese), burner.
Method: Mix the protein and soda-lime together and heat in the test-tube. Ammonia is given-off. Test the gas by smell, turning moist red litmus-paper blue, and producing a dense white cloud of ammonium chloride when the stopper of a bottle of concentrated hydrochloric acid is placed near it.

Ammonia (NH_3), contains nitrogen which must have been derived from the protein.

Experiment: To show the presence of phosphorus in protein.
Apparatus: Casein powder (or cheese), crucible, fusion mixture, test-tubes, 50% nitric acid, ammonium molybdate solution, *burner,* tripod, pipe-clay triangle.
Method: Mix the casein and twice its bulk of fusion mixture in the crucible and heat, gently at first, then strongly. Allow to cool, and

extract with nitric acid. To the extract add ammonium molybdate solution and boil. A yellow precipitate indicates the presence of phosphorus in the protein.

Experiment: To demonstrate the presence of sulphur in protein.
Apparatus: Egg albumen, 40% sodium hydroxide solution, lead acetate solution, test-tubes, burner.
Method: Boil a little of the egg albumen in the sodium hydroxide solution, and add a few drops of lead acetate. This gives a black precipitate of lead sulphide (PbS) if sulphur is present.

Experiment: To test various food substances for carbohydrates, proteins and fats.
Apparatus: Flour, butter, cheese, castor-oil seed, potato, milk, iodine solution, Benedict's (or Fehling's solution), osmic acid, Millon's reagent, test-tubes, burner, pestle and mortar.
Method: Grind up each of the foods separately with a little water in a pestle and mortar. Divide the extracts into four and test each separately for starch, reducing sugars, protein and fats, as previously described. Tabulate your results.

Mineral salts
Experiment: To calculate the water content, dry weight, and ash content of various organic substances.
Apparatus: Grass, leaves, beans (or other seeds), liver, and any other organic material, crucible, oven, burner, pipe-clay triangle, stand, balance.
Method: Weigh a crucible and partly fill it with finely cut-up organic material. Re-weigh. Place in an oven at 100°C, (or a steam oven), removing it and weighing it at intervals, after cooling, until the weight is constant. All the water is now driven-off and the water content and dry weight can be calculated by difference. The organic matter is now burnt off leaving the mineral salts. The burning is carried out by heating gently, then more strongly,

ultimately burning directly with the flame. Cool and weigh repeatedly until the weight is constant. The organic matter content and ash content can now be calculated by difference. Tabulate your results and notice the differences in the composition of the various materials.

Experiment: To investigate the elements and ions present in plant ash.
Apparatus: Plant ash (from previous experiment), platinum wire, cobalt glass, concentrated hydrochloric acid, dilute nitric acid, potassium ferrocyanide solution, ammonium molybdate solution, concentrated nitric acid, barium chloride solution, silver nitrate solution, 0.5% diphenylamine solution, concentrated sulphuric acid, burner.
Method: Moisten a loop of platinum wire in concentrated hydrochloric acid and dip it into the plant ash. Place this in a colourless flame and examine with and without the cobalt glass. A yellow flame without the glass indicates the presence of sodium, a lilac flame with the glass indicates potassium, and a brick-red flame calcium. (The latter two may mask each other). The cobalt glass cuts out the yellow light of the sodium.

Dissolve some of the ash in dilute nitric acid and divide the solution into three parts. Test each separately by adding silver nitrate solution—a white precipitate indicates chloride ions; concentrated nitric acid and ammonium molybdate solution—a yellow precipitate, or colour, indicates phosphate ions; and potassium ferrocyanide solution—a blue colour, or precipitate indicates iron.

Shake some of the ash thoroughly with water and decant off the liquid. To this add a little concentrated sulphuric acid and a little diphenylamine solution. A blue colour shows the presence of nitrate ions.

Finally test for sulphate ions by shaking the plant ash in concentrated hydrochloric acid. Decant off the liquid and dilute it to about twice its volume. Add barium chloride solution. A white precipitate shows that sulphate ions are present.

Some of these colours and precipitates may be very slight.

Plant pigments

Experiment: To extract chlorophyll from leaves, and separate the pigments by paper chromatography and column chromatography.
Apparatus: Herbaceous leaves (nettle leaves are very suitable), beaker, Buchner funnel and flask, filter pump, filter paper, acetone, petroleum ether, 200 ml. separating funnel, evaporating basin, carbon disulphide, glass tube ($\frac{1}{2}-\frac{5}{8}$ in. internal diameter, and 6 in. long), glass wool, heavy magnesium oxide, ethyl alcohol (ethanol).
Method: Dry the nettle leaves in air for several days and powder them. Mix about 3 g. of the powder with 40 ml. of 80% acetone. Shake well, and when the whole solution is well-coloured filter through the Buchner funnel. Pour the filtrate into the separating funnel and add 40 ml. of petroleum ether. Mix thoroughly by rotating the funnel. Now add 70 ml. of distilled water and mix again. Allow to settle, and the pigments have passed into the ether. Separate.

Place a circle of filter paper on the evaporating basin, and place 1 ml. of the petroleum ether extract on the centre in successive spots, drying each spot before applying the next. When all the extract is added finally dry. Add drops of carbon disulphide to the spot, and as the carbon bisulphide is spread through the paper, successive circles of chlorophyll a, chlorophyll b, and xanthophyll develop. Carotene is not often distinguished this way.
N.B. Carbon bisulphide is best handled in a fume cupboard.

Heat one end of the glass tube, so as to practically close it, and plug this end with a piece of glass wool. Pack the tube evenly and firmly with the magnesium oxide powder. (The method of packing is important, and should be done by inserting the powder a little at a time.) Set the tube up in a Buchner funnel as shown in Figure 5. Pour the remainder of the petroleum ether extract

from the initial experiment through the tube while suction is being applied. Turn off the pump while there is still a little extract left in the tube. Now suck through a mixture of

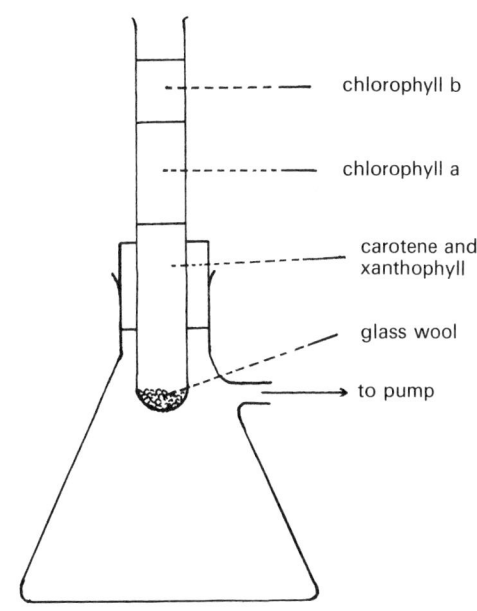

Fig. 5. Separation of chlorophyll pigments by column chromatography.

50 ml. petroleum ether, 2.5 ml. acetone, and 0.25 ml. ethyl alcohol. The pigments will develop as shown in the diagram.
N.B. Do not handle glass wool—use a forceps.

Experiment: To extract various **plant pigments and examine the effect of acid and alkali on them.**
Apparatus: Blue, red or purple flowers, white flowers, beetroot, test-tubes, burner, sodium hydroxide solution, dilute hydrochloric acid, ammonium hydroxide.
Method: Boil the coloured material in water to extract the pigments. Divide the extracts into two and add sodium hydroxide to one of each pair, and hydrochloric acid to the other. What changes in colour do you see?

Dip the petals of the white flower into ammonium hydroxide. Is there any change in colour?

3 Enzymes

Amylases

Experiment: To show the action of amylase on starch.

Apparatus: Starch, rubber bung, test-tubes, iodine solution, Benedict's solution, beaker, watch, burner.

Method: Chew on a rubber bung to encourage the flow of saliva. Wash out the mouth in a little warm water and spit out into a beaker. (Saliva contains the amylase ptyalin.) Make a 0.5% solution of starch by boiling in water. Allow to cool. Divide the starch solution equally into sixteen test-tubes and arrange them in pairs. To each add a few drops of the saliva extract and test one of each pair for reducing sugars and starch at 0, ½, 1, 2, 5, 10, 20, and 30 minute intervals. Tabulate your results. How long does it take the amylase to change the starch into sugar under these conditions?

Experiment: To show the effect of heat on the rate of enzyme action.

Apparatus: Test-tubes, thermometers, beakers, rubber bung, starch, iodine, solution, tripods, gauze, watch, glasses, dropper.

Method: Make a 0.5% starch solution and allow it to cool. Make a solution of ptyalin as described in the previous experiment. Pour equal amounts of starch solution into seven test-tubes and place these into beakers of water, ice and water, or ice. Warm if necessary to maintain one of each of the beakers at the following temperatures, 0°C, 20°C, 30°C, 40°C, 50°C, and 60°C. When the temperature is steady, add a few drops of saliva mixture. Take drops of the solutions out of the tubes

into the watch glasses at 2 minute intervals and test for starch. Notice the time at which there is no positive starch reaction. Graph the temperature against time as shown in Figure 6. What is the optimum temperature for the action of this enzyme?

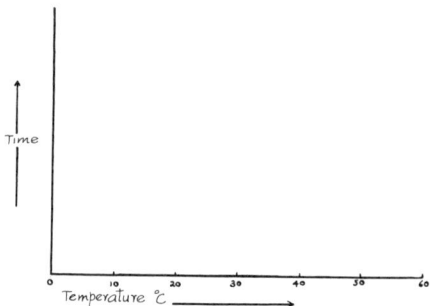
Fig. 6. Graph to show the effect of temperature on enzyme action.
This diagram is to illustrate the axes used in the experiment, and not to show any particular set of results.

Experiment: To examine the effect of pH on amylase action.

Apparatus: Buffer solution tablets, rubber bung, test-tubes, starch, watch glass, pipette, iodine solution.

Method: Make a 1% solution of starch by boiling. Allow to cool. Collect a saliva sample as previously described. Half fill the test-tubes with buffer solutions of pH 10, 9, 8, 7, 6, 5, 4, 3, and 2, and add an equal volume of starch solution. Shake well. Add a few drops of saliva to each tube. Shake. Test samples from each tube for the presence of starch every two minutes by placing a few drops in a watch glass and adding iodine solution. Graph time against pH, and decide which is the optimum pH for this enzyme.

Buffer solutions are used to keep the pH constant throughout the experiment.

Lipases

Experiment: To examine the effect of lipases on fats.

Apparatus: Test-tubes, cream (or milk), bile salts, pancreatin solution, phenol red solution, water-bath, dilute sodium hydroxide solution, thermometer, pipette, watch.

Method: About quarter fill three test-tubes with cream and add a few drops of phenol red, until the cream is pink. If it is not pink adjust the acidity by adding sodium hydroxide solution a drop at a time until it is pink. To the first tube add an equal volume of water, to the second, the same volume of water with a little bile salt, and to the third an equal volume of pancreatin solution and a little bile salt. Place in a water-bath at about 37°C. Record the time at which any of the solutions give a yellow colour. What conclusions about fat digestion can be drawn from this experiment?

Fats are digested to fatty acids and glycerol, so that a change in pH to the acid side indicates the formation of acids by digestion.

Phenol red is red in neutral or alkaline solution and yellow in acid.

Proteases

Experiment: To examine the digestion of protein by pepsin, and the effect of pH on the reaction.

Apparatus: Hard-boiled egg-white, test-tubes, water-bath, thermometer, tripod, burner, pepsin solution, watch, buffer solutions.

Method: Cut the egg-white into $\frac{1}{4}$ in. cubes. Half fill one tube with water, and another with 50:50 water and pepsin solution. Half fill a series of tubes with buffer solutions at pH 2, 3, 4, 5, 6, 7, 8. Quarter fill another series of tubes with the same buffer solutions, but now fill each to half full with pepsin solution. Put a piece of egg-white in each. Place the tubes in a water-bath at about 37°C. Record the times at which the egg-whites are dissolved. Graph your results as previously. What conditions are necessary for pepsin to digest protein?

Experiment: Using chromatography, show that proteins are digested to amino-acids.

Apparatus: White of hard-boiled egg, pancreatin solutin, test-tubes, filter paper, butanol (butyl alcohol), glacial acetic acid,

ninhydrin reagent, bungs, flat dish (Petri dish), forceps, burner.

Method: Place the egg-white in pancreatin solution and leave in a warm place until digested. This may take a day or longer. Cut a piece of filter paper so that it fits into a test-tube without touching the sides or bottom. Pin it, or stick it to the cork. Place a spot of the digested material near one end of the strip, as shown in Figure 7. This is done by placing on a spot, and allowing it to dry, and repeating about ten times. While this is being done, make a solution of 4 parts (by volume) butanol, 1 part glacial acetic acid, and 1 part water. Pour a few ml. of this mixture into the tube and cork it. This saturates the atmosphere in the tube with the solvent vapour. Place the strip in the tube as shown. Watch the solvent rise up the paper, and when it reaches within 2 cm. of the top remove it. Hang the paper to dry for 24 hours. Pour a little ninhydrin in the flat dish and briefly immerse the paper in it. Dry the paper over a small flame (do not burn!). A series of purple spots appear, each indicating a different amino acid.

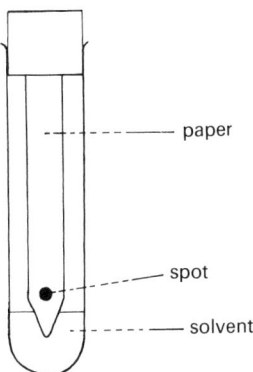

Fig. 7. Method of paper chromatography.

Ninhydrin is poisonous—do not inhale the fumes. Use a fume-cupboard if possible.

Ninhydrin gives a purple colour with certain amino-acids.

Handle the filter paper with a forceps to keep it clean. Amino-acids from the fingers will develop on the paper.

Catalase

Experiment: To examine the effect of catalase derived from various plant and animal material.

Apparatus: Liver, potato, horse-radish, (and any other living material), pestle and mortar, sand, guaiacum extract, splints, burner, 30% hydrogen peroxide solution, watch.

Method: Cut the living material in small pieces and grind some of this in a pestle and mortar with sand. Boil some further pieces. Pour about 3 ml. of hydrogen peroxide into each of three test-tubes and add a few drops of guaiacum extract. Place the material into separate test-tubes. Which colours the solution most quickly? Test any gas evolved with a glowing splint. Identify the gas. Which tissue has the most efficient catalase system, and under what conditions does it function most efficiently?

Catalase reduces hydrogen peroxide to oxygen and water. The oxygen released turns guaiacum blue. The blue colour may be masked by the colour of the material in the tube.

4 The Cell

The individual tissues will be dealt with under the appropriate sections.

In connection with the following exercises, examine electron-microscope photographs of cells and their inclusions.

Animal Cells

Apparatus: Microscope, methylene blue solution, cover-slips and slides, pipette, 70% ethyl alcohol, tooth-pick or splint of wood.
Method: Sterilize the tooth-pick by dipping in alcohol. With it scrape the inside of the cheek. Mount the scrapings in water on the slide and cover it with the cover-slip. Irrigate with methylene blue and examine under low, then high power of the microscope. These are simple, flat cells (Figure 8). Draw a few of them.

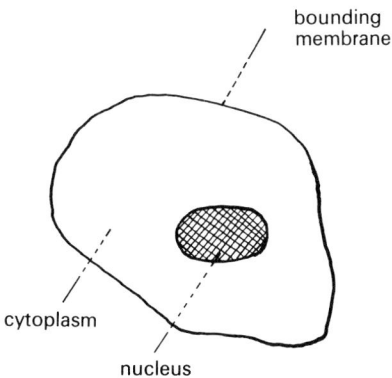

Fig. 8. Typical animal cell. Epithelial cell from inside human mouth.

Examine prepared slides of frog's blood. Notice the same features as in the cells of the cheek epithelium. Draw a few cells.

Examine a prepared slide of liver. Draw a few of these cells. Notice that there are numerous rod-like bodies scattered throughout the cytoplasm. These are the *mitochondria*. Compare their appearance under the optical microscope with their appearance in electron microscope photographs.

Plant cells

Apparatus: Microscope, onion, cover-slips, slides, dropper fine forceps, scalpel.

Cut an onion into fairly large pieces, and the individual fleshy scale-leaves are easily separated. Bend one of the leaves over your finger with the concave surface upwards, and cut across the fine outer skin (epidermis). Hold the free edge of the cut surface in the forceps and pull off the skin. Mount this in water on a slide and examine under the microscope. (See Figure 9.)

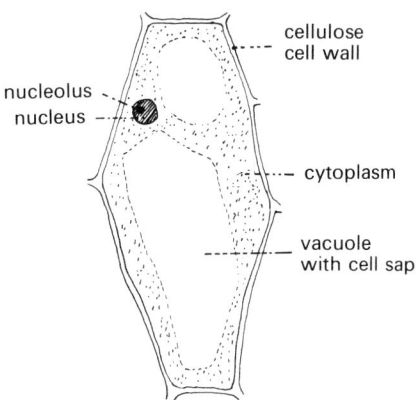

Fig. 9. Cell of Onion epidermis.

Mount a leaf of *Elodea* in water on a slide and examine under the microscope. The leaves of this little water-plant are only two cells thick, so that if you examine the edge of the leaf some of the cells should be clear enough for you to draw. Notice the cellulose cell-wall, the large central vacuole, and the *chloroplasts*. (Figure 10.)

Plant cells frequently have various inclusions in them. Some of these are *plastids* the commonest of which are chloroplasts which have been seen in *Elodea*. The colour of cells is often due to pigmented plastids called *chromoplasts*.

Apparatus: Microscope, cover-slips, slides, dropper, tomato, scalpel.

Method: Peel the tomato and take a small

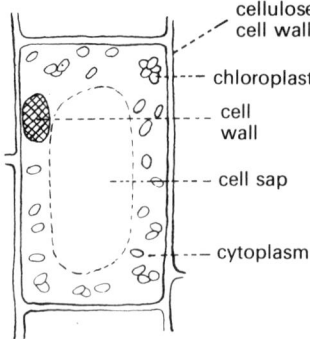

Fig. 10. Single cell from the leaf of Elodea.

amount of the tissue beneath the skin. Mount it in water and examine it under the microscope (see Figure 11). The small orange-red bodies in the cells are chromoplasts, giving the fruit its colour. Draw a few of the cells.

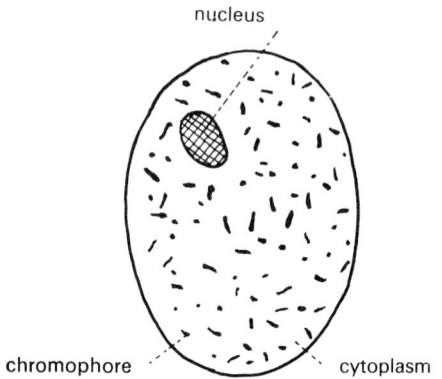

Fig. 11. Cell from the pulp of a Tomato.

Green plants are characterised, in a great many cases, by storing starch as a food reserve. This can be seen as characteristically shaped grains.

Apparatus: Microscope, cover-slips and slides, dropper, razor, scalpel, iodine solution, potato, rice.

Method: Soak the rice. (This is not essential, but makes it easier to cut.) Cut thin sections of the rice and potato and mount

on separate slides. Observe and draw the starch grains (Figure 12). Irrigate the slide with iodine solution and observe the staining of the starch grains.

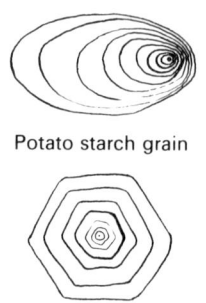

Potato starch grain

Maize starch grain

Fig. 12. Starch grains—Potato and Maize.

Other storage material occurs in plant cells, e.g. the carbohydrate *inulin* in *Dahlia* tubers (Figure 13), and protein stored as *aleurone grains* in the endosperm of Castor Oil *(Ricinus)* seeds (Figure 14). These can be examined as previously described.

Fig. 13. Inulin crystals in the cells of Dahlia root-tuber.

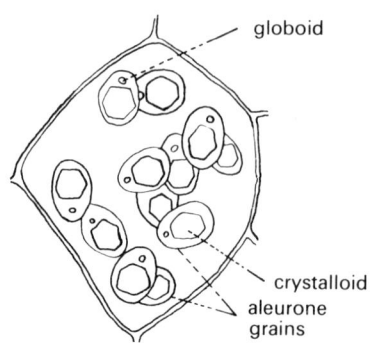

Fig. 14. Aleurone in a cell from the endosperm of Castor oil seed.

Animal tissues

The elementary organization of individual cells to form a tissue, with subsequent division of labour, and specialization is well-illustrated by *Hydra*.

Examine and draw a prepared transverse section of *Hydra*. Notice that the cells with different functions have different structures, in shape, and internal organization (Figure 15).

with walls thickened with cellulose.
Apparatus: Dead-nettle *(Lamium)* stem, or celery petiole, or nettle *(Urtica)* stem, razor, cover-slips, slides, dropper, watch-glass, microscope, needles.
Method: Cut transverse sections of the material, mount, and examine microscopically (Figure 16). Look at the corners of the stems and observe the thickened cells.

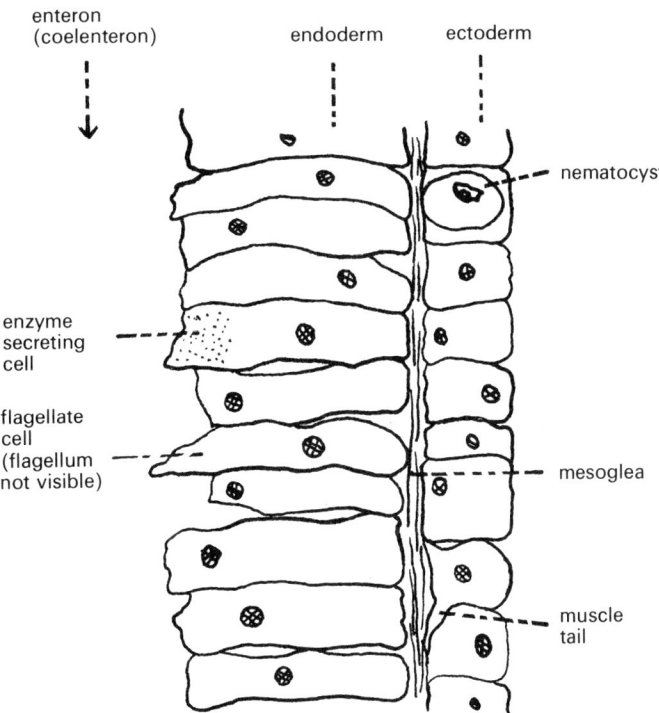

Fig. 15. High power L.S. through wall of Hydra.

Plant tissues

Parenchyma—This is an undifferentiated tissue, frequently concerned with storage.
Apparatus: Elder *(Sambucus)* stem, razor, cover-slips, slides, dropper, watch-glass, microscope.
Method: Cut some thin sections of elder stem; mount them on a slide and examine under the microscope. Look at the pith. Notice the large, undifferentiated cells with relatively thin walls. This is parenchyma.
Collenchyma—This is a supporting tissue

Take a small piece of the material on a slide in a drop of water. Tease it out with the needles into the individual fibres. Examine the elongated collenchyma cells.

Sclerenchyma—This is a supporting tissue which is thickened with *lignin*. Small masses of rounded sclerenchyma is seen in pear fruit. These are *sclerides* or stone cells (Figure 17). If the cells are elongated with pointed ends and a thin lumen, they are *fibres* (Figure 18). These are found widely in plants, especially in the stem cortex in bundles, but may be intermingled with any other tissue, e.g. xylem and phloem.

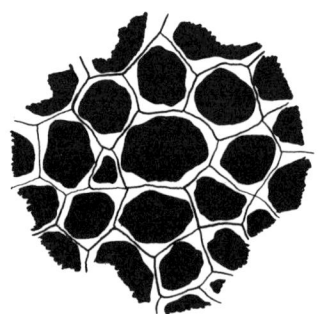

Fig. 16. Transverse section through typical collenchyma.

Apparatus: Pear fruit, hop (or other suitable) stem, slides, cover-slips, microscope, aniline hydrochloride (sulphate), 1 % phloroglucinol solution (or other lignin stain).

Method: Crush a small piece of the flesh of the pear on a slide. Irrigate with aniline hydrochloride (or other stain). Observe unde the microscope. Macerate a piece of the cortex of the hop stem, and stain as before. Draw the lignified cells.

Aniline hydrochloride stains lignin yellow, and phloroglucinol stains it red.

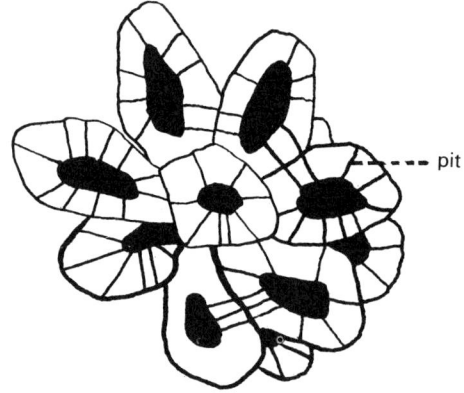

Fig. 17. Stone cells from Pear fruit.

Xylem and Phloem—The conducting tissues. Examine prepared slides of sunflower *(Helianthus)* stem in transverse and longitudinal section, paying particular attention to the xylem. Notice the wide

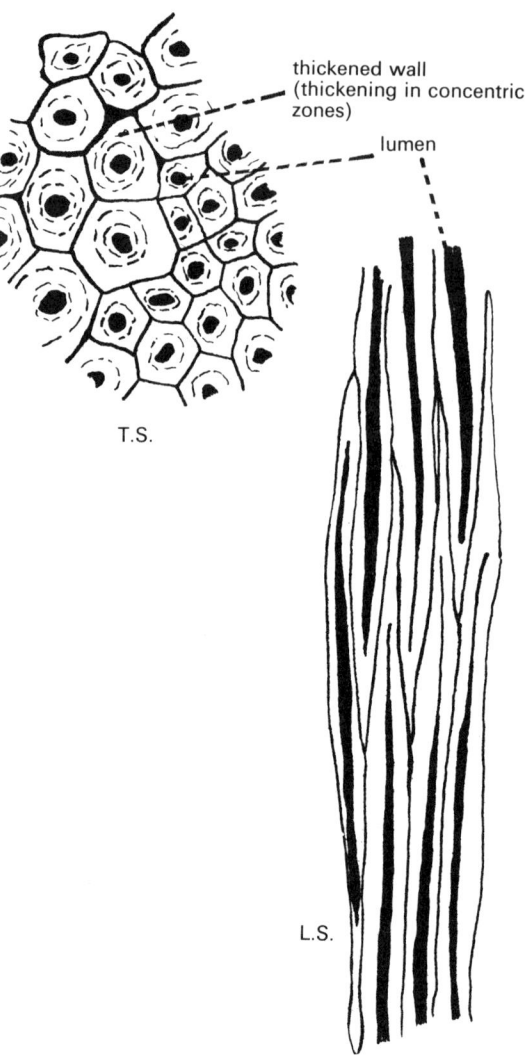

thickened wall (thickening in concentric zones)

lumen

T.S.

L.S.

Fig. 18. Fibres.

vessels, the narrower *tracheids* and some fibres. Notice, in longitudinal section. the spiral, annular (ring) and scalariform (ladder) thickening of the conducting elements, and the presence of *pits* in the walls. Draw a few representative cells.

Similarly examine prepared slides of *Cucurbita* (marrow), especially the phloem, where the *sieve-tubes* and *companion cells* are particularly clear. Draw a few representative cells. Notice that there is also parenchyma in the phloem.

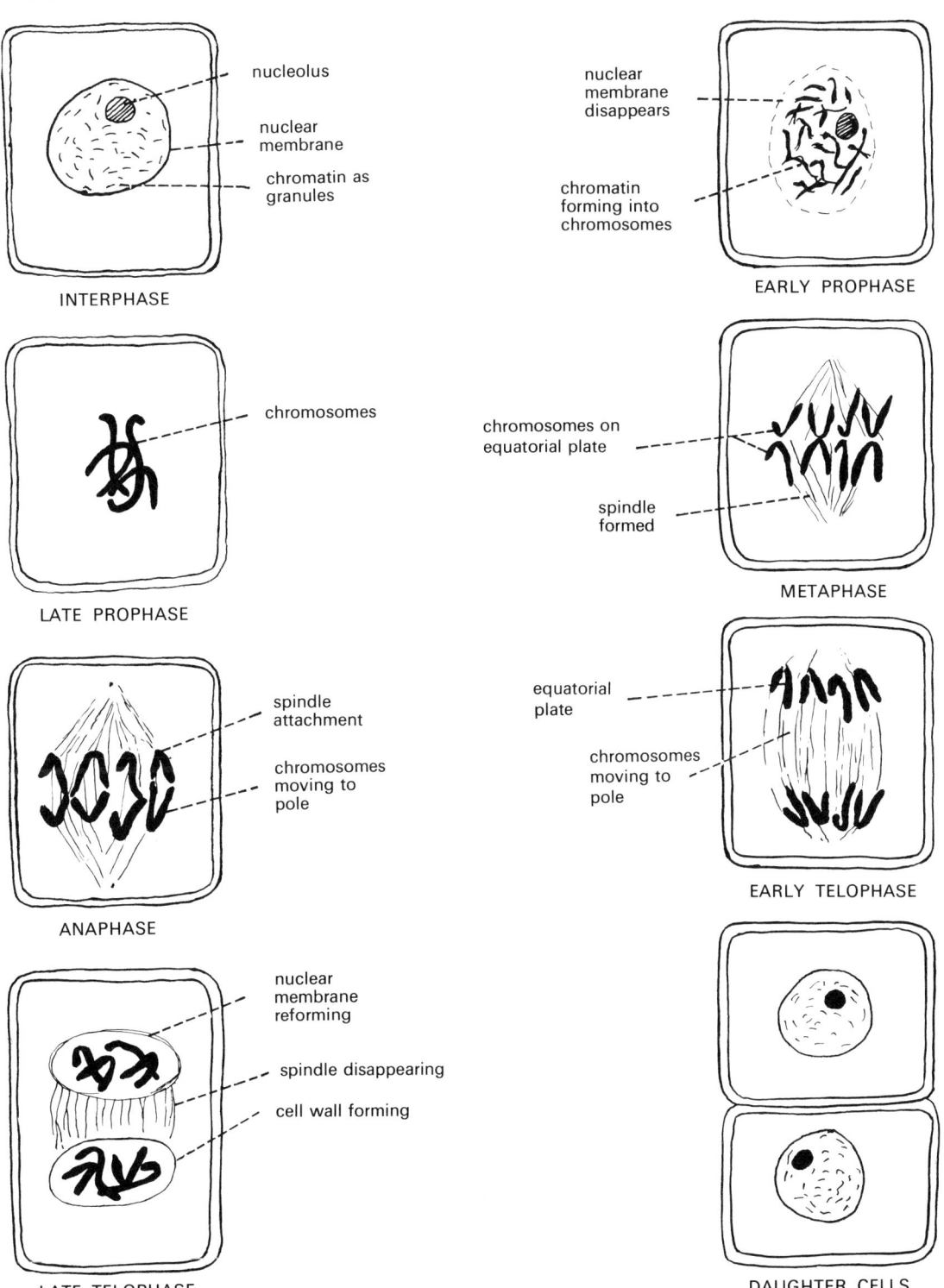

Fig. 20. Mitosis in a plant cell. (This is not any particular cell. The number of chromosomes is usually larger.)

Protoplasmic movement

Apparatus: Microscope, cover-slips, slides, forceps, dropper, *Tradescantia* flower.
Method: Carefully remove a stamen of the flower, mount it on a slide in water, and observe the movement of the protoplasm under the microscope. If *Tradescantia* is not available, *Elodea* leaves, or filaments of *Oedogonium* will do.

Mitosis in plants

Examine prepared slides of longitudinal sections through root-tips (usually onion or broad bean *(Vicia)*). Draw as many stages of mitosis as are visible (Figure 19 and 20).

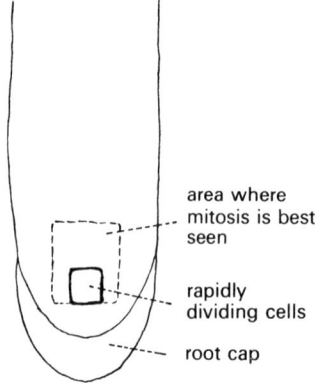

area where mitosis is best seen

rapidly dividing cells

root cap

Fig. 19. Diagram of L.S. of root to show zone of mitosis.

Apparatus: Flower buds, *(Lilium* or *Tradescantia* are good), slides, forceps, needle, aceto-carmine, filter paper.
Method: Carefully dissect out the young pollen-sacs from the flower bud. Place them on a slide in a drop of water. Place a cover-slip on them and press firmly to squash. Irrigate the slide with aceto-carmine. Leave for a few moments and examine under the microscope.

Aceto-carmine stains chromosomes red. Look for the various stages of mitosis and draw them. You may have to repeat this several times, with pollen-sacs of different ages (size), before you see all stages.

Mitosis in animals

Examine prepared transverse sections through the ovary of female *Ascaris*. This will contain developing eggs which should show the stages of mitosis. Draw the stages. What are the similarities and differences between mitosis in plants and animals?

Remember that the cells are three-dimensional, but under the microscope you see only two dimensions, so that the mitotic figures may appear to be different from the usual drawings.

5 Respiration

For experimental purposes respiration of carbohydrates is summarized in the equation:

$$C_6H_{12}O_6 + 6O_2 = 6CO_2 + 6H_2O + Energy.$$

This is not the full story, but it is only convenient in a school laboratory to show that these stages take place. Anaerobic respiration can also take place; this involves the release of energy without utilising oxygen, but as well as a variety of other substances, carbon dioxide is released.

Absorption of oxygen

Experiment: To show that oxygen is absorbed during respiration.

Apparatus: Four short pieces of wide ($\frac{5}{8}$ in.) glass tubing, 6 1-holed bungs, and 2 solid bungs to fit the tubing, 2 bent glass tubes, germinating peas, 20% potassium hydroxide solution, light oil, gauze, clamps, rubber tubing, glass tubing.

Method: Prepare two sets of apparatus as shown in Figure 21, with one tube containing potassium hydroxide solution, and the other oil. Push the gauze firmly into the top of one of the tubes and place the peas on it. Replace the bungs. Leave the apparatus for a few days having marked the original levels of the liquids. Remembering that potassium hydroxide absorbs carbon dioxide, and oil does not, account for your results.

The constant-pressure respirometer

Apparatus: Two large bottles (500 ml.–1 litre), manometer tube, 2 two-holed bungs to fit the bottles, glass tubing, clips, rubber tubing, gauze, soda-lime, millimeter scale.

Method: Set up the apparatus as shown in Figure 22, leaving the clips open for at least 10 minutes to allow the apparatus to come into equilibrium. This is a sensitive piece of apparatus, and must be used in a constant environment (away from heat, draughts, etc.). As the pressure on both sides of the manometer is constant, any variation in the manometer reading is a measurement of the change in volume of the internal atmosphere. Remember that soda-lime absorbs carbon dioxide.

You can design your own experiments using the apparatus, but here are a few suggestions: (a) compare the respiration rate

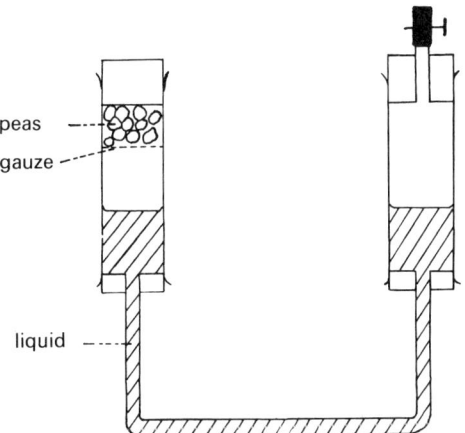

Fig. 21. To show that oxygen is absorbed during respiration.

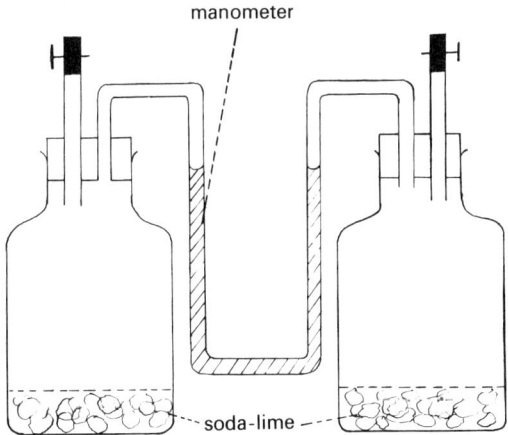

Fig. 22. Constant pressure respirometer.

of different organisms (remember to take weight into account); (b) compare the rate of respiration of the same organism under different conditions (the two bottles can be placed in a water-bath); (c) compare the respiration rate of warm-blooded and cold-blooded animals.

Production of carbon dioxide

Experiment: To show that there is a change in the composition of the air during breathing.
Apparatus: Two 250 ml. conical flasks, 2 two-holed bungs to fit the flasks, Y-piece, rubber tubing, glass tubing, lime-water.
Method: Set up the apparatus as shown in the diagram (Figure 23). Place the end of the Y-piece in your mouth, and breathe normally through it. Notice the flow of air through the flasks. What happens to the lime-water?

Lime-water turns milky in the presence of carbon dioxide due to the formation of insoluble calcium carbonate.

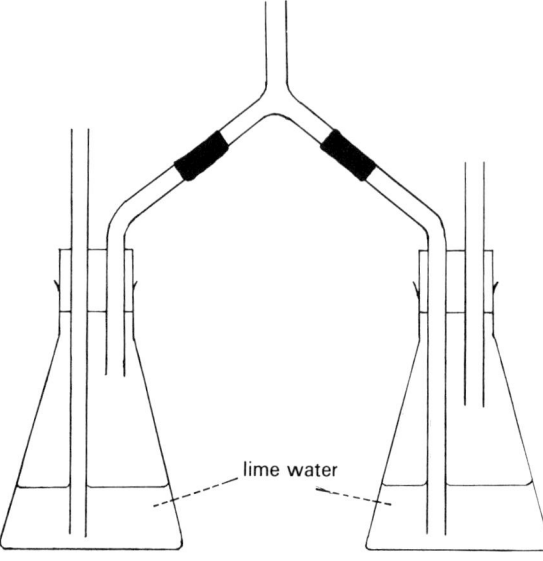

Fig. 23. To show the change in the composition of inhaled and exhaled air.

Experiment: To show that carbon dioxide is evolved by a variety of living organisms.
Apparatus: Two conical flasks (250 ml.), 2 bungs to fit the flasks, muslin, cotton, lime-water, soil, germinating peas, worms, and any other living organisms, beaker, 1% formaldehyde solution.
Method: The basic apparatus is stated above, but you will want this for each material if the experiments are to be run simultaneously. Kill one half of the organisms by heating, and surface sterilise with 1% formaldehyde (except soil). Set up the apparatus as shown in Figure 24, with a control containing dead material. Place in the dark to eliminate any complications of photosynthesis (if necessary) and leave for 2–3 days. Compare the colours of the lime-water in each pair of flasks. What does this show?

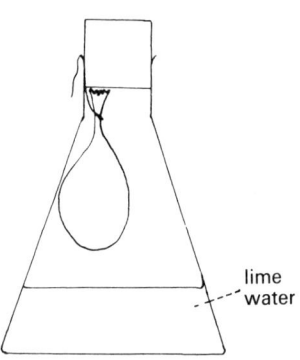

Fig. 24. To show the evolution of carbon dioxide by a variety of organisms.

Experiment: To show the evolution of carbon dioxide by small aquatic organisms.
Apparatus: Boiling-tube, 0.04% cresol red solution, small aquatic organism, e.g. minnow.
Method: About half-fill the boiling-tube with water and add a few drops of cresol red solution. Place the organism in the liquid, and leave for about 10 minutes. The cresol red changes colour due to the water becoming more acid. The increase in acidity is caused by an increase in the dissolved carbon dioxide.

Experiment: To measure the amount of carbon dioxide produced by living organisms.
Apparatus: Four flasks or bottles, 4 two-holed bungs to fit them, rubber tubing, aspirator,

clip, pipette, burette, dropper, phenol red solution, lime-water, 5% sodium hydroxide solution, sodium hydroxide solution of exactly 0.40 g/l.

Method: N.B. 1 ml. of the 0.40 g/l. sodium hydroxide solution is equivalent to 0.0004 g. of carbon dioxide.

Set up the apparatus as shown in the diagram (Figure 25). First use water instead of the solutions and run the apparatus to get used to the aspirator. When you are satisfied that it is running suitably (the bottle should empty in about 20 min.), pour 5% sodium hydroxide solution into flask 1 (to remove the carbon dioxide from the atmosphere), lime-water in flask 2 (to test that the air is carbon dioxide-free, and the test organism in flask 3. Pour 100 ml. of distilled water into flask 4 and add a few drops of phenol red solution. Add the 0.40 g/l. sodium hydroxide solution *carefully,* shaking the while, until the solution is pink. Connect the flasks as shown, and draw air through the apparatus, using the aspirator, for a definite time (about 10 min.). Note any colour change in flask 4. Disconnect flask 4 and add the 0.40 g/l. sodium hydroxide solution to it from a burette, until the solution returns to its original colour. Calculate the amount of carbon dioxide evolved from the data given above.

Various living material can be placed in flask 3, the animal organism can be replaced by a potted plant in a bell-jar, covered to

prevent photosynthesis. The temperature of flask 3 can be varied by placing it in a water-bath.

In any comparative study, the weights of the organisms must be taken into consideration.

Release of energy

Experiment: To show that heat is evolved from living organisms, in the presence of air.

Apparatus: Two 'Thermos' flasks, cotton wool, germinating peas, 2 thermometers ($-10°C$ to $110°C$), 1% formaldehyde solution, beaker, clamps.

Method: Divide the peas into two equal groups. Kill one group by dipping into boiling water for a few minutes. When the peas are cooled, surface-sterilise them by dipping them in the formaldehyde solution. This prevents the growth of bacteria which will cause the release of energy, and upset the experiment. Set up the apparatus as shown in Figure 26. Compare the two thermometers initially, and again at regular intervals (about two hours, if this is possible).

The cotton-wool allows the entry of air, but heat-insulates the flasks sufficiently to give a result. The flasks are upside-down so that the thermometers can be read without removing them. Thermometers should actually be read with the bulb downwards, but the error is slight, and at least compensated for in part as the readings are comparative rather than absolute.

This can be replaced by a plant under a bell-jar

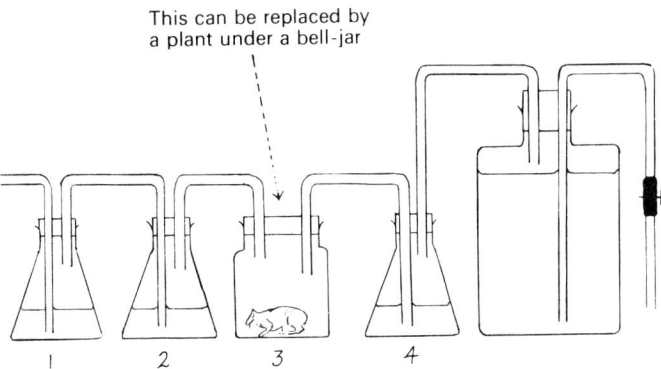

1   2   3   4

Fig. 25. Measurement of the amount of carbon-dioxide produced by a living organism.

Experiment: To show that energy is evolved during anaerobic respiration.

Apparatus: 10% glucose solution, 2–3 g. dried yeast, 2 'Thermos' flasks, glass tubing, rubber tubing, beaker, 2 thermometers, 2 two-holed bungs to fit the flasks.

Method: Set up the apparatus as shown in the diagram (Figure 27), placing the yeast in one of the flasks and leaving the other as a control. Take the initial temperature in each flask, and take readings at regular intervals (2 hours, if possible) for the next 2–3 days. What conclusions do you draw?

The beaker of water acts as an air trap. The yeast gives-off carbon dioxide displacing the air above the solution in the flask, so that the conditions become anaerobic quickly.

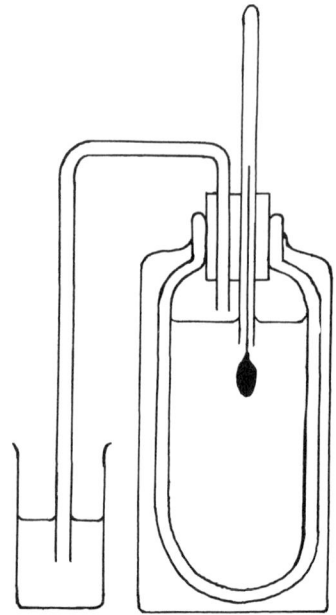

Fig. 27. To show that heat is evolved during anaerobic respiration.

Anaerobic respiration

Experiment: To show that carbon dioxide is produced during anaerobic respiration.

Apparatus: Two test-tubes, mercury, clamps, small trough, 20 well-soaked peas, lime-water, small pipette with the nozzle bent to introduce the lime-water into the tubes, beaker, 1% formaldehyde solution.

Method: Kill half the peas by immersing them in boiling water for a few minutes, and surface sterilize them with formaldehyde having removed the seed-coats from all of the peas. (This prevents air being introduced into the tubes.) Completely fill the tubes with mercury, and invert them in a trough of mercury. Introduce the living and dead peas into separate tubes in equal numbers. (This is quite easy—push the peas one by one into the open end of the tube, and they will float upwards.) Leave in a warm place for about two days. What happens to the mercury levels? Using the pipette introduce lime-water into each of the tubes. Again observe the mercury levels, and the colour of the lime-water. Account for your results.

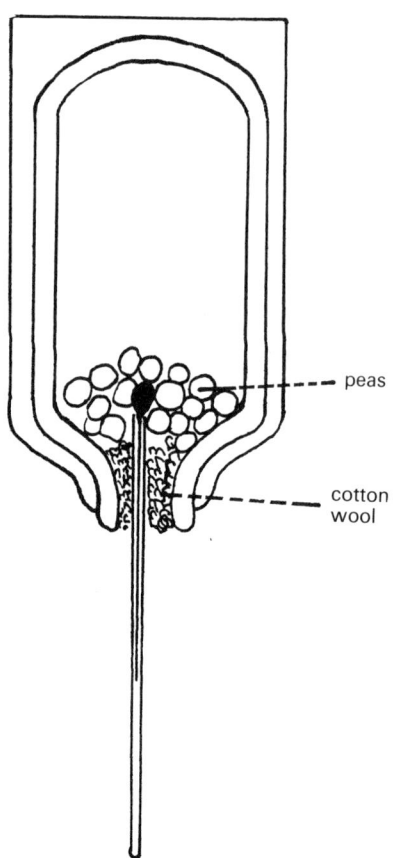

Fig. 26. To show evolution of heat by germinating peas.

Experiment: To show some of the products
of anaerobic respiration.
Apparatus: 10% glucose solution, yeast,
2 1 litre (or bigger) flat-bottomed flasks,
2 one-holed bungs to fit the flasks, test-tubes,
cotton-wool, lime-water, iodine in potassium
iodide solution, burner, filter-funnel, filter
paper, glass tubing.
Method: Set up the apparatus as shown in
Figure 28, putting 15–20 g. of yeast in the
flask. Set up a similar apparatus, but without
the yeast, as a control.

Leave for a few days in a warm place.
Observe what happens in the flask and
lime-water. What does this indicate.

After a few days, filter some of the liquid
in the flask into a test-tube. Add the same
volume of water to the filtrate. Add a few
drops of iodine solution and then the sodium
hydroxide solution, until the solution
becomes slightly yellow. Shake thoroughly.
If the colour disappears, add a few drops of
iodine until it is restored. A yellow
precipitate with a pungent, vaguely anaesthetic,
smell (iodoform) indicates the presence of
ethyl alcohol (ethanol) in the original
solution. The recation may not take place
without gentle warming.

You may like to distil-off the alcohol
fraction, using a Leibig condenser.

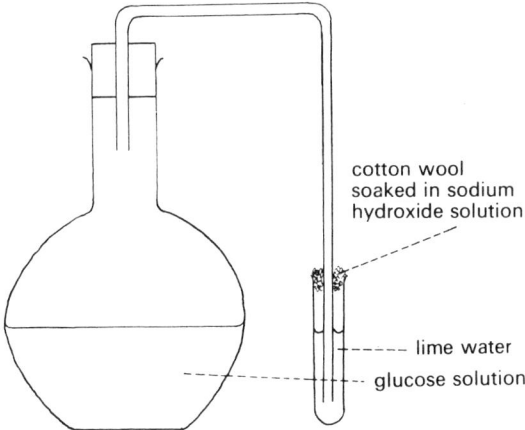

cotton wool
soaked in sodium
hydroxide solution

lime water

glucose solution

Fig. 28. To show the products of anaerobic
respiration.

6 Dissection of the Rat

EXTERNAL FEATURES

Before beginning to dissect the rat, make a thorough examination of the external features. This is frequently skimped, and a question on it set in the examination comes as a shock.

The head

Examine the head from the front and side, and draw, noticing the points shown in the diagrams of Figure 29.

Notice especially the arrangement of the whiskers *(vibrissae)*, and their length in relation to the width of the body. These are important sensory organs of touch, enabling the animal to judge the size of an opening into which it places its head.

There is a *nictitating membrane* (third eye-lid) in the anterior corner of the eye. If the specimen is fresh, this membrane may be drawn across the eye, using a fine forceps. Notice the colour of the eye. Albino rats have no eye pigment, and the eyes are pink.

Using the handle of a steel scalpel, force the mouth open. Be sure not to break anything! Examine the teeth and tongue. The dental formula is

$$I\frac{1}{1}, \ C\frac{0}{0}, \ PM\frac{0}{0}, \ M\frac{3}{3},$$

so that there is a space *(diastema)* immediately behind the incisors. Notice the

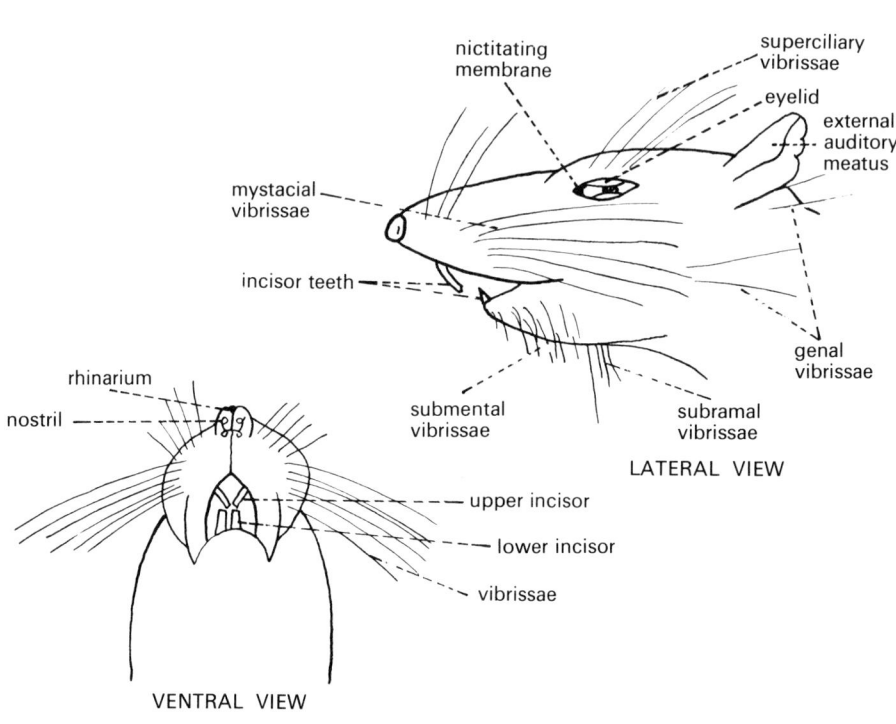

Fig. 29. Head of the Rat.

surface of the molars, and the lack of enamel on one surface of the incisors. This enables the teeth to be sharpened by constant gnawing. This is all that can be seen before the dissection. The teeth will be examined in detail later.

Thorax and abdomen

Place the animal on its back, but do not pin it out, as this will damage the limbs. Notice the relative proportions of the head, thorax and abdomen, and the length of the tail in relation to the total body-length. The tail is a balancing organ held out horizontally behind the animal when it is moving at speed.

The thorax is supported internally by the rib-cage, which can be felt underneath the skin. The abdomen is not ventrally supported by bone and is therefore soft to the touch. The structures visible on the male and female are shown in the diagram (Figure 30). The hair will have to be moved on the female to expose the nipples.

The limbs

Draw the under surface of the fore and hind feet on one side of the body. Notice the reduction of the thumb *(pollex)* on the front foot (Figure 31).

THE SKELETON

Examine a prepared mounted skeleton and notice the relative sizes of the bones. When you come to draw the individual bones, make sure that the drawings are to scale. Use a ruler to get the proportions right. Unless they are drawn carefully, illustrations of bones can look nothing like the original. Cut down the shading to a minimum.

The vertebral column

Count the number of vertebrae in each zone of the column. The tail is made up of a large number (about 30) relatively simple caudal vertebrae. Notice that the bones from each zone grade into those of the next. When drawing individual vertebrae, try to chose

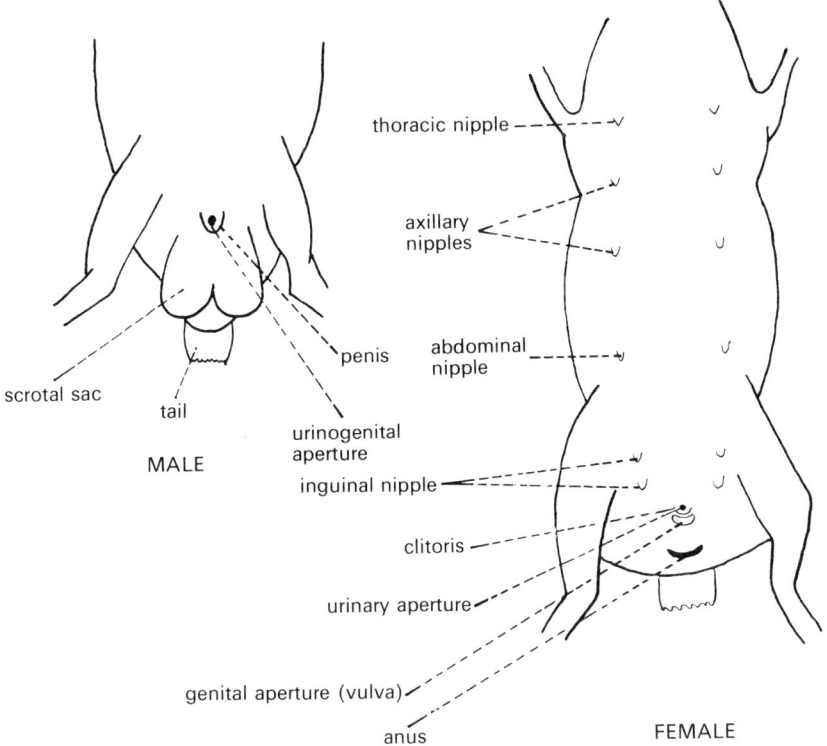

Fig. 30. Ventral view of the Rat.

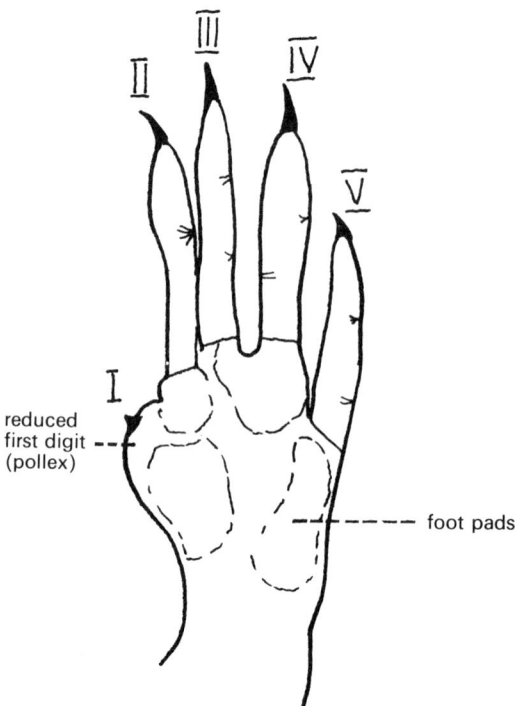

reduced
first digit - - -
(pollex)

- - - - foot pads

Fig. 31. Ventral view of left fore foot.

one from the middle of the zone. Draw an
anterior and lateral view of each bone.
See Figures 32(a) and (b).

The following are the special points to look
for on the individual bones.

Atlas: 1. The absence of a centrum; 2. The
transverse ligament—this may be destroyed
during the preparation of the specimen. The
space below it is occupied by the odontoid
process of the axis; 3. The vertebro-arterial
canals; 4. The large transverse processes;
5. The large facets on the anterior surface for
articulation with the skull, and the smaller
ones on the posterior surface for articulation
with the axis; 6. The small neural spine.

Axis: 1. *The odontoid process,* which fits
into the atlas and allows lateral movement
of the skull; 2. Posterior zygopophyses only;
3. Vertebro-arterial canals; 4. Small centrum,
which can be seen clearly only from the
posterior.

Typical cervical: 1. Centrum; 2. Neural
arch; 3. Posterior and anterior zygopophyses;

4. Vertebro-arterial canals; 5. *Costo-transverse
processes*. These are compound structures
formed by the fusion of the transverse
processes and vestigial cervical ribs.

Thoracic: 1. Large neural spine;
2. Demi-facets at either end. The demi-facets
of adjacent vertebrae form a socket for
the articulation of the head *(capitulum)* of
the rib; 3. Anterior and posterior
zygopophyses; 4. Facets at the end of
transverse processes for the articulation of
the tuberculae of the ribs.

Lumbar: 1. A large heavy bone with
well-developed centrum; 2. Large transverse
processes, with *metapophyses* and *anapophyses*
for the attachment of the abdominal muscles;
3. An *intervertebral notch* below the
anapophyses through which pass the spinal
nerves.

Sacrum: 1. This is made up of two sacral
and two caudal vertebrae fused together;
2. The sacral vertebrae have large transverse
processes which form an immovable joint
with the ilia of the pelvic girdle; 3. The
neural spines become smaller posteriorly;
4. The anterior zygopophyses of successive
bones are fused. The extent of the fusion
depends on the age of the rat; 5. Intervertebral
notches.

Caudal: 1. The gradual reduction of the
zygopophyses posteriorly; 2. The reduction
of the neural arch and spine until the end
10 bones consist only of a centrum.

Rib cage

Examine the complete rib cage on a mounted
specimen. Notice the attachment of the ribs
to the sternal ribs, and the articulation of
the latter with the sternum. The 11th, 12th,
and 13th ribs are not attached to the sternum,
and are called floating ribs.

Skull

Examine the skull as a complete unit, and
draw the views shown in the diagram
(Figure 34). You are not expected to know
the names of the individual bones of the
skull, but concentrate more on the relative

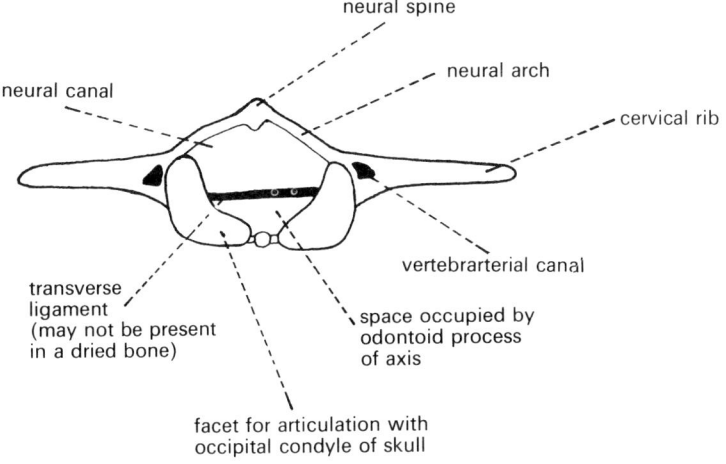

ATLAS: Anterior View

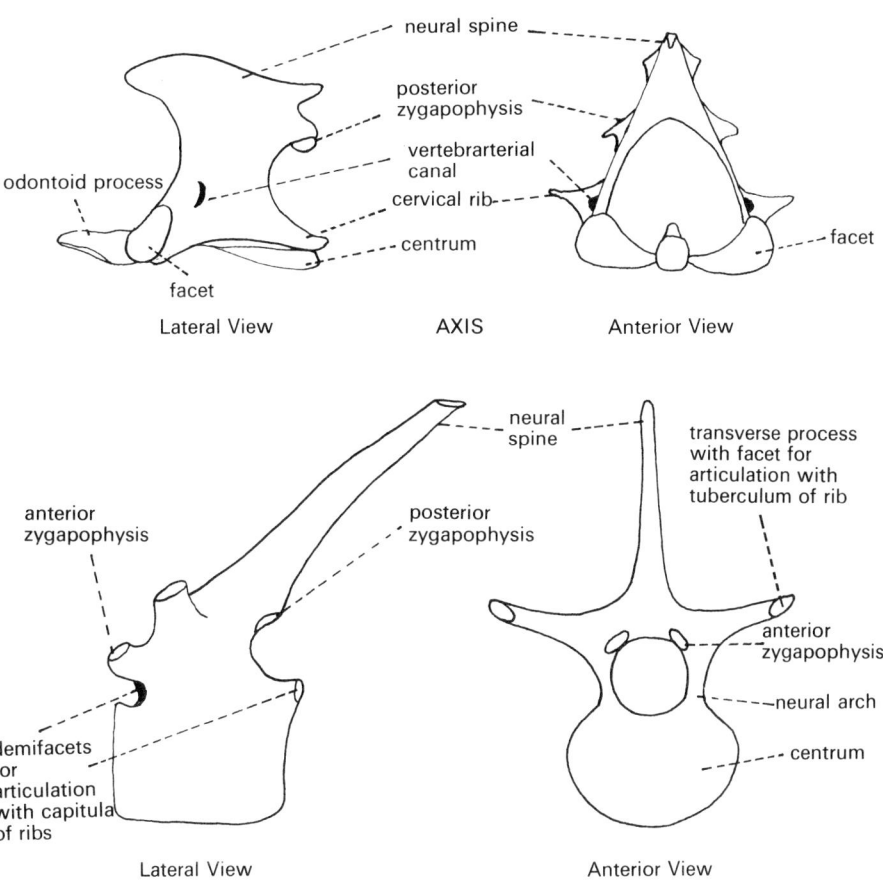

Fig. 32a. Vertebrae of the Rabbit.

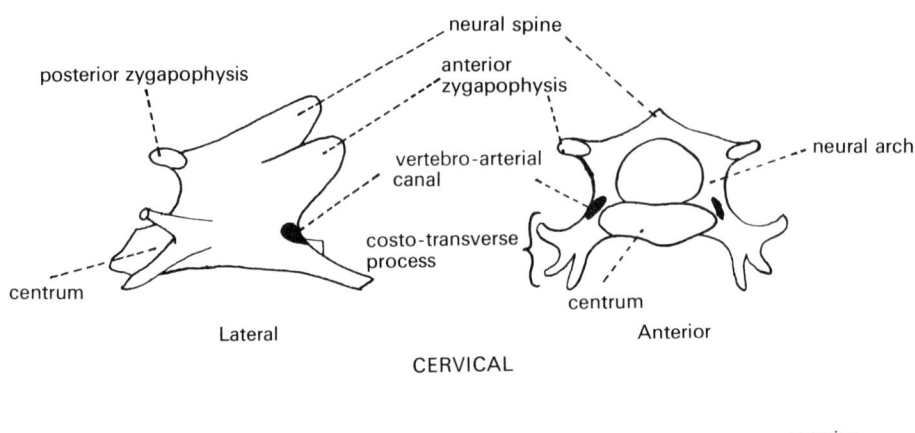

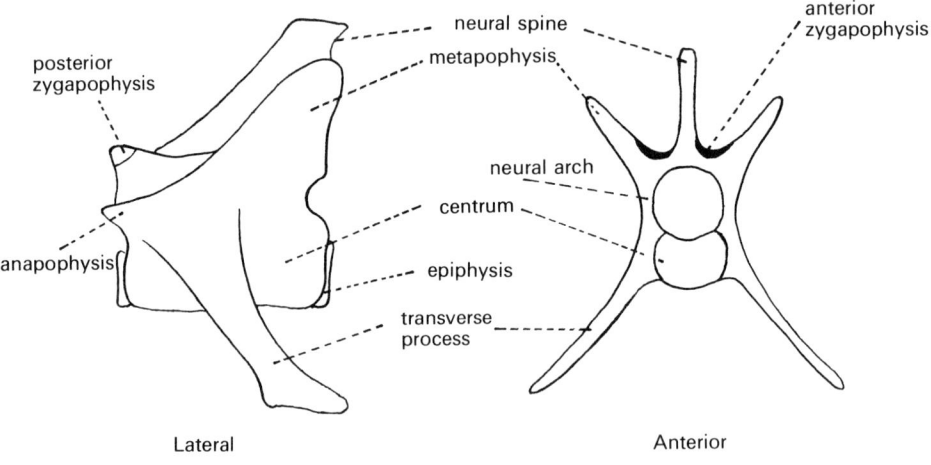

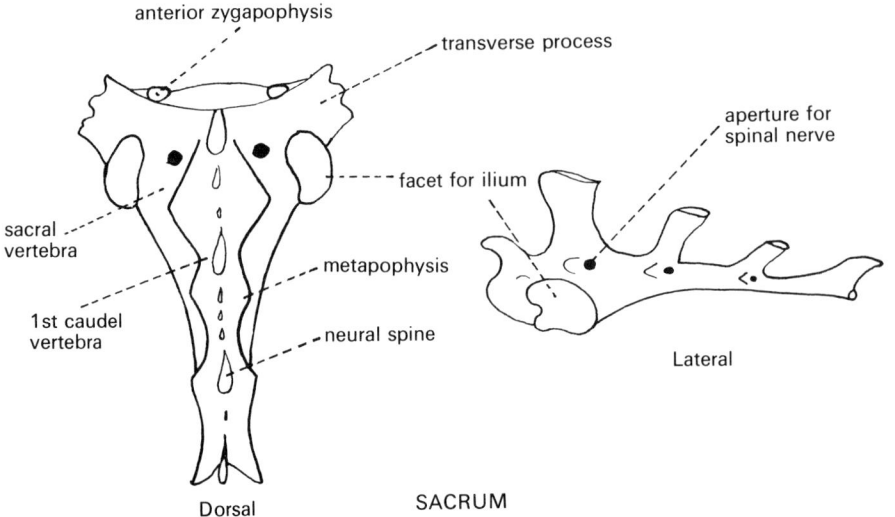

Fig. 32b. Vertebrae of the Rabbit.

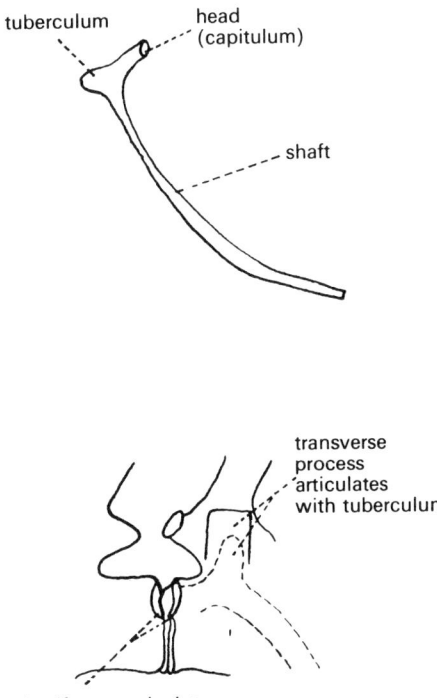

Fig. 33. Detail of articulation of rib with thoracic vertebra.

sizes of the different parts and their functions. The cranium is relatively small, compared with the total volume of the skull. This gives an indication of the size of the brain. The nose containing the nasal scrolls and chemo-receptors, is big, suggesting the relative importance of this sense-organ. From the position of the eye-orbits, can you say if the rat has binocular vision?. From the articulation of the jaw, and the points of attachment of the muscles, draw conclusions as to the way the rat chews its food. The *tympanic bullae* contain the bones and canals of the ear—notice the *Eustachian canal.* There are several openings in the bones of the skull that allow the passage of nerves and blood vessels—notice particularly the *foramen magnum* through which passes the spinal cord to the brain.

Pectoral girdle and fore-limb

Examine the complete girdle and the articulation of the humerus (Figure 35). This is essentially a universal (ball-and-socket) joint, but it is not a very substantial one when compared with the corresponding joint of the femur with the pelvic girdle. Does this give any indication of the relative importance of these two limbs in movement? Draw one half girdle with the attached limb and the separate bones.

Notice the smooth shiny surfaces on the extremities of the limb bones forming articulating surfaces with the cavities in the

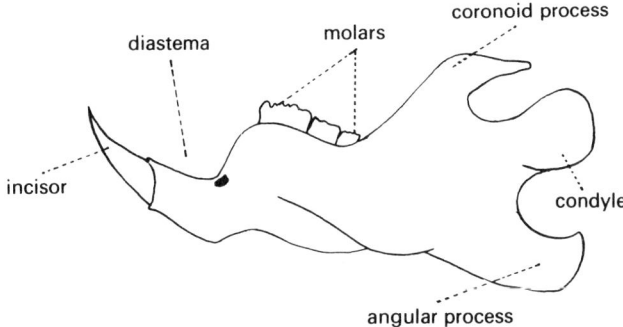

Fig. 34b. Lateral view of the lower jaw of the Rat.

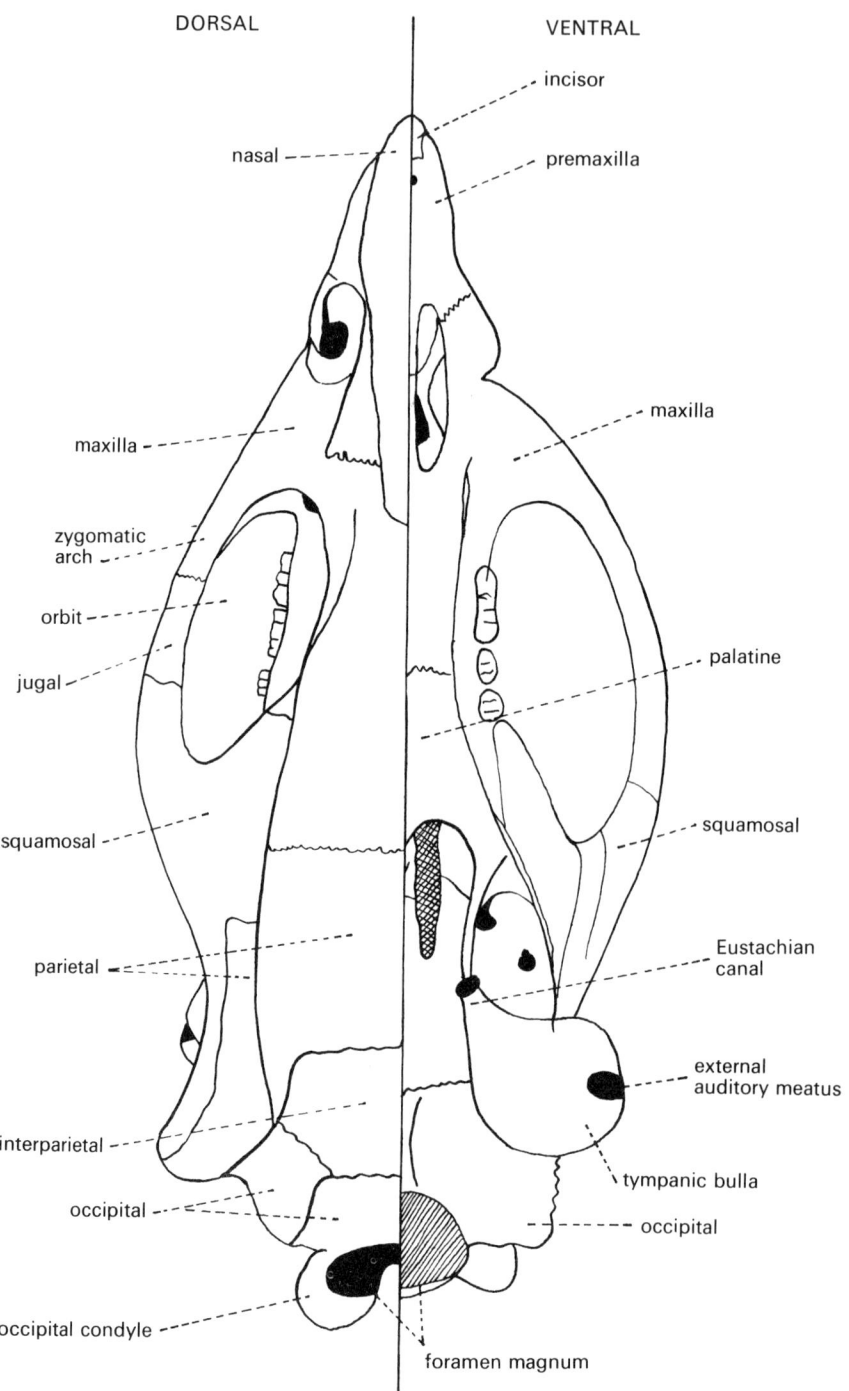

DORSAL   VENTRAL

incisor

nasal   premaxilla

maxilla   maxilla

zygomatic
arch

orbit   palatine

jugal

squamosal   squamosal

Eustachian
canal

parietal

external
auditory meatus

interparietal

tympanic bulla

occipital   occipital

occipital condyle

foramen magnum

Fig. 34a. The skull of the Rat.

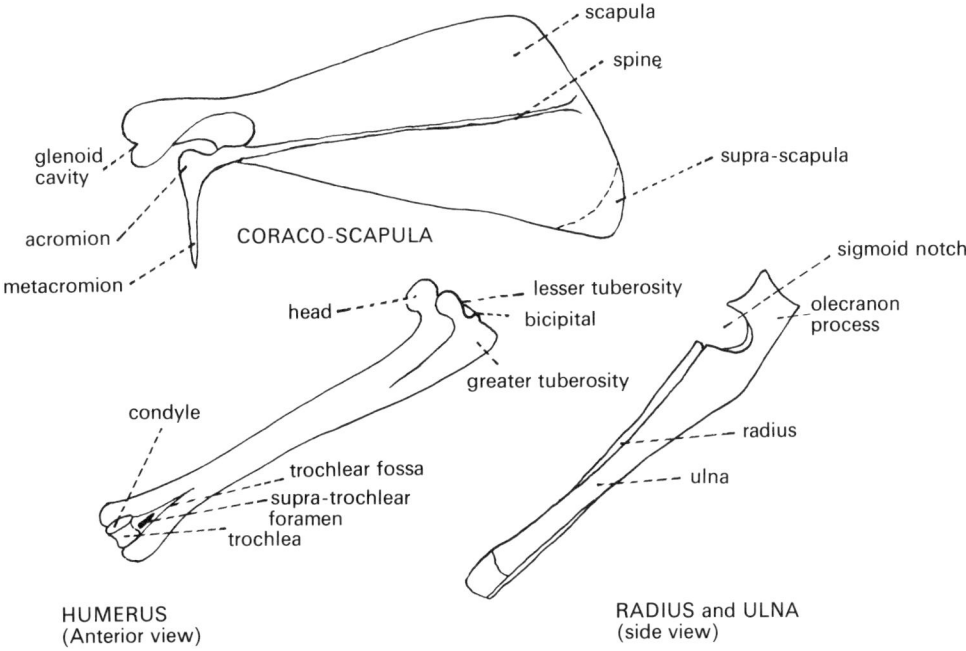

scapula

spine

glenoid cavity

supra-scapula

acromion

CORACO-SCAPULA

metacromion

sigmoid notch

lesser tuberosity

head

bicipital

olecranon process

greater tuberosity

condyle

radius

ulna

trochlear fossa

supra-trochlear foramen

trochlea

HUMERUS
(Anterior view)

RADIUS and ULNA
(side view)

Fig. 35. Left pectoral girdle and limb bones of the Rabbit.

corresponding bones. The projections e.g. the tuberosities and deltoid ridge are for the attachment of muscles.

In your examination of the front foot, look for the fusion of the wrist bones, and their number. Fusion, and consequent reduction in number leads to reduced flexibility, but additional strength. Observe the relative lengths of the toes. Is there any suggestion here of how the rat used its fore-limb in locomotion?

Pelvic girdle and hind limb

The dominant feature of these bones is the fusion, of the two halves of the pelvic girdle and of the fibula with the tibia (Figure 36). Notice the size of the bones, and compare them with the corresponding bones of the fore-limb and girdle. Again there are ridges for muscle attachment, and smooth surfaces for articulation. The hind foot is much larger than the fore-foot, and once again observe the ankle bones. Which toes are dominant in movement? Can you follow the line of

greatest pressure through the foot to the articulation of the limb with the pelvic girdle?

INTERNAL ANATOMY

Initial stages

Using dissecting spikes (awls), pin the animal back downwards on the dissecting board. Push the awls through the feet, so as to slightly stretch the specimen. The awls should enter the board at 45°. Now remove the skin. These initial stages are very important, and are frequently carried out too quickly and superficially, ruining the dissection. Lift a small piece of skin near the centre of the abdomen (be careful not to lift the underlying muscle) and make a small lateral cut with scissors (about $\frac{1}{4}$ in.). Now cut the skin up the medial line to about $\frac{3}{4}$ in. under the jaw, and backwards towards the anus. (Be careful not to damage the scrotal sacs of the male.) Now make a lateral incision down the upper part of the hind legs, and out

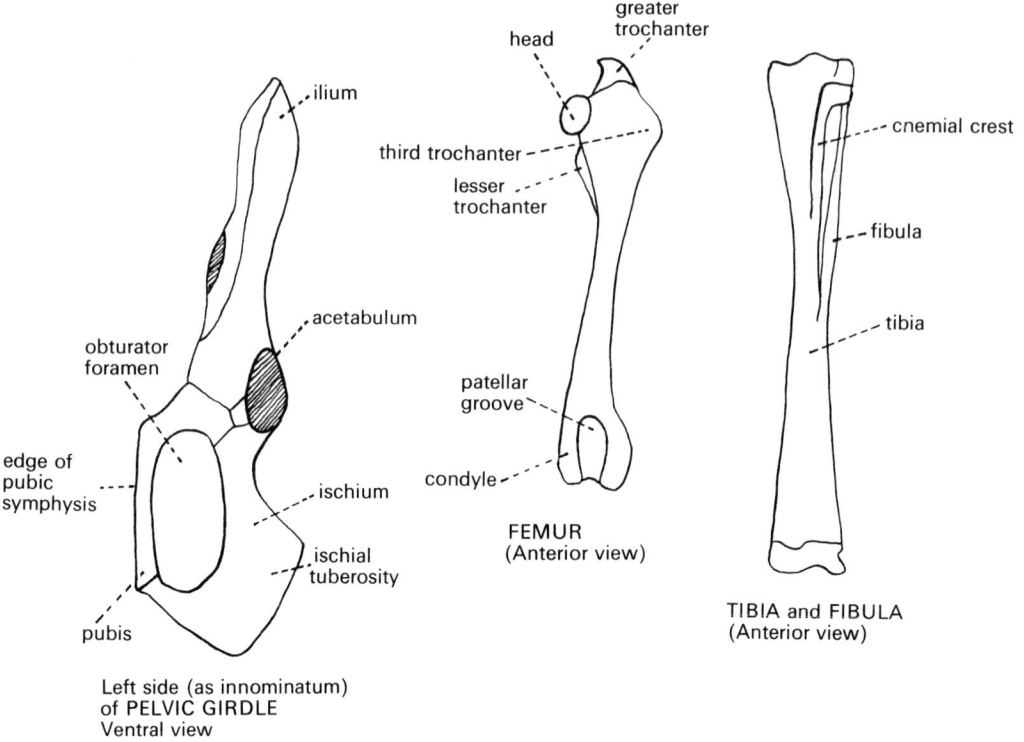

Fig. 36. Pelvic girdle and hind limbs of Rabbit.

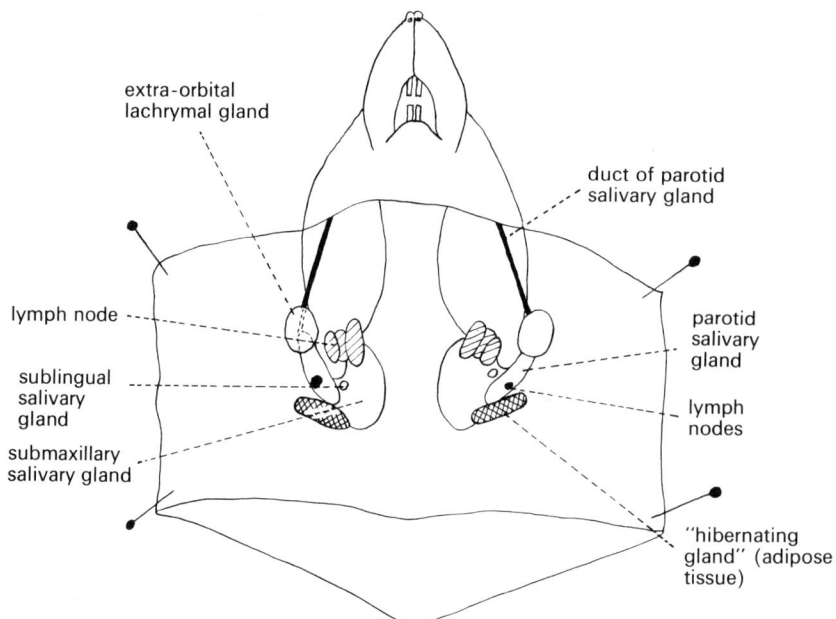

Fig. 37. The Rat: Ventral surface of neck after removal of skin.

towards the fore-legs (blood vessels near the surface here!). Peel back the skin, using your fingers and the handle of a scalpel. Use firm, but gentle pressure. Pin back the flaps of skin.

Under the skin of the female the *mammary glands* may be visible, looking like patches of fat. Notice the *salivary glands* and *tear glands* in the region of neck, and draw them (Figure 37).

Abdominal viscera

Now cut into the abdominal muscle to open the coelom. This should be done with scissors. Keep the point of the scissors facing upwards so as not to damage the under-lying organs. Make a median incision up to the xiphoid cartilage of the sternum and down to the pubic region, taking care not to damage the penis and the scrotal sacs (Figure 38). Carry the incision along the line of the ribs, and along the top of the hind limbs. Fold back the flaps of muscle to expose the organs of the abdomen. There may be a large amount of fat present—large yellow masses—especially two masses originating in the scrotum. Gently remove these from the organs and fold them outwards towards you. There will be other masses of fat on the abdominal wall, around the kidneys, but these can be dealt with later. Draw the organs as in Figure 38.

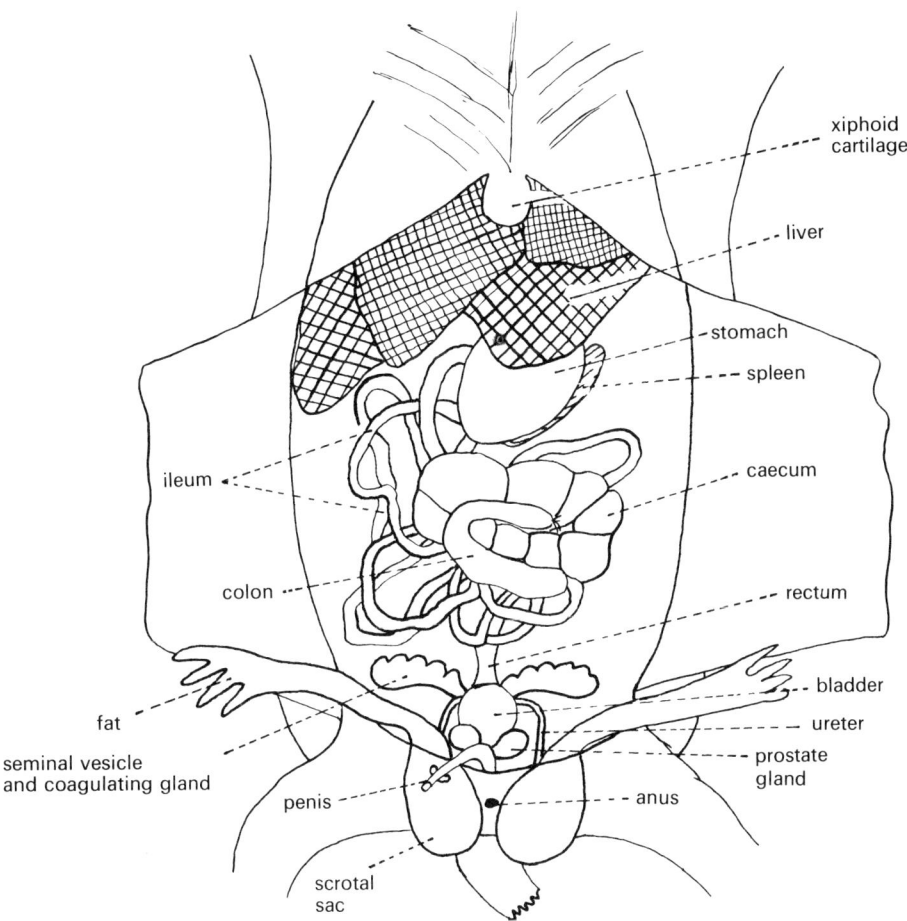

Fig. 38. Male Rat showing the abdominal viscera *in situ*. There are pockets of fat between the viscera and the abdominal wall.

Notice: 1. The *liver* which is lobed and dark red, lying under the diaphragm: 2. The *stomach*—light-coloured, and lying slightly to the left (your right); 3. To the left, and slightly below the stomach is the dark red *spleen*. (It looks rather like liver.) The fine, coiled part of the intestine is the *ileum*, which is light coloured; 4. Dominating the centre of the dissection are the darker, wider *colon* and *caecum;* 5. The *bladder, uterus, penis* and associated glands are low in the pubic region, and may still be covered by uncut muscle. In a preserved specimen, the bladder will be smaller and harder than in a fresh one.

Digestive system, hepatic portal vein, and arteries of the digestive system.

Lift the whole digestive system free from the coelomic cavity. This must be done very carefully with a minimum of cutting. Some fat and mesentery may have to be removed, but do *not* cut any blood vessels. Turn the whole intestine over to the animal's left. Lift the liver back towards the head, and, if possible, pin back the lobes. Do not stick the pins into the thorax. The pins may rip the liver of a preserved specimen. Turn the duodenum over to the right and pin it out. Pin out the rest of the intestine as far as possible to expose it as shown in the diagram (Figure 39). This will display the intestine, the hepatic portal vein and its branches to the various parts of the digestive system.

Notice: 1. The lobing of the *liver;* 2. The *bile duct* and the absence of the gall bladder; 3. The *stomach* with the lighter coloured cardiac region (to the left) and the darker pyloric region; 4. The pink-yellow (or pink

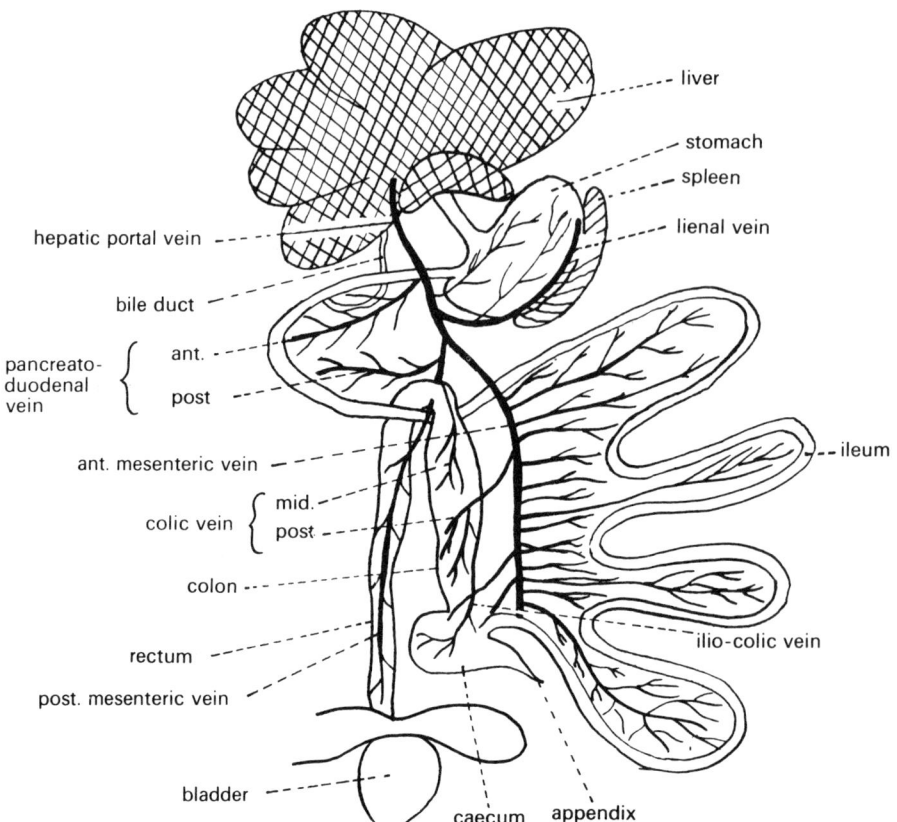

Fig. 39. The hepatic portal view of the Rat.

in a fresh specimen) *pancreas* in the loop of the duodenum; 5. The sac-like *caecum* terminating in the *appendix;* 6. The saccate *colon;* 7. The *rectum,* distinguished by the presence of pellets of faeces (Figure 41); 8. The fairly short *hepatic portal vein* running close to the bile duct and oesophagus and through the intestinal mesentery as the *anterior mesenteric vein.* This gives off branches to all parts of the intestine as shown in the diagram (Figure 39).

During this dissection you may have to remove fat from the mesenteries and you may damage blood-vessels and nerves.

Turn the whole of the intestine over to the right, to expose the arteries running to it. The *dorsal aorta* (light-coloured) and the *vena cava* (wider, with blood visible through

it) should be visible immediately, but they are frequently embedded in fat. Remove this with great caution, and be careful not to cut the ureter which runs close to the longitudinal blood vessels. The three arteries to the intestine are the *coeliac, anterior mesenteric* and *posterior mesenteric* (Figure 40).

Notice the *solar plexus* immediately posterior to the coeliac artery. Dissect this out.

Remove the intestine from the abdominal cavity. Cut through the oesophagus near to the stomach, and through the rectum. Holding the oesophagus, make your cuts as close as possible to the intestine. In a fresh specimen, the larger blood vessels may bleed and need ligaturing.

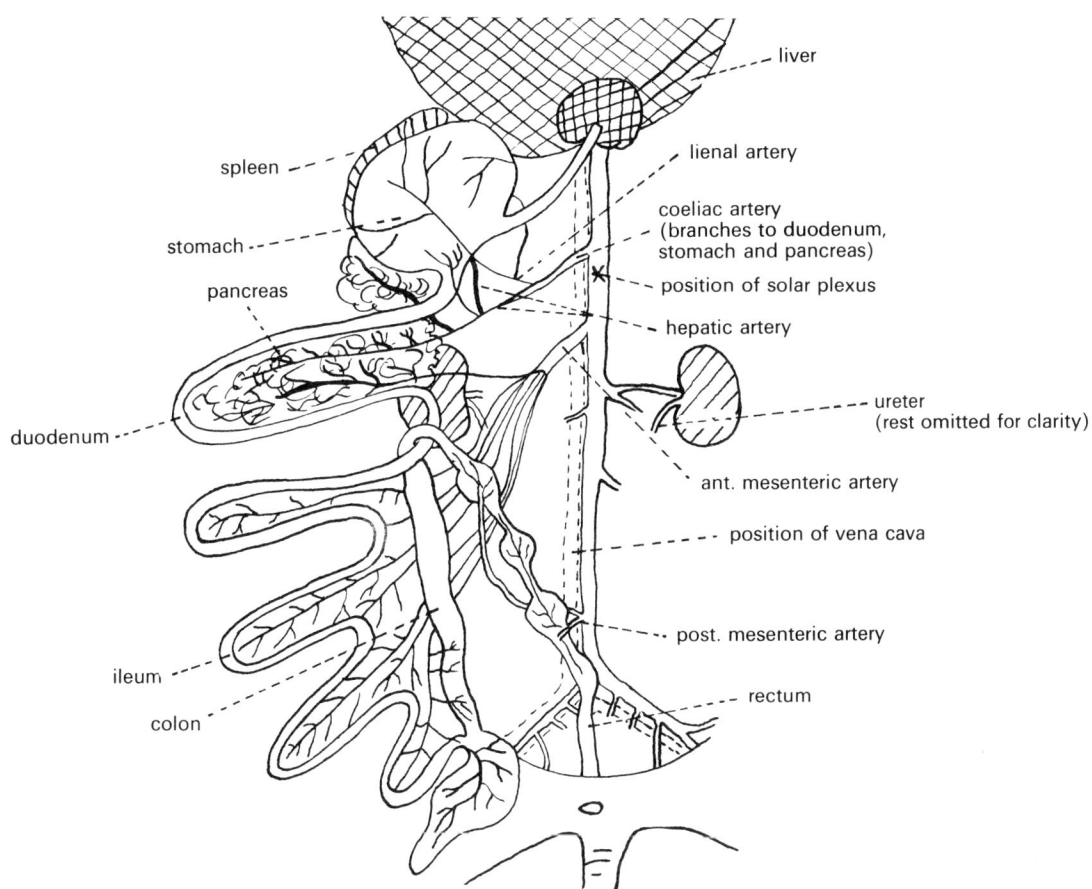

Fig. 40. Arteries to the alimentary canal of the Rat.

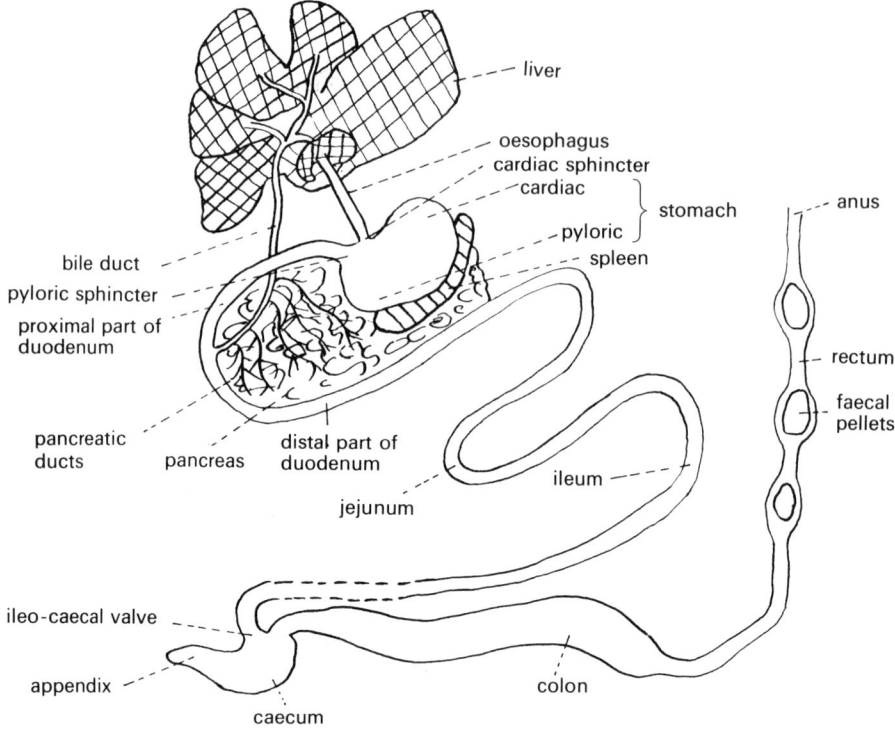

Fig. 41. Alimentary canal of the Rat displayed.

Display the intestine as shown in the diagram (Figure 41). Notice the relative lengths of the various parts of the intestine. Does this suggest anything about the diet of the rat and the water content of the food?

Cut off the stomach and open it along the posterior surface. Observe the condition of the food. Wash out the food and examine the inside of the stomach. Look for the folding of the mucosa. Which part is most muscular? Similarly examine the small intestine. Cut out the caecum, slit it longitudinally and examine the inner surface. The colon should be examined likewise. This should give a broad idea of the macrostructure.

Look at prepared slides of the stomach, duodenum, and large intestine. Do not draw them yet, as the histology of these organs will be studied later.

The urinogenital system

The intestine is now removed and the reproductive and renal systems are partially exposed on the wall of the abdomen. *In both sexes* remove the fat and peritoneum from around the kidneys which are red, bean-shaped bodies lying close to the abdominal wall. Notice the relative positions of the two kidneys. Dissect out the *adrenal bodies* which lie anterior to the kidneys. Clean up any blood vessels *(renal artery, renal vein, spermatic artery,* or *utero-ovarian artery and vein)*. Trace the *ureter* down to the *bladder*. The blood vessels of the genitalia lie over the ureter. A lot of fat may have to be removed at this stage.

Cut one kidney longitudinally, and draw the cut surface (Figure 42). It may be more convenient to use a pig's or sheep's kidney to see the internal structure.

Parts of the reproductive system lie obscured by the pelvic girdle. This has now to be split and pulled apart. The girdle lies underneath the urethra, so that its bisection is rather difficult. Remove the muscle over the middle of the girdle to expose the pubic

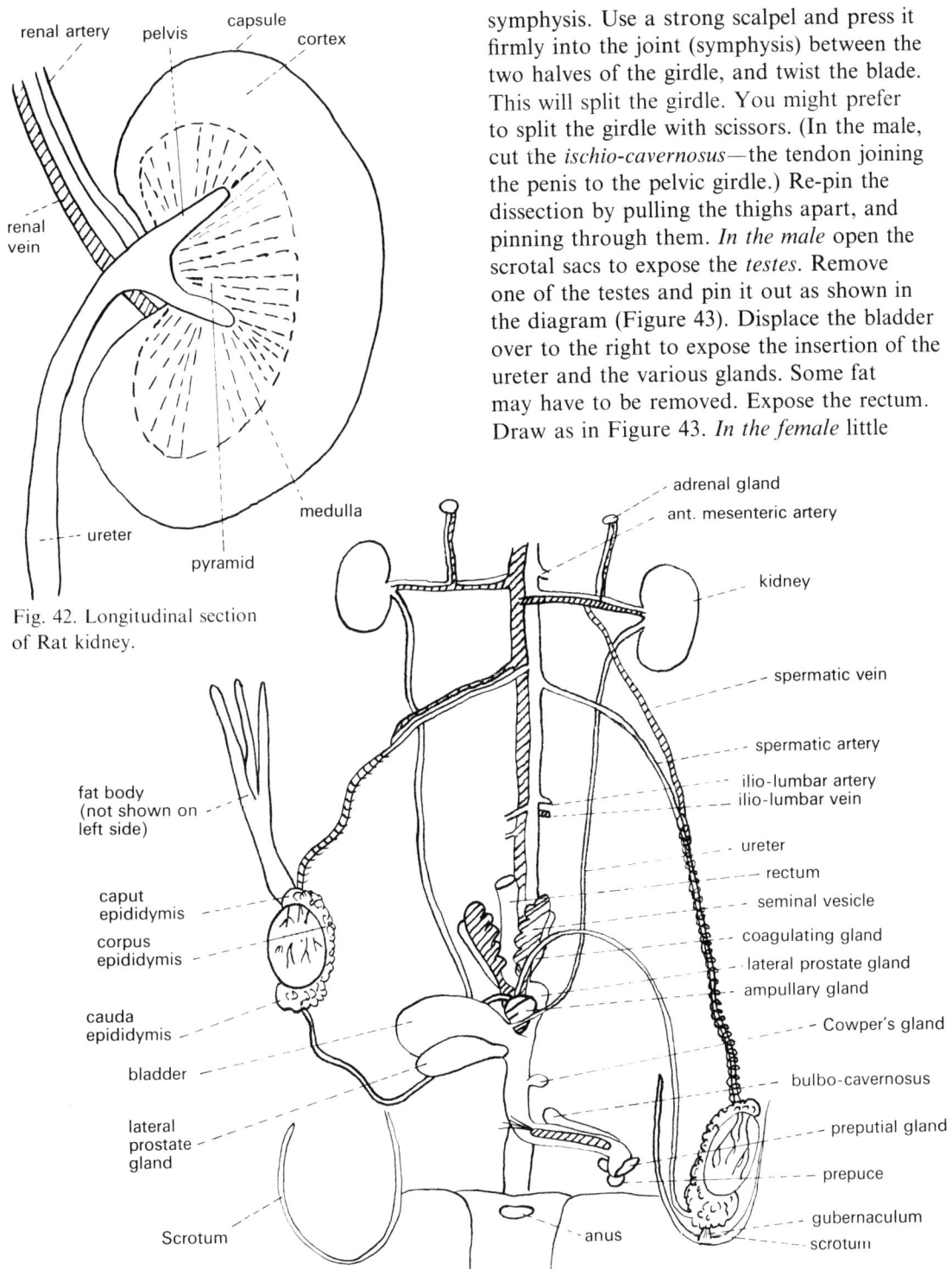

symphysis. Use a strong scalpel and press it firmly into the joint (symphysis) between the two halves of the girdle, and twist the blade. This will split the girdle. You might prefer to split the girdle with scissors. (In the male, cut the *ischio-cavernosus*—the tendon joining the penis to the pelvic girdle.) Re-pin the dissection by pulling the thighs apart, and pinning through them. *In the male* open the scrotal sacs to expose the *testes*. Remove one of the testes and pin it out as shown in the diagram (Figure 43). Displace the bladder over to the right to expose the insertion of the ureter and the various glands. Some fat may have to be removed. Expose the rectum. Draw as in Figure 43. *In the female* little

renal artery   pelvis   capsule   cortex

renal vein

ureter

medulla

pyramid

Fig. 42. Longitudinal section of Rat kidney.

adrenal gland

ant. mesenteric artery

kidney

spermatic vein

spermatic artery

ilio-lumbar artery
ilio-lumbar vein

ureter

rectum

seminal vesicle

coagulating gland

lateral prostate gland

ampullary gland

Cowper's gland

bulbo-cavernosus

preputial gland

prepuce

gubernaculum

scrotum

fat body (not shown on left side)

caput epididymis

corpus epididymis

cauda epididymis

bladder

lateral prostate gland

Scrotum

anus

Fig. 43. Urinogenital system of the male Rat. The cut edges of the pelvic girdle are omitted, and the lower part of the dissection stretched laterally.

further dissection is necessary. The ovaries may have to be cleared of fat. Examine the ovaries for *Graafian follicles* which appear as bulges on the surface. The uteri may contain embryos, if they do open the uterus and examine the embryos and their attachment by the placentae. Expose the rectum and separate the urethra and the vagina. Display the system and draw (Figure 44).

Examine prepared slides of the kidneys, ovary and testes. Do not draw them yet.

Abdominal blood vessels

In the male move the bladder, rectum, glands, etc. towards you and pin down. *In the female* cut the utero-ovarian vein on one side, and deflect the ovary towards you to expose the femoral artery and vein. Treat the bladder and rectum as in the male. Draw (see Figure 45).

The position of the blood vessels is variable, and there may be some fat to be removed.

Thorax

Cut across the rib-cage just in front of the diaphragm, and then along both sides through all the ribs except the first. Make sure you do not damage any of the nerves and blood vessels especially near the arm-pits. Lift the thoracic wall over the animal's head and draw the organs *in situ* (Figure 46). Remove the wall of the thorax.

Remove the *thymus gland.* It is better to lift off the pieces, rather than cut, as you will be working very close to the blood vessels and nerves. Lift and cut off the pericardium over the heart. This will expose the heart. Now follow the veins and arteries forward and backward from the heart, having deflected the heart over to the right. Be

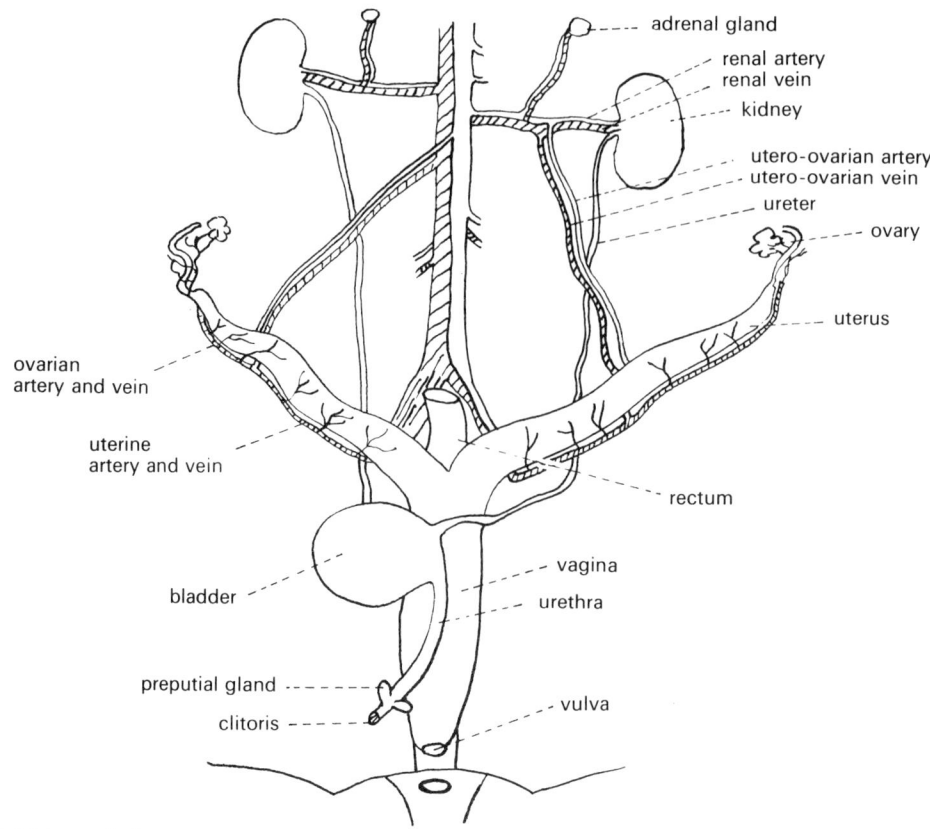

Fig. 44. Urinogenital system of the female Rat. The cut edges of the pelvic girdle are omitted.

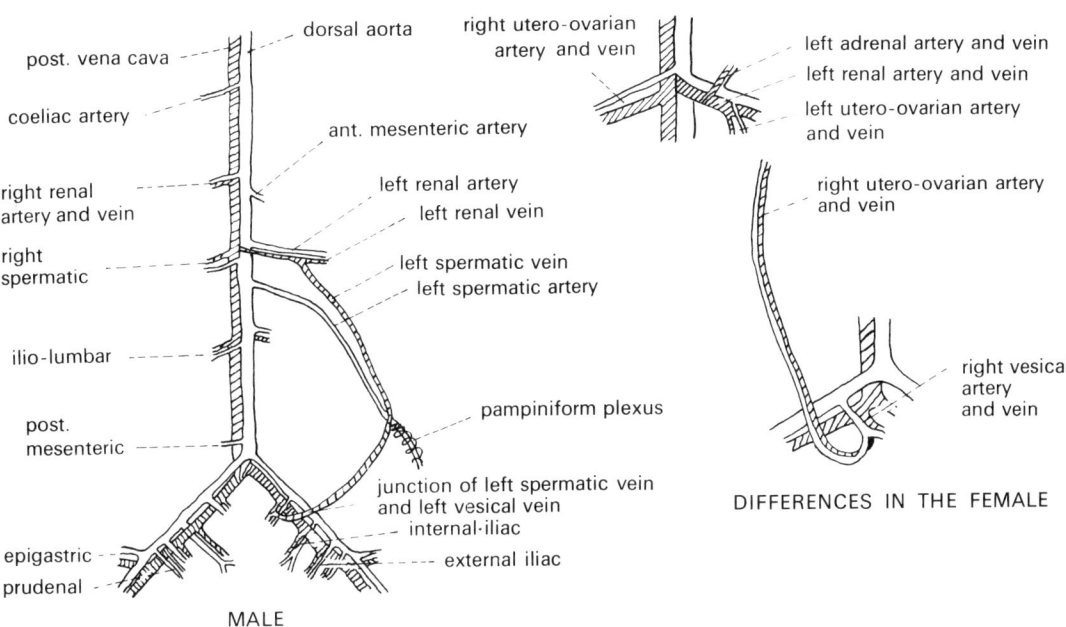

Fig. 45. Blood vessels in the abdomen of the Rat.

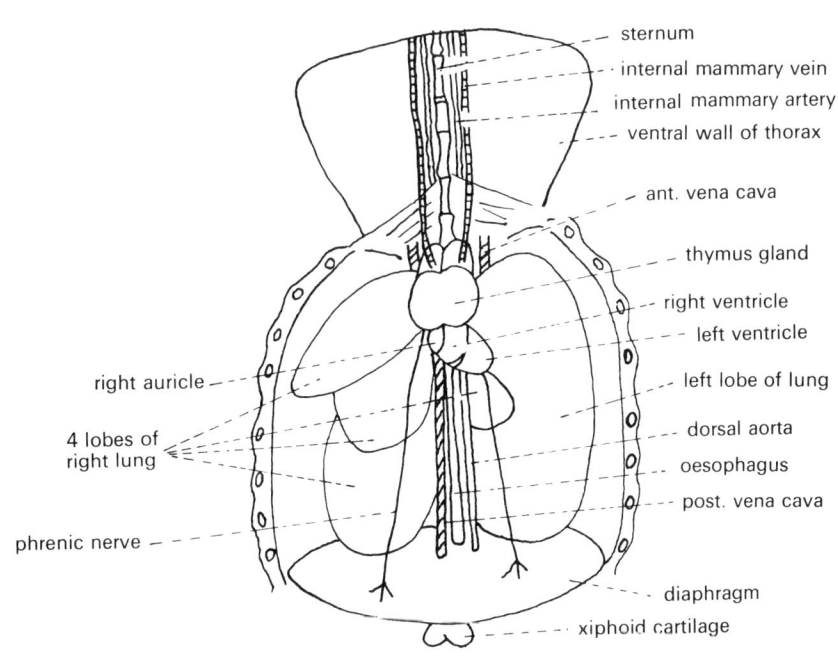

Fig. 46. Thorax of the Rat with ventral wall lifted forward.

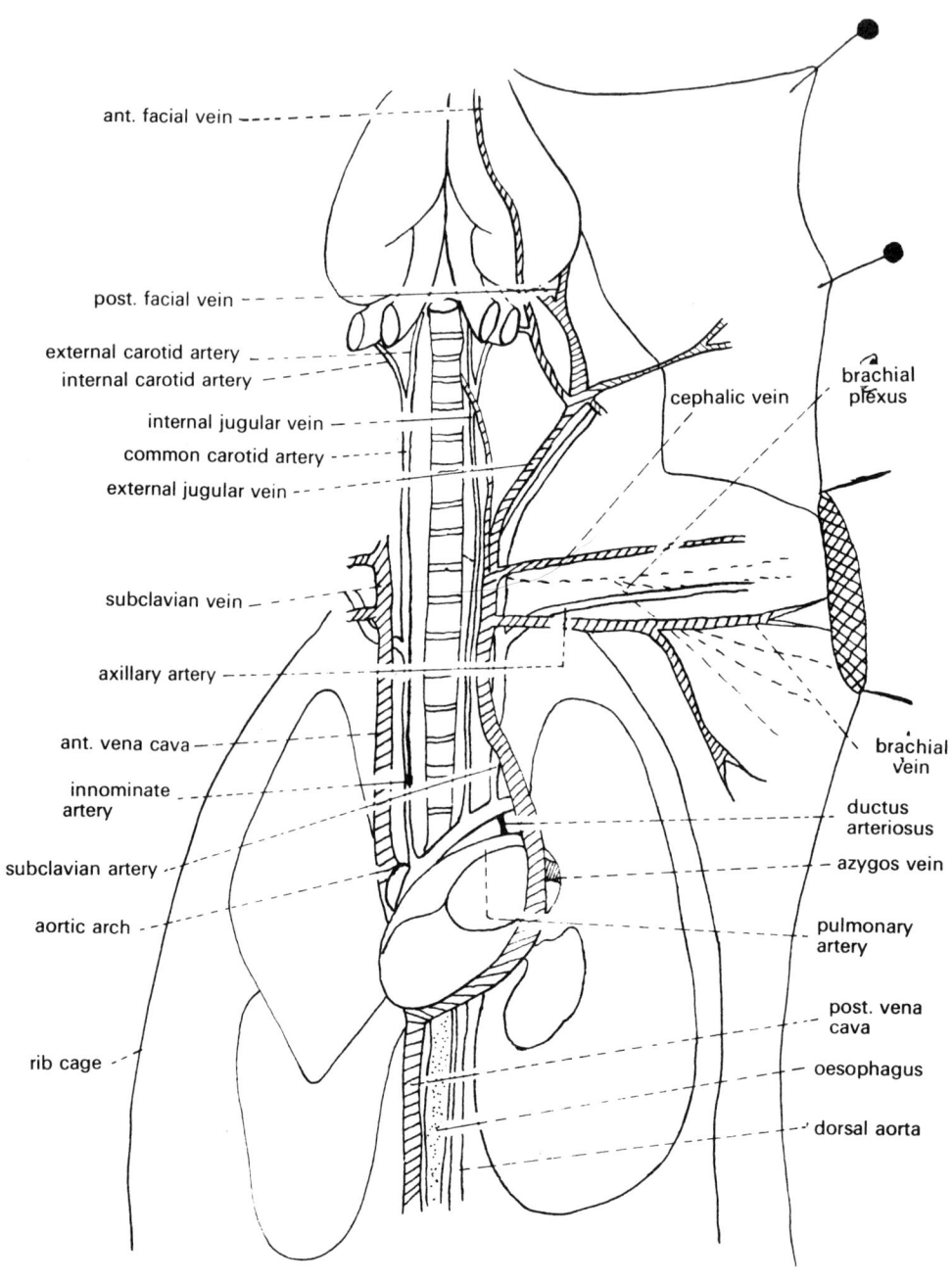

Fig. 47. The blood vessels of the thorax of the Rat. Nerves are shown as dotted lines. Detail is shown on the left side only.

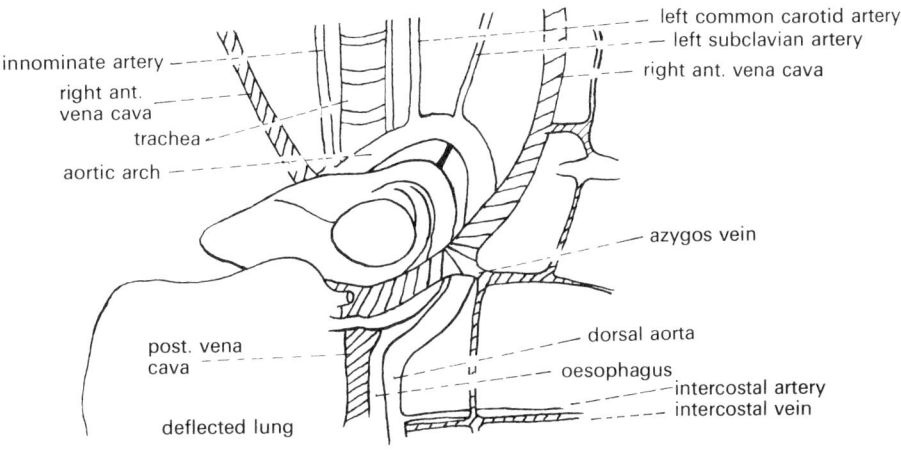

innominate artery

right ant. vena cava

trachea

aortic arch

left common carotid artery

left subclavian artery

right ant. vena cava

azygos vein

post. vena cava

deflected lung

dorsal aorta

oesophagus

intercostal artery

intercostal vein

Fig. 48. The blood vessels of the Rat around the heart, with left lung deflected, to show vessels running to the dorsal wall of the thorax.

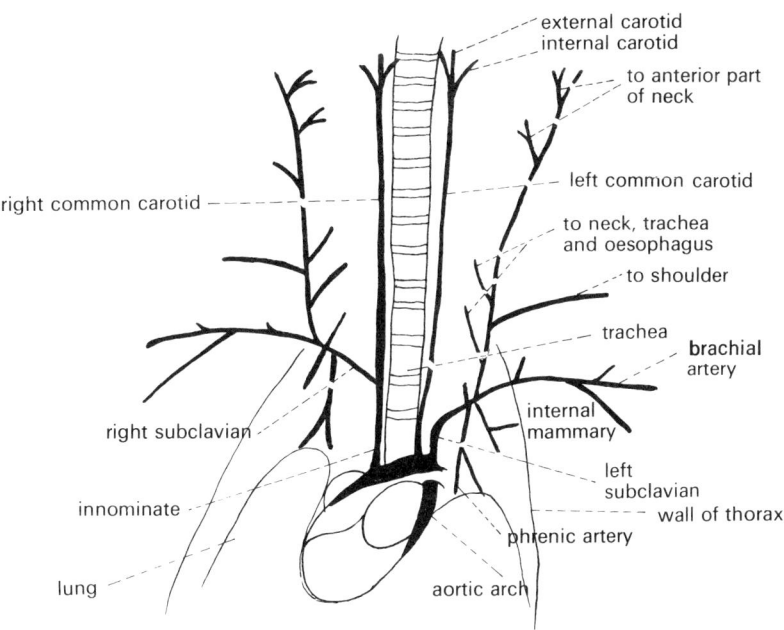

external carotid

internal carotid

to anterior part of neck

left common carotid

right common carotid

to neck, trachea and oesophagus

to shoulder

trachea

brachial artery

right subclavian

internal mammary

innominate

left subclavian

wall of thorax

phrenic artery

lung

aortic arch

Fig. 49. Arteries of thorax of the Rat.

careful not to cut the *ductus arteriosus* and the loop of the *Xth cranial nerve* (left recurrent laryngeal nerve) passing under it. There are nerves running close to the *trachea* and *right common carotid artery*. This is a dissection demanding great care—do not try and do it too quickly. Draw (Figure 47).

Lift the left lung and heart over to the right. This exposes the vessels on the dorsal inner surface of the thorax. Draw (see Figure 48).

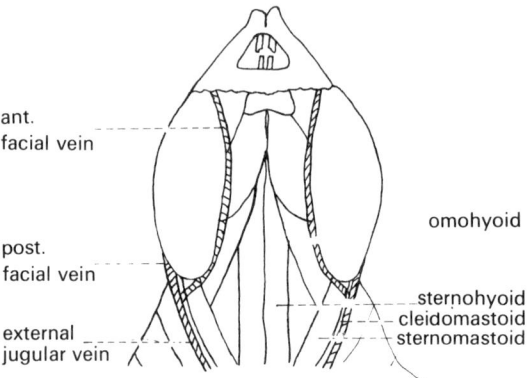

Fig. 50. Simplified diagram of the neck muscles of the Rat. (Only those to be removed have been named).

Replace the heart and lungs and remove the veins to expose the arteries which lie near them. Draw the complete arterial system (Figure 49).

The neck

A new specimen is necessary to carry out the dissection satisfactorily, as some of the blood vessels and nerves will have been damaged in the dissection of the thorax. Skin the neck and thorax. Stretch the neck by placing a piece of wood under it, and pinning the head through the cheeks. Remove the ventral wall of the thorax as before. Clean away lymph nodes, fat and salivary glands to expose the muscles. Running centrally from under the chin to the *manubrium* (top of the sternum) are the *sternohyoid muscles,* cut through them near the manubrium and near the hyoid. This will expose the *trachea, thyroid gland* and *larynx.* Remove the *sternomastoid, omohyoid,* and *cleidomastoid muscles.* (See Figure 50).

Clean up the various tissues to show the veins arteries and nerves. In this dissection, concentrate on exposing the nerves, and remove any organs that obscure them

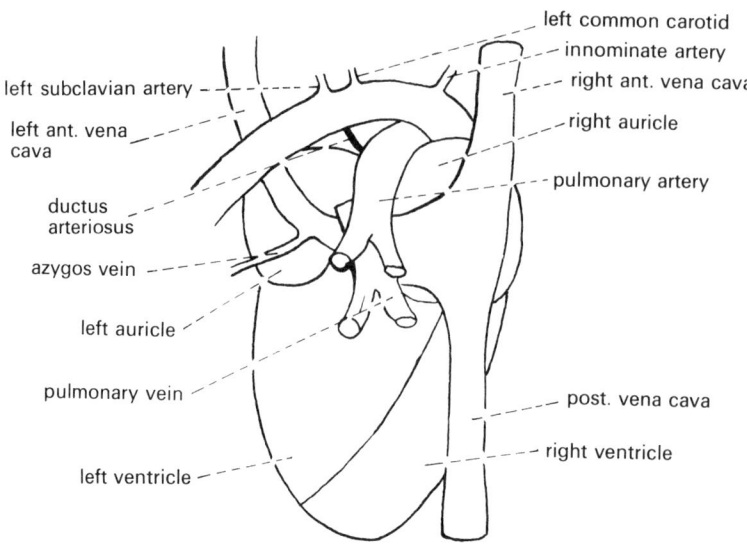

Fig. 52. Heart with attached blood vessels.

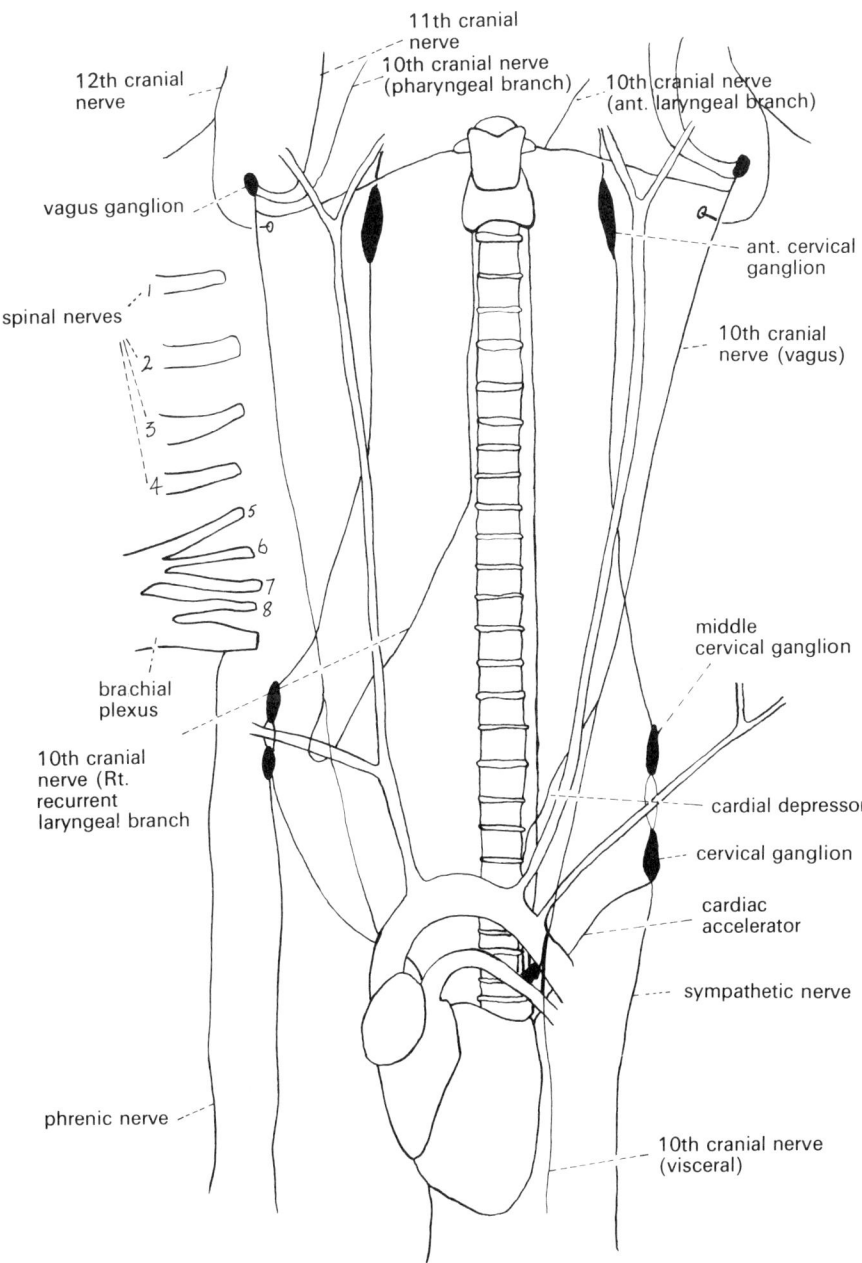

Fig. 51. The nerves of the neck of the Rat. (For clarity, the nerves are drawn separated from associated structures and only nerves are labelled.)

(Figure 51). The nerves are very close to the blood vessels in places, and may be shown more clearly by removing the veins.

The heart

Remove the heart by cutting through the main blood vessels, and draw it from the dorsal surface (Figure 52).

Use a sheep's heart for the dissection because a rat's heart is too small to see the details clearly (Figure 53).

Pin the heart in a dish of water, ventral surface upper-most. Remove the walls of the auricles and wash out any blood.

In the *right auricle* notice the opening of the *venae cavae,* the *septum auricularum*—a thin division between the two auricles and the *auriculo-ventricular opening.*

In the *left auricle* is the opening of the *pulmonary vein* and the opening of the *left auriculor-ventricular aperture.*

Remove the wall of the ventricle, and notice the relative sizes and thicknesses of the walls of both auricles and ventricles.

In the right ventricle you will see the *semilunar valves* at the base of the *pulmonary artery* (part of which has been removed), the *moderator band* running across it diagonally, and the *tricuspid valve* separating the right ventricle from the right auricle. It is held in place by the *chordae tendinea* attached to the ventricle wall by the *muscular papillae.*

The left ventricle contains the *biscuspid (mitral) valve,* separating it from the left auricle and attached to the wall of the ventricle by chordae tendinae and muscular papillae. The base of the aorta, opening into it is closed by semi-lunar valves.

The brain

If your specimen is preserved you can proceed directly with the dissection of the brain, but if it is fresh, the tissues will have to be hardened, as described below. In either case it is better to soften the skull

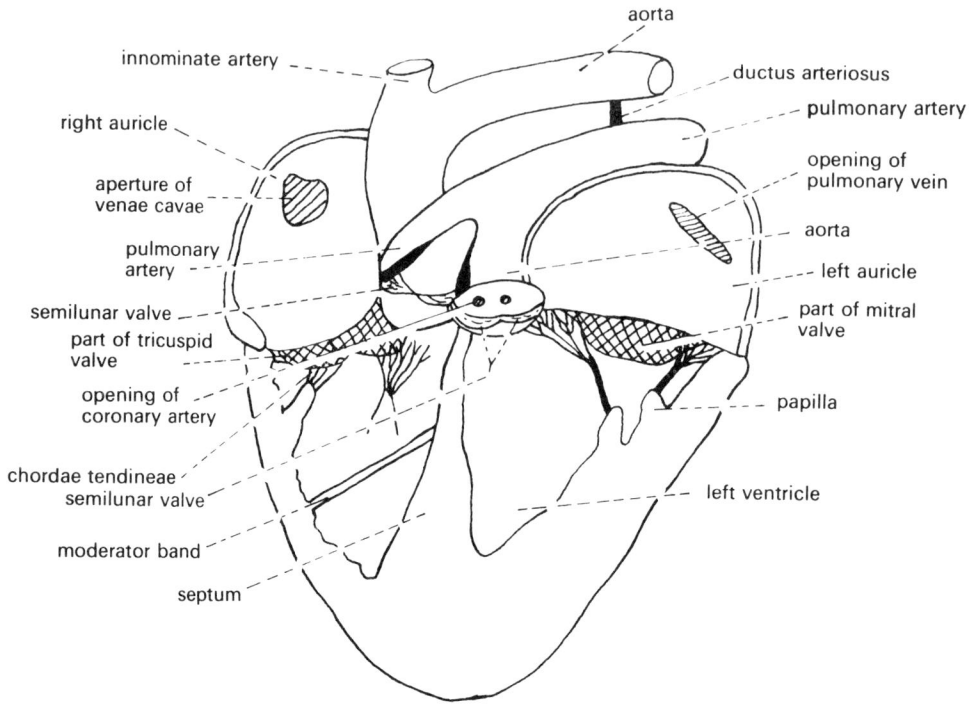

Fig. 53. Dissection of Sheep's heart.

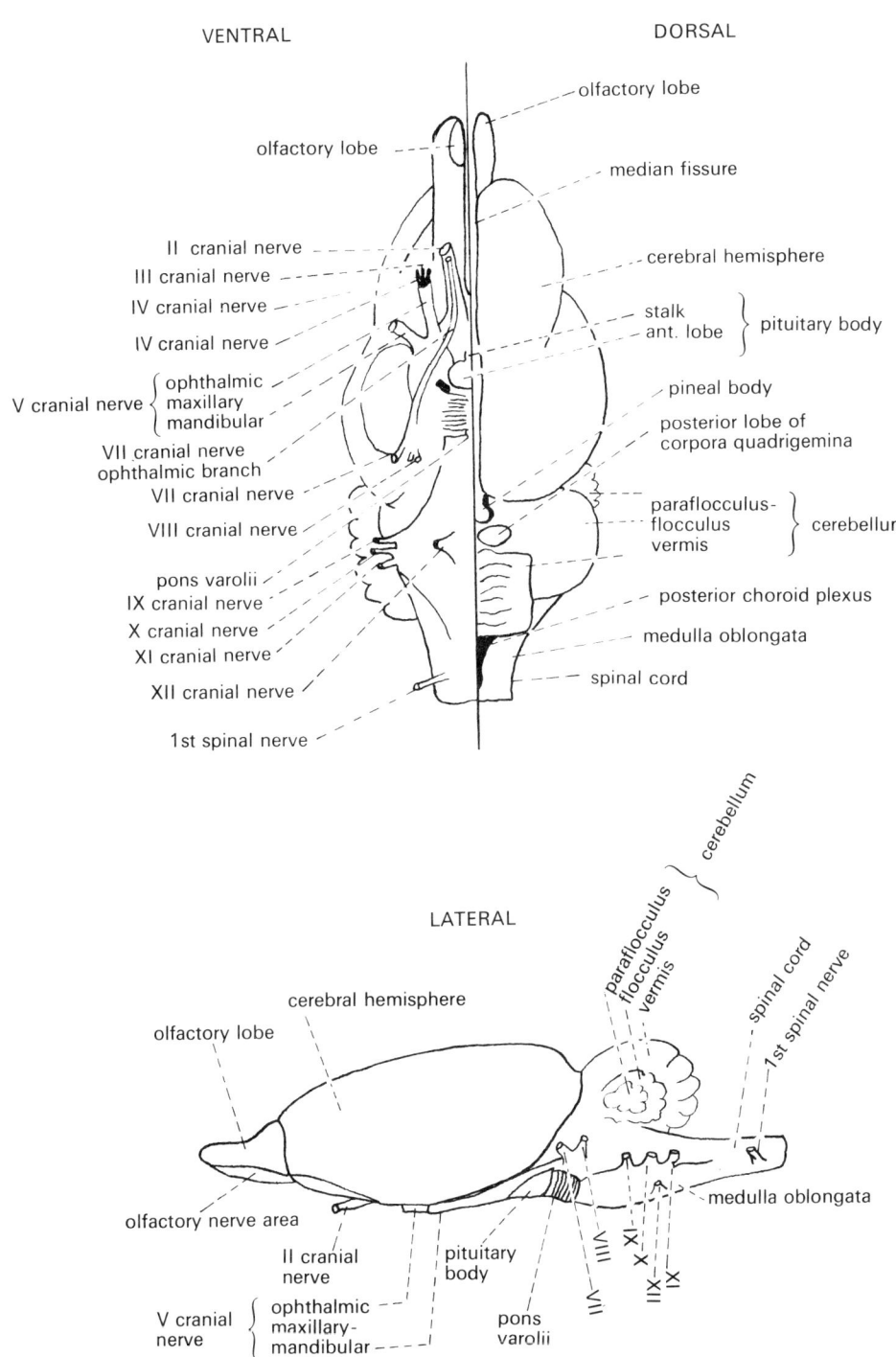

Fig. 54. Brain of the Rat.

first of all, but it is possible to remove the skull of a preserved specimen directly.

Cut the head off the carcass about 1 in. from the base of the skull. Clean it to expose the median longitudinal fissure. Near the base of the cranium make a small hole and insert the point of scissors. Cut forward along the fissure, and through the bone at the back of the skull. From this incision work to either side, chipping away the bone. Split the neural arches of the vertebrae and remove them, exposing the *medulla oblongata*. Work carefully to either side of the median incision until the brain is fully exposed. A great deal of patience is necessary so as not to do any damage. The brain tissue is very delicate. You may have difficulty around the eye orbit, and in freeing the lateral part

(paralocculi) of the *cerebellum*. When the brain appears free, gently lift the spinal cord, and, working forward, cut the cranial nerves as near the bone as you can. The brain should now be free and can be lifted out, or shaken from the skull. The *pia mater* may have to be removed. Draw as in figure 54.

To soften the skull, remove the head and skin it. Then place it in 10% hydrochloric acid for 3–4 days. Wash thoroughly before handling.

To harden the brain tissue, make a small hole in it and immerse it in 50% ethanol for 3–4 days.

Now cut the brain medianly just a little off centre and draw the median saggittal plane (Figure 55).

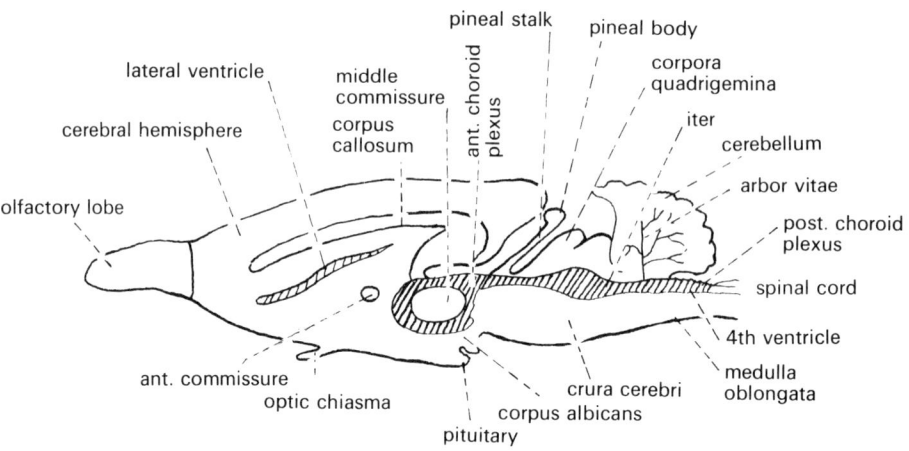

Fig. 55. Median sagittal section of the brain of the Rat.

7 Histology and Physiology of the Mammal

In this section any tissue to be examined microscopically will be a prepared slide, unless otherwise stated.

DIGESTIVE SYSTEM

Dentition

Examine the skulls of a typical carnivore (dog, cat), herbivore, (sheep), and rodent (rat, rabbit).

The carnivore has a complete dentition (all types of teeth are represented). Write out the dental formula of the example you have chosen. The relative sizes of the teeth are important when considering the diet, particularly the dominance of the *canines*, and, in the dog, the large *carnassials*. The teeth are spiked rather than ridged related to the need to grasp and crush the food rather than grind it. Examine the individual teeth. Notice the *closed* pulp cavity. Draw the skull from the side (Figure 56).

The *herbivore* has an incomplete dentition (some of the types of teeth are not represented). Give the dental formula. The canines are absent, leaving the *diastema*, and the sheep has no incisors in the upper jaw, the incisors of the lower jaw articulating against a horny pad in place of the incisors of the upper jaw. The ridges of the molars and premolars are low and sharp, the teeth of a jaw have the appearance of a file formed by the close proximity of the separate surfaces. This common surface fits closely into the corresponding surface of the other jaw forming an efficient grinding mechanism. Draw the skull from the side. Notice the *open* pulp cavity in the individual teeth.

The *rodent's* dentition is similar to the herbivore's insofar as the presence of the diastema and the "filing" molars and premolars are concerned. It differs in the pair of extremely large incisors in both jaws.

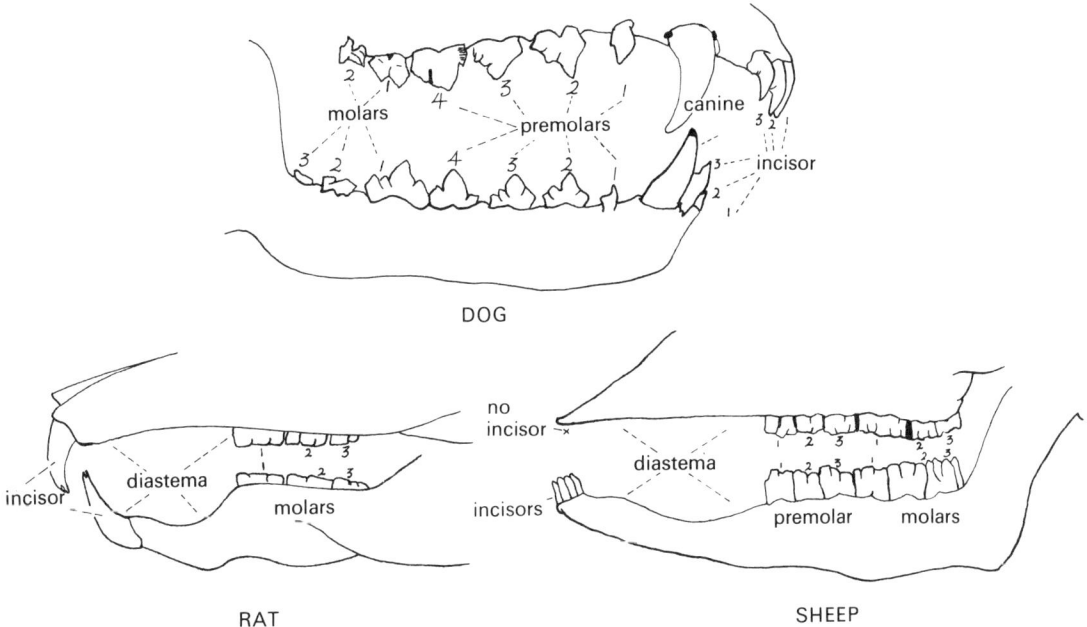

Fig. 56. Dentition of various animals.

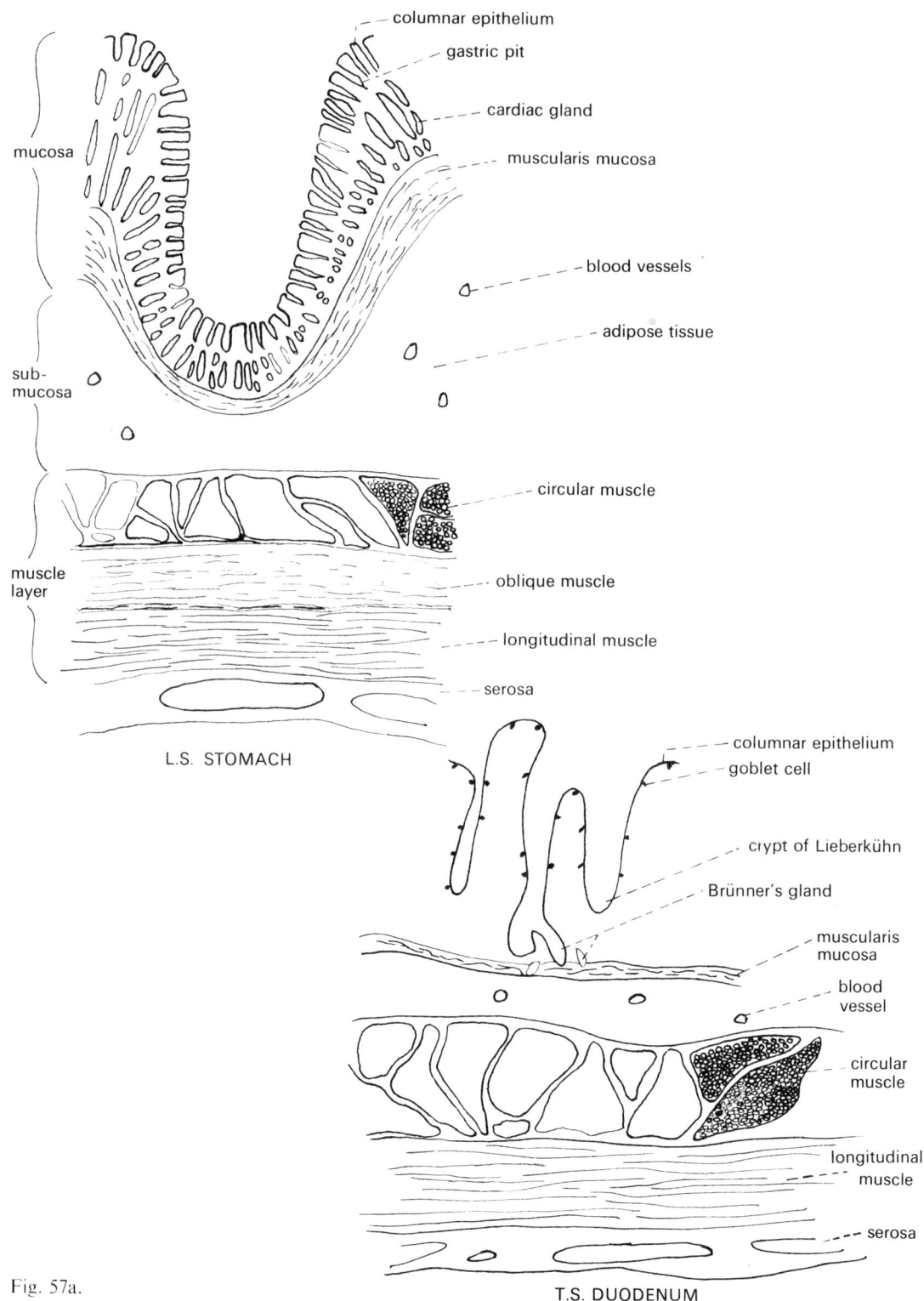

columnar epithelium
gastric pit
cardiac gland
muscularis mucosa

mucosa

blood vessels

adipose tissue

sub-
mucosa

circular muscle

muscle
layer

oblique muscle

longitudinal muscle

serosa

L.S. STOMACH

columnar epithelium
goblet cell

crypt of Lieberkühn

Brünner's gland

muscularis
mucosa

blood
vessel

circular
muscle

longitudinal
muscle

serosa

Fig. 57a.

T.S. DUODENUM

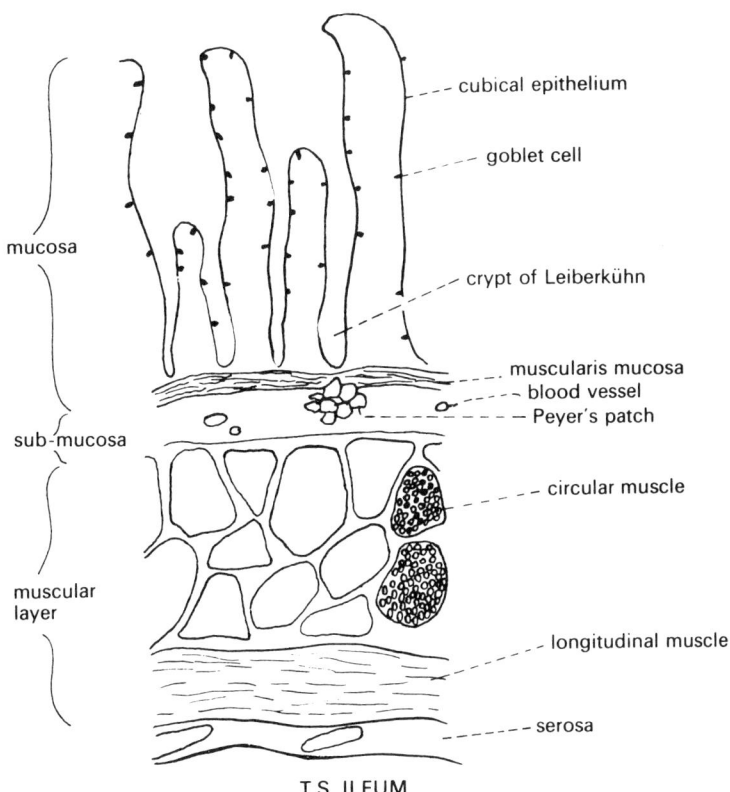

cubical epithelium

goblet cell

crypt of Leiberkühn

muscularis mucosa
blood vessel
Peyer's patch

circular muscle

longitudinal muscle

serosa

mucosa

sub-mucosa

muscular
layer

T.S. ILEUM

Fig. 57b.

These are used for gnawing—the typical method of feeding of rodents.

Examine the cut surface of a tooth in longitudinal section.

Histology

Draw plans, and portions from microscope slides viewed under high-power of transverse sections of the stomach, duodenum, ileum and colon. Remember that longitudinal muscle is cut transversely in transverse section, (i.e. appears end-on) but circular muscle is cut along its length and appears as fibres. The reverse is true in longitudinal section. Notice the relative thicknesses of the various tissues in these sections, and the presence or absence of the various glands. Make a table of the similarities and differences.

Enzymes

These have been dealt with previously.

SKELETON AND MUSCLES

Bone

Examine a transverse section of bone under high and low power. Notice the central *Haversian canal* and the concentric rings of *lacunae* with the fine canals running from them (Figure 58). Between the rings are further rings of dead bone-material, mainly calcium phosphate and calcium carbonate.

Cut a bone across transversely. Draw (see Figure 58). Notice the soft pulp and the hard material around the periphery of the bone. This gives greatest strength, particularly resistance to bending, at the same time keeping the bone light, and conserving material.

Experiment: To examine the constituents of bone.
Apparatus: Bones, beaker, dilute nitric acid, burner, test-tubes, platinum wire, ammonium

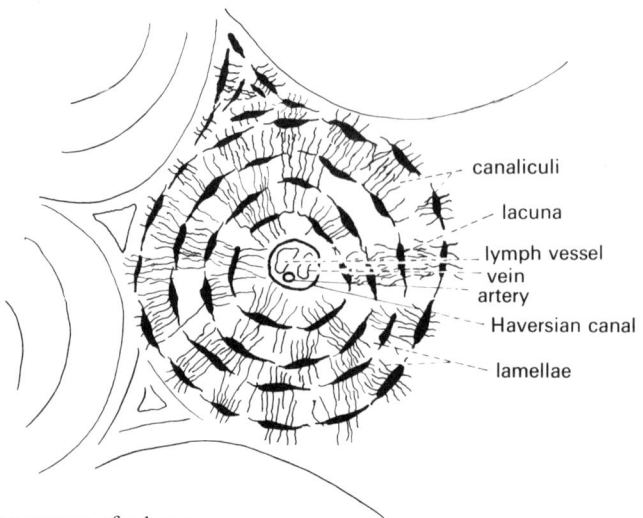

Fig. 58. An Haversian system of a bone.

molybdate solution, tripod, pipe-clay triangle, crucible, 50% hydrochloric acid solution.

Method: In a test-tube, dissolve a few small pieces of bone in the dilute nitric acid. Pour off the supernatent liquid, and test it for calcium and phosphorus.

Weigh a small bone, and place it in hydrochloric acid for a few days. This will dissolve the calcium salts. Warm gently to dry and reweigh. Calculate the weight of the calcium salts by difference. The bone left after acid-treatment is largely collagen. If this is put in a crucible, and heated strongly, it will burn completely away.

The collagen of the bone will form gelatine. This can be shown by boiling bones for about 4 hours in a minimum amount of water, decanting the liquid and allowing it to cool. It sets as a jelly due to the presence of gelatine.

Muscle

Experiment: To examine the effect of a stimulus on muscle contraction.

Apparatus: Frog, clamps and stands, cotton, pivoted lever with a fine wire to make a smoked surface, revolving drum with smoked surface, three $1\frac{1}{2}$-volt batteries (accumulators), ammeter, variable resistance (rheostat), key, fine copper wire, Ringer's solution.

Method: Kill a frog, and dissect out the gastrocnemius muscle ("calf" muscle) under Ringer's solutions, by cutting the Achilles'

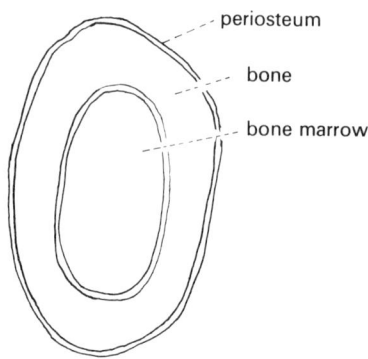

Fig. 59. Section through bone.

tendon as near to the ankle as possible, and the end of the bone to which the top of the muscle is attached. Set up the apparatus as shown in Figure 60. This must be done quickly, keeping the muscle moist. Connect up the electrical circuit with fine wire, fixing the ends of the wire firmly on to the muscle. Set the resistance at its highest value. (If the muscle contracts, use a higher resistance, or remove a battery from the circuit.) Gradually reduce the resistance until the muscle contracts. What is the ammeter reading? Notice the height of the peak drawn on the smoked drum. Remove the key and allow the muscle to rest for 1 minute. Repeat the experiment using different resistances. Does the strength of the current effect the degree of contraction?

Re-set the resistance at its maximum value for contraction. Press the key up-and-down quickly. Does the muscle contract and relax in the same way? What is the minimum time between successive shocks for the muscle to respond?

Finally keep the key depressed and observe what happens to the muscle. Can you account for your result? Copy and label your records from the smoked drum into your results.

Experiment: To demonstrate muscle fatigue.
Apparatus: Stop-watch.
Method: Work in pairs, one of the pair record, and the other do the exercise; then change over. Extend one arm out fully from the shoulder, and keeping it extended, move it upright and back to shoulder level as quickly as possible. Continue for 4 minutes. Rest for 1 minute, and repeat. Record the number of movements in each successive minute. Graph your results and account for them.

Draw slides of striped muscle, unstriped muscle, hyaline cartilage (e.g. from the trachea), white fibro-cartilage, (e.g. from intervertebral disks), yellow elastic cartilage (e.g. from the ear), areolar tissue (e.g. from under the skin as in Figure 61).

The differences between the various types of muscle and connective tissues can be seen in the diagram. Under your drawings tabulate the differences between the two types of muscle, and the connective tissues.

CIRCULATORY SYSTEM

Blood vessels and capillaries

Examine prepared slides of the aorta, artery and vein. Notice the relative thickness of the walls of the blood vessels (Figure 62).

Tie a handkerchief firmly around the arm just above the elbow. The surface veins stand out due to the stopping of the flow of blood back up through the arm. Lumps appear in the veins, showing the position of the valves. If the arm is stroked upwards (away from the hand) the lumps disappear as the blood is forced through the valves, but they become exaggerated if the arm is stroked in the opposite direction. Tighten the handkerchief. This stops the flow of blood through the arteries which are deeper than the veins, and the arm becomes white and numb. Do not leave the handkerchief tied too long.

Experiment: To examine the capillary circulation, and the effect of various substances on the capillaries.
Apparatus: Live frog or small fish, board on

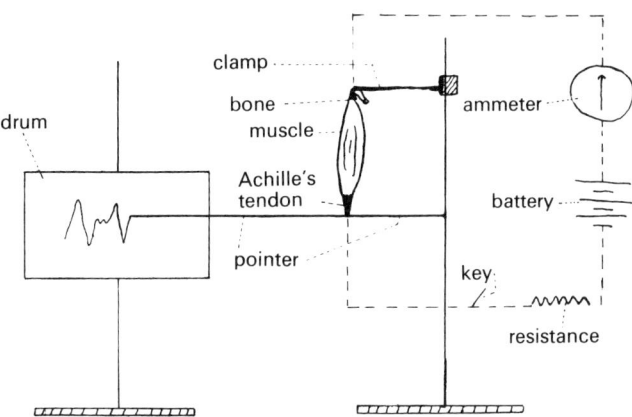

Fig. 60. Apparatus to demonstrate muscular contraction.

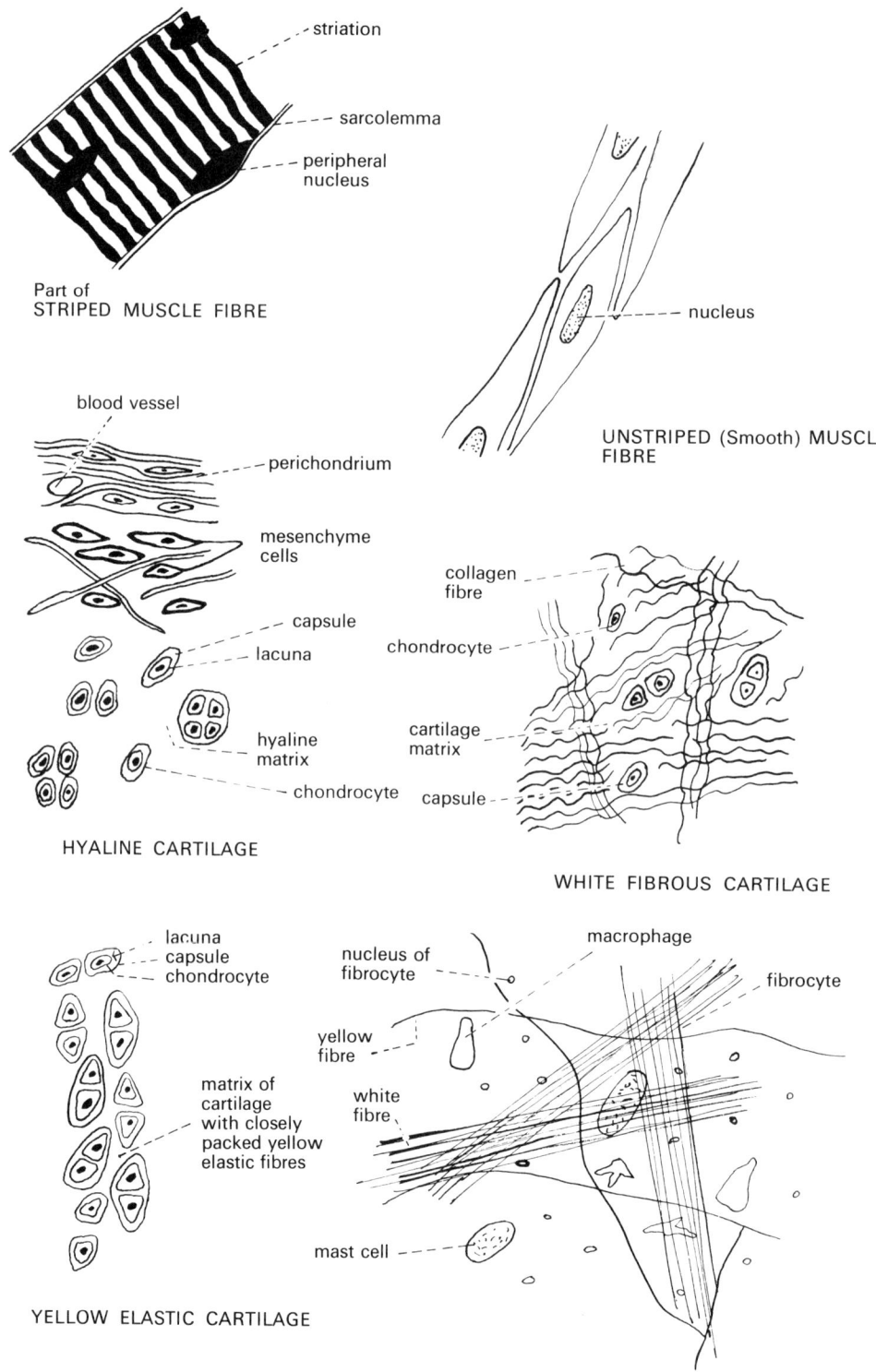

striation

sarcolemma

peripheral
nucleus

Part of
STRIPED MUSCLE FIBRE

nucleus

UNSTRIPED (Smooth) MUSCLE
FIBRE

blood vessel

perichondrium

mesenchyme
cells

capsule

lacuna

hyaline
matrix

chondrocyte

HYALINE CARTILAGE

collagen
fibre

chondrocyte

cartilage
matrix

capsule

WHITE FIBROUS CARTILAGE

lacuna
capsule
chondrocyte

matrix of
cartilage
with closely
packed yellow
elastic fibres

YELLOW ELASTIC CARTILAGE

macrophage

nucleus of
fibrocyte

fibrocyte

yellow
fibre

white
fibre

mast cell

AREOLAR TISSUE

Fig. 61.

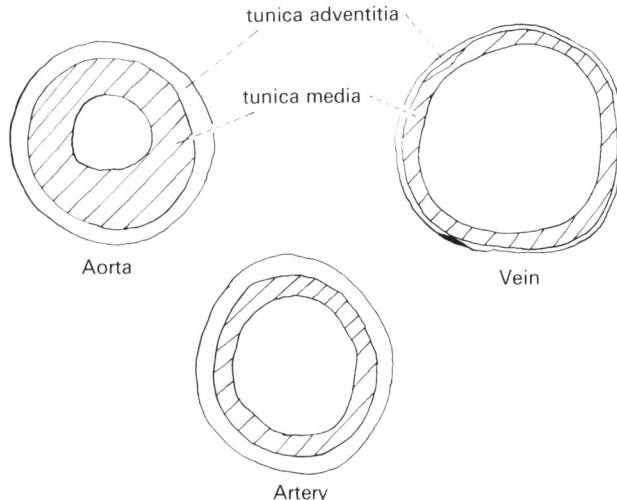

Fig. 62. Diagram to show the relative thickness of the walls of blood vessels.

which to pin the animal, with a hole about $\frac{3}{8}$ in. diam. near one side, pins, damp cotton-wool, cotton, microscope, pipette, 0·0001% solutions of sodium nitrite, adrenaline chloride, histamine acid phosphate, lactic acid, 0·00005% acetylcholine, 95% ethanol

Method: Cover the body of the animal with damp cotton-wool and tie it *firmly* to the board. Stretch the web of the foot (or tail) over the hole in the board, and pin it near to the bones. This will not hurt the animal as there are very few nerve-endings in this part of the body, but it is essential that the animal is firmly anchored so that it cannot wriggle and rip the flesh held by the pins. When the experiment is finished, release the pins through the foot (or tail) first. Examine the web under the microscope and the capillaries will be seen clearly. Compare the size of the capillaries with the diameter of the red blood-cells, and estimate the rate of the blood flow.

Add a few drops of one of the solutions to the web and observe the effect on the rate of flow of the blood, and on the diameter of the capillaries. Wash the web well and allow it to recover before adding the next solution. Note the various effects.

Blood

Examine a prepared slide of human blood, or make a blood smear in the following way: Tie a piece of string tightly around your finger just above the first joint, compressing the blood up to the nail. Sterilize the surface of the finger by wiping it with 70% ethanol. With a *sharp, sterilized* needle puncture the skin between the tournequet and the nail. A few drops of blood will exude. Let them drop on to a clean slide and smear the drop by firmly pulling the edge of another slide over it. Wave the slide in the air to dry it quickly.

Under the microscope notice the predominance of the red blood-corpuscles *(erythrocytes)* (see Figure 63). In surface view, the centre will appear light due to focussing on a concave surface. The various white blood-cells will show best in a stained preparation, but it is doubtful if you will identify the different types with a student microscope. The platelets appear as small deeply-stained grains.

Experiment: To examine the properties of haemoglobin.

Apparatus: Blood (obtained from an abbatoir), large beaker, test-tubes, oxygen source, carbon dioxide source, glass-tubing, rubber tubing, 10% sodium citrate solution, 10% potassium hydroxide solution, burner, spectrometer, 20 volume hydrogen peroxide solution, guaiacum tincture.

Method: Pour some blood into a test-tube and add about $\frac{1}{10}$th its volume of the sodium citrate solution to prevent clotting. Bubble oxygen through it. What happens to the colour of the blood? Account for any change in colour. Now pass carbon dioxide through the same tube, note any further change in colour, and account for it.

Dilute some blood to about 1 part in 50 with water. Add a little potassium hydroxide solution and warm *(not boil)*. Notice the change in colour from red to brown. This is caused by the breakdown of the haemoglobin into its component haematin and globulin.

The change from haemoglobin to oxyhaemoglobin can be seen by examining blood spectroscopically. Dilute some blood. This has to be done by trial and error until the absorption bands of the spectrum are visible. Using white light, observe the absorption bands. Shake the blood to oxygenate it, and note the change in the absorption spectrum. This change indicates a change in the original composition of the haemoglobin.

Half fill a test-tube with water and add a few drops of blood (not enough to colour it strongly). Now add a little guaiacum tincture and hydrogen peroxide solution. Note any colour change and account for it.

Clotting of blood

Experiment: To examine the clotting properties of blood and the blood constituents.
Apparatus: Blood, beaker, test-tubes, crucible, tripod, pipe-clay triangle, 10% sodium citrate solution, glass beads, Millon's reagent, glacial acetic acid, silver nitrate solution, dilute nitric acid, ammonium molybdate solution, Fehling's solution, dilute hydrochloric acid, potassium ferrocyanide solution, slides, microscope.
Method: Leave some fresh blood to stand in a beaker (or large test-tube) for some time, and observe the various changes that take place. Finally a clot is formed, leaving a supernatent yellowish serum. To a similar blood sample add $\frac{1}{10}$th its volume of 10% sodium citrate solution. What happens? Suggest a practical application for this procedure.

Place a drop of blood on a slide, but do not spread it. Leave it still and observe it under the microscope. Notice the strands of *fibrin* forming, the platelets becoming aggregated together, and the red corpuscles clumping to form *rouleaux*.

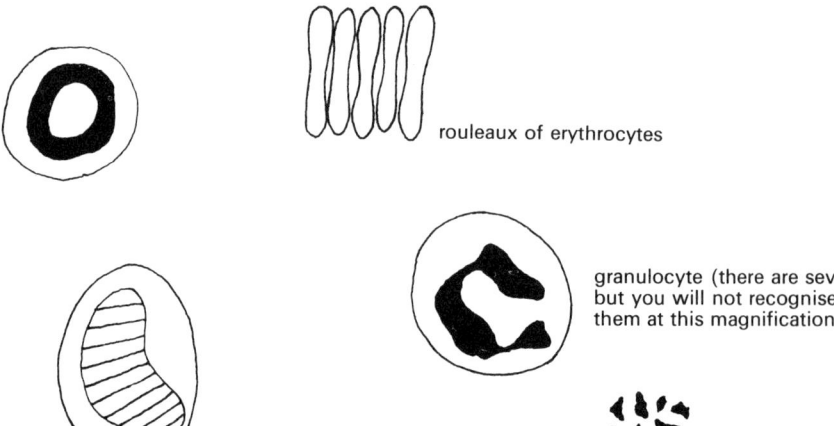

rouleaux of erythrocytes

granulocyte (there are several types but you will not recognise them at this magnification.)

lymphocyte

platelets

Fig. 63. Various blood cells.

Place some blood in a beaker and add sufficient glass beads to cover the bottom to a depth of about $\frac{1}{2}$ in. Shake vigorously. This collects the fibrin on the beads, and the residual blood does not clot on standing. Collect some of the fibrin and test it for protein.

Dilute some of the defibrinated blood 1:6 with water and boil. Add an equal volume of water and 3 drops of glacial acetic acid. Boil again. A precipitate forms which should be allowed to settle. Pour off the supernatent liquid. Heat the precipitate in the crucible, cool, and dissolve the residue in dilute hydrochloric acid. Test for iron.

Evaporate the supernatent liquid from the blood to about a quarter its bulk. Divide it into four portions and test for reducing sugars, sucrose, chloride, and phosphate.

Heart beat

Kill a frog, and quickly dissect it by opening the body cavity from the ventral side to expose the heart. Remove the tissues from around the heart, and observe the rhythmic contractions of the auricles and ventricle. The frog's heart has two auricles and one ventricle.

Experiment: To show the effect of various factors on heartbeat.
Apparatus: Live *Daphnia*, microscope, slide, stop-watch, cotton-wool, pipettes, 0·5% ethanol, tea, coffee, mineral water (contains dissolved carbon dioxide), D-amphetamine sulphate solution, chlorpromazine solution, beaker, burner, thermometer.
Method: Daphnia (the 'water flea') is a small crustacean, which is transparent. Consequently its heart can be seen clearly and its pulse-rate measured by timing with a stop-watch.

Place a few strands of cotton-wool on a slide and place a drop of water containing *Daphnia* on it. The cotton-wool fibres inhibit the movement of the *Daphnia*, making it easier to see.

Measure the pulse rate over three successive minutes, and average the results. To different

specimens, add a drop of the various solutions, and measure the pulse rate again. Which substances are stimulants, and which depressants?

Warm some water and take drops from it at different temperatures. Place a few *Daphnia* in the drop as before and measure the rate of heart-beat. By the time you come to observe the *Daphnia* the water will not be at the original temperature, but the experiment will give some idea of the effect of temperature on heart-beat.

Experiment: To examine the effect of exercise on pulse-rate.
Apparatus: Stop-watch.
Method: Work in pairs. Take each others pulse-rate when sitting relaxed. Make this measurement over five successive minutes. Then take some vigorous exercise. (Run 100 yd. as hard as you can!) Measure the pulse-rate again over successive minutes. How long does it take to return to normal? Graph your results.

Remember to take the pulse with the first three fingers over the artery that runs through the wrist above the thumb.

SKIN

Examine and draw prepared slides of the skin. Notice in particular the structure of the *hair* and the *follicle*, the structure and position of the *sebaceous glands* and the *sweat* glands (see Figure 64). The epidermis is a good example of a *stratified epithelium* in section. You will have to move the slide about a lot before seeing good examples of all these structures.

Experiment: To show the excretory function of the skin.
Apparatus: Rubber finger-stall, burner, platinum wire, very dilute potassium permanganate solution, test-tubes.
Method: Wear the finger-stall for about an hour so that it accumulates sweat inside it. Remove the stall and test the sweat by the flame test. What colour is the flame? What

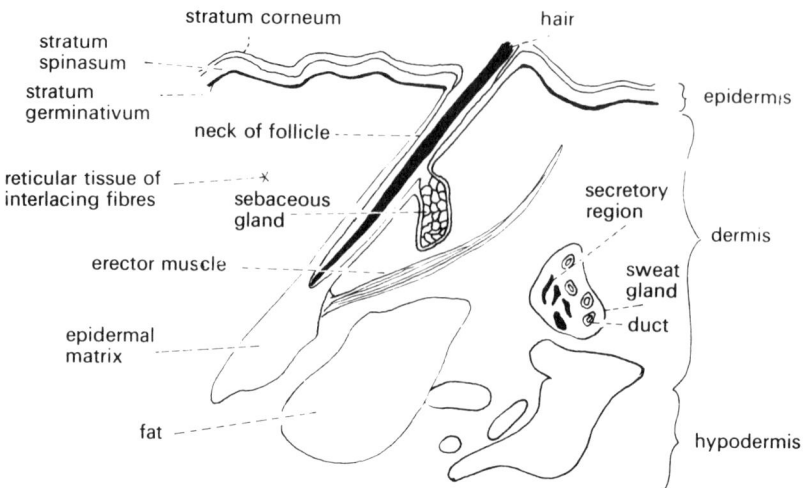

Fig. 64. Vertical section through the skin.

element is present? Wash out the finger-stall with the potassium permanganate solution and pour the liquid back into the test-tube, comparing the colour of the solution with that of the original. A browning of the solution indicates the presence of organic matter.

EXCRETION

The kidney

Examine and draw prepared slides of a longitudinal section through the kidney (Figure 65). This section is clearly recognised by the two distinct regions (*cortex* and *medulla*). The cortex is easily distinguished, even on superficial observation, by the *Bowman's capsules,* each containing a *glomerulus,* and the medulla by the ducts converging on the pyramid.

Experiment: To test urine for various chemicals.
Apparatus: Urine, test-tubes, burner, filter funnel, filter paper, Fehling's solution, concentrated hydrochloric acid, silver nitrate solution, barium chloride solution, ammonium hydroxide, dilute acetic acid, potassium oxalate solution, concentrated nitric acid, ammonium molybdate solution,

microscope, slides, evaporating basin, acetone, urea crystals.
Method: Place a few ml. of urine in a test-tube and test for reducing sugars with Fehling's solution.

Evaporate some urine nearly to dryness and allow it to crystallize. Dissolve the crystals in acetone and filter. Crystallize the acetone extract by putting a few drops on a slide and allowing the acetone to evaporate. Compare the structure of these crystals with the urea crystals and identify them.

Test a sample of urine for chloride and sulphate.

Heat some urine with ammonium hydroxide solution until a white precipitate is formed (calcium phosphate). Filter, and dissolve the residue in dilute acetic acid. Divide the solution into two and test one part for calcium by adding oxalic acid (forming a white precipitate of calcium oxalate), and the other for phosphate with nitric acid and ammonium molybdate solution.

The lungs

Examine and draw a prepared section through the lungs (see Figure 66). Under low power notice the extremely spongy nature of the organ, produced by the *alveoli.*

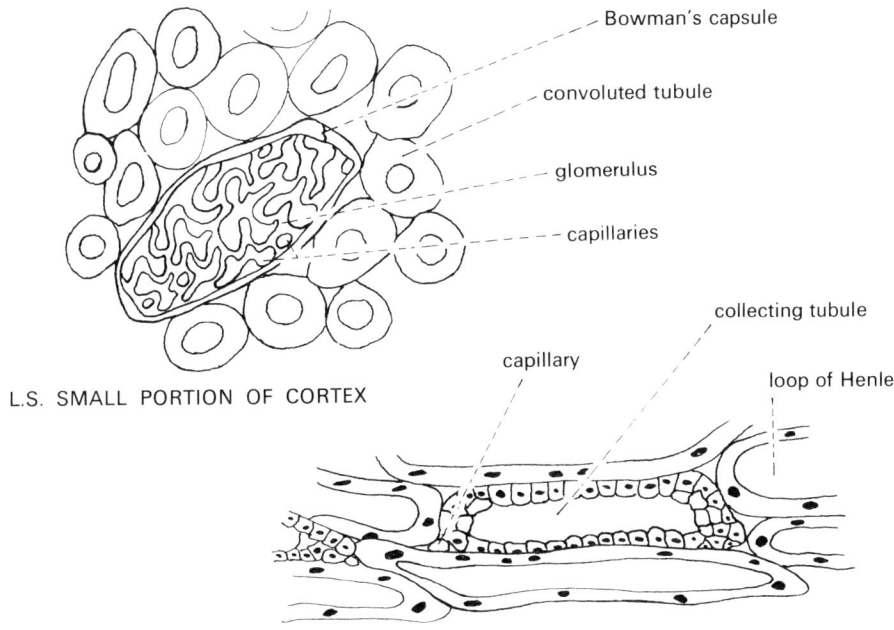

Fig. 65. Sections through kidney.

Experiment: To show that exhaled air contains carbon dioxide and water.
Apparatus: Test-tubes, glass tubing, watch-glass, lime water, anhydrous copper sulphate.

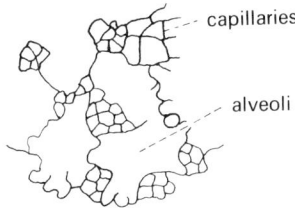

Fig. 66. Section through lung.

Method: Breathe out through the glass tube into a test-tube of lime water which goes milky, showing that the expelled breath contains carbon dioxide. Breathe out on to a dry watch-glass and notice the condensed vapour. Breathe out on to some anhydrous copper sulphate. This goes blue indicating the presence of water in the breath.

CO-ORDINATION AND CONTROL

(Hormones will be dealt with in the next section.)
The experimental studies in this section are not confined to the mammal, but it seems more reasonable to bring all the experiments on the subject under one heading, rather than have them dispersed as isolated pieces of information on animals that are not studied in so much detail.

Nervous system

Examine and draw prepared slides of a medullated nerve fibre, a transverse section through a nerve and a transverse section through the spinal cord (Figure 67). The medullated nerve will probably be stained with osmic acid, showing the medullary sheath in black (fat). Notice the *nodes of Ranvier* and the nuclei of the *Schwann cells*.

Unconditioned reflex

Experiment: To demonstrate spinal reflexes.
Apparatus: Live frog, dilute hydrochloric acid, dissecting instruments.

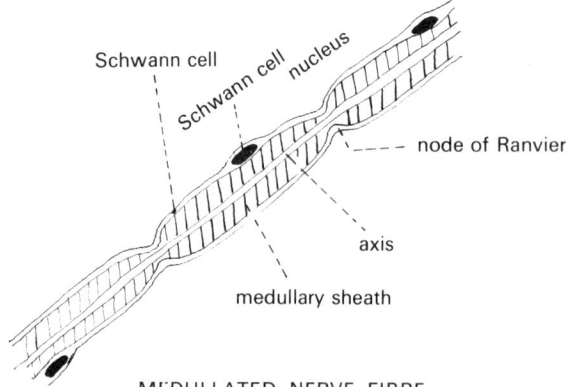

MEDULLATED NERVE FIBRE

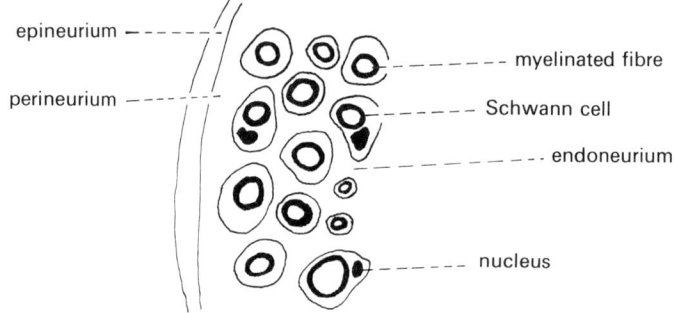

SMALL PIECE OF TRANSVERSE SECTION OF
A NERVE

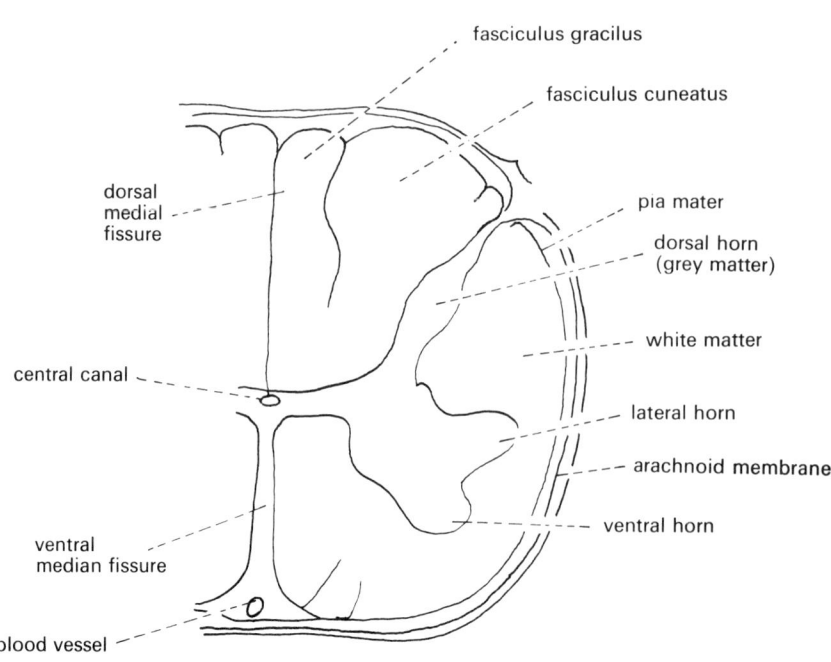

Fig. 67. Histology of nervous system.   T.S. SPINAL CORD

Method: Anaesthetize the frog, and cut the head off immediately behind the skull. This is best done with a large scissors. Allow the animal to recover from the anaesthetic and the shock (this will take about an hour) and squeeze the toes with a forceps. What reaction takes place? Now place a drop of dilute acid on the leg and observe the reaction.

Remember that any reaction cannot be controlled by the brain, and must therefore be a reflex.

Experiment: To demonstrate reflex actions.
Method: For these experiments work in pairs. One of the pair sit with the legs crossed with one knee over the other. Tap the cross leg just below the knee-cap with the edge of the hand. What happens? Can this reaction be consciously stopped? You will have to give a sharp tap, and may not hit the right spot first time.

One of the pair should now close his eyes for a few minutes. When he opens them his partner should observe the movements of the iris. Repeat the experiment, but open one eye first. When the iris has become adapted to the light, open the other eye and observe both eyes. Can these reactions be controlled?

The eye

Dissection of the eye
Use a fresh bull's or sheep's eye.

Draw the eye from the side in such a way as to show the muscles and the optic nerve. Now pin it down with the pupil upwards.

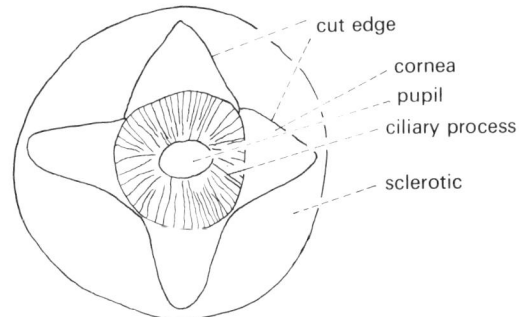

Fig. 68. Surface of eye with cornea deflected.

This can be done by pushing the pins through the muscles or the fatty tissue. Make an incision in the centre of the *cornea* and from this point make four radial cuts to the edge of the cornea (Figure 68). Notice the consistency of the *aqueous humour.* Fold back the four flaps to expose the *iris* and the *ciliary process.*

Now make similar incisions in the iris to expose the *lens,* held in position by the *suspensory ligaments* (Figure 69).

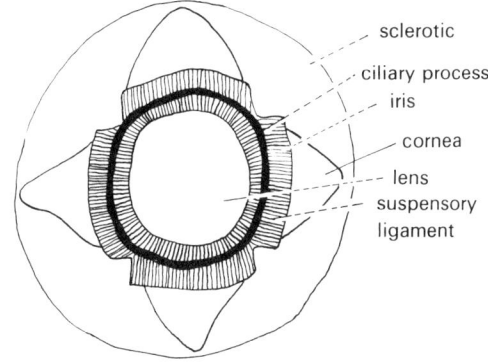

Fig. 69. Surface of eye with cornea and iris deflected.

The lens can now be squeezed out, and with it the *vitreous humour.* Notice the consistency of the latter. Continue the incisions through the rest of the eye-ball, and as far as possible, flatten it out to expose the *retina* and *blind spot.* Notice the colour of the retina (what substance gives it this colour?), the *blood vessels,* and the thickness of the *choroid* and *sclerotic* at the cut edge of the eye-ball. The following are a few simple experiments which do not require complicated apparatus.

Hold a pencil in each hand about 2 ft. apart. Close one eye and bring the points of the pencils together slowly so that they touch. Repeat the experiment using both eyes. Which way is easier? This shows the importance of binocular vision.

Look at a small bright light in a dark room then turn it out. After the light is out you will see a very bright image of the lamp for a very short time (less than a second)—the positive after image, then an intensely black area—the negative after image.

Look at a brightly coloured red shape for about a minute, then look at a piece of white paper immediately afterwards. You will see a green shape. This phenomenon is called successive colour contrast.

Close your eyes and firmly press on the lids. Do you see anything? Any image produced cannot be caused by light, but by pressure. So that the nerves of the eyes are capable of transmitting 'sensations' other than light, to the brain, but these are translated into light images by the brain.

On a sheet of paper draw a cross (to the left) and a dot (to the right) about $2\frac{1}{2}$ in. apart, and $\frac{3}{8}$ in. high. Close the left eye and focus the right eye on the cross. Gradually move the paper towards or away from the eye. Notice the point at which the spot becomes invisible. At this point the spot is focussed on the blind spot of the eye.

The inversion of the image by the eye is shown by the following experiment.

Make a pin-hole in a piece of card and hold it about 1 in. from one eye. Now bring the head of the pin between the hole in the card and the eye, and observe that the image of the pin-head is inverted. The pin-head is so close to the eye that the lens forms an erect image of the pin's shadow on the retina, and, as the brain interprets the image produced by the retina as inverted, the shadow appears inverted.

The ear

Examine prepared slides (or photographs) of sections through the ear, and compare them with the text-book diagrams. Notice the overall and relative sizes of the various parts.

Experiment: To show that sound waves can be conducted through the head bones.
Apparatus: Watch, cotton-wool.
Method: Plug the ears with cotton-wool so that the watch cannot be heard. Then place the watch on various parts of the head, and hold it between the teeth. Can it be heard now? The sound waves are conducted through the bones of the skull and not through the air to the ear-drum.

Taste and smell

Experiment: To distinguish between taste and smell.
Apparatus: Any strong and mild tasting substances, e.g. onion, apple, potato, carrot, oil of cloves (diluted!), oil of peppermint, radish, distilled water.
Method: Work in pairs. Blindfold your partner and close his nose with a clip. Then give him the substances in succession to identify. He must wash out his mouth after each test. Repeat the experiment still blindfold, but with the clip removed. How do the two sets of results compare?

Experiment: To identify the areas of the tongue sensitive to sweet, sour, salt, and bitter.
Apparatus: Solution of sucrose (sweet), vinegar (sour), sodium chloride (salt) and quinine sulphate (bitter), glass rods.
Method: Work in pairs, and blindfold your partner. With the glass rods place the solutions, one at a time, at random on various parts of the tongue. Note the position on the tongue, the solution used, and whether it was tasted or not. If sufficient trials are carried out, a pattern of the sensitive areas of the tongue will emerge.

Experiment: To estimate the sensitivity of the tongue to various substances.
Apparatus: Sucrose, acetic acid (not vinegar, as the concentration cannot be known accurately), sodium chloride, quinine sulphate in the following dilutions: 1%, 0.1%, 0.05%, 0.025%, 0.01%, 0.001%, 0.0001%, glass rods.
Method: Work as in the previous experiments. Treat each one of the series separately, and apply them to the tongue at random, recording whether they can be tasted. This will give the limits of sensitivity for each substance in each part of the tongue.

Sense-organs of the skin

Experiment: To map the various sensitive spots on the skin.

Apparatus: Blunt needle, fine needle, bristle, hot water, ice.

Method: Work in pairs. Blindfold one of the partners. Heat the blunt needle in the hot water, and touch various points on the skin surface selected, e.g. the back of the hand. Mark the points which are positive for heat, as distinct from pressure. Repeat for cold spots, using the blunt needle cooled in ice, for touch using a bristle, and pain, using a sharp needle.

Repeat the whole experiment on different parts of the body, e.g. finger tips, ear lobes, shoulder, palm of the hand.

In the following experiments, make sure that the conditions, other than those being tested are constant throughout the experiment.

Phototaxis

N.B. (1) The intensity of the light varies inversely as the *square* of the distance of the light source.

(2) All these experiments should be carried out in a darkened room.

Experiment: To examine the reaction of fruit flies *(Drosophila melanogaster)* to light.

Apparatus: Fruit flies, boiling tube, cotton-wool, bench lamp, coloured screens, clamp and stand.

Method: Place about 12 fruit flies in the boiling tube and plug it with cotton-wool. Clamp the tube so that the light shines horizontally on the end of the tube. Allow the flies to settle down. Fix the light source 25 cm. from the end of the tube. Observe the reaction of the flies. Repeat, moving the light away from the tube, giving the flies time to react at each distance. Record your results. What is the minimum distance at which a reaction takes place? Repeat the experiment using different coloured lights.

Experiment: To examine the reaction of blowfly larvae *(Calliphora erythrocephala)* to light.

Apparatus: Blowfly larvae, 2 light sources, 2 pieces of cardboard,.

Method: Place the blowfly larvae at random on one piece of cardboard, and arrange the light source and the second card so that the first card is partially shaded. Observe the reaction of the larvae. Vary the light intensity by moving the light source. What happens to the larvae?

Arrange the two lamps so that they shine at right-angles across the card, but have one light farther from the card than the other, to give beams of different intensity. Place the larvae where the beams cross. In what direction do they move? Vary the distances of the two lights, and repeat the experiment. What conclusions do you draw?

Experiment: To examine the reaction of woodlice *(Oniscus* sp.*)* to light.

Apparatus: Woodlice, wide glass tube (2·5 cm. by 25–30 cm.), card, lamp, strip of damp filter paper, cotton-wool.

Method: Place the damp filter paper along the bottom of the tube which is lying horizontally. Arrange the card vertically at the centre of the tube, so as to shade one half. Illuminate one half of the tube, and place about 12 woodlice in the illuminated end. Stop both ends with cotton-wool. Observe the reaction of the woodlice, and the time taken for a reaction to take place. How long does it take for all the woodlice to congregate at one end of the tube? Vary the intensity of the light and repeat the experiment.

Geotaxis

Experiment: To examine the effect of gravity on woodlice and cabbage white butterfly larvae *(Pieris* sp.*)*.

Apparatus: Wide glass tube (2·5 cm. by 25–30 cm.) damp filter paper, woodlice, caterpillars, cotton-wool.

Method: Slope the tube at about 30° from

the horizontal, and introduce the woodlice at the higher end. Close the tube with cotton-wool. Observe the movement of the woodlice. Gently tip the tube so that it slopes in the opposite direction. What happens?

Repeat the experiment using the caterpillars.

Kinesis

Experiment: To demonstrate chemokinesis in *Paramecium.*

Apparatus: Paramecium culture, coverslips, slides, pipette, 0.05% acetic acid solution, 0.5% sodium chloride solution, carbonic acid, ('soda water' will do).

Method: Support one cover slip on the slide by coverslips at either edge. Place the culture of *Paramecium* under the central coverslip and allow the animals to become scattered at random. Using the pipette, introduce a drop of acetic acid under one side of the coverslip, and observe the reorientation of the paramecia.

Repeat the experiment using the salt solution and the carbonic acid.

Experiment: To demonstrate thermokinesis in *Paramecium.*

Apparatus: Paramecium culture, hand lens, two Petri dishes, coverslips and slides.

Method: Set up the apparatus as shown in the diagram, one Petri dish containing hot water, the other cold (Figure 70). Place a drop of water containing paramecia under the central coverslip and observe the final position adopted by the animals.

Experiment: To demonstrate hygrokinesis in woodlice.

Apparatus: Two small lipless beakers, 2

ground glass plates, 2 crucibles to fit into the beakers, metal gauze, concentrated sulphuric acid, woodlice.

Method: Set up the apparatus as shown in Figure 71. Introduce equal numbers of woodlice (about 10) into each beaker, and observe the rate of movement. What effect does humidity have on it?

The relative humidity above the sulphuric acid will be roughly 0%, and above the water 100%.

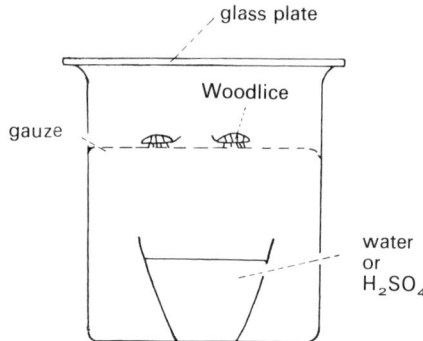

Fig. 71. Apparatus to demonstrate hygrokinesis in Woodlice.

There are a great many more experiments which demonstrate various taxes and kineses, using different animals. If you have time, devise some further experiments of your own.

GROWTH

Change in form and weight

Plot a graph of your age against weight. The figures will probably be available from the school authorities. What shape is the curve? Over what period did your weight increase most per unit time? To determine the increase in weight over a shorter time, weigh yourself weekly, and graph the result. In doing this

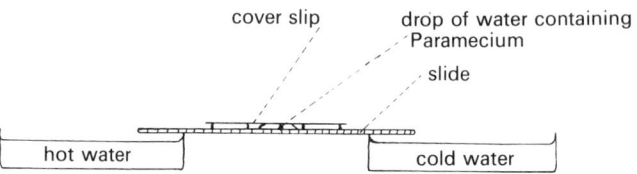

Fig. 70. The apparatus required to show thermokinesis by Paramecium.

make sure to use the same balance, and make the weighings at the same time of day, preferably before a meal. (The difference in weight before and after a meal is not an increase in weight!)

Look through some old photographs of yourself at different ages. Measure the length of your head, and the length from your shoulders to your feet in the photographs. Divide the head length by the other. How does this ratio vary from age to age?

Experiment: To examine the change in weight and form of a mouse.
Apparatus: Balance, young mouse (about 2 weeks old), ruler, cage etc. to keep mouse.
Method: Weigh the mouse at weekly intervals. This is conveniently done in a weighed box. At the same time measure the length of the tail (l_1) and the length of the trunk, from the base of the tail to the base of the neck (l_2). Tabulate your results. In your table include, $\log l_1$, $\log l_2$, and l_2^3. Calculate the *allometric coefficient* for the tail length and body length, i.e. $\log l_1/\log l_2$. If this is greater than 1 the tail is elongating more quickly than the body.

Graph the same pairs of figures, this should give a straight line. If the gradient is greater than 45°, the tail is elongating more quickly than the body, i.e. the tail length is showing a positive allometry.

Graph body weight against l_2^3. If weight and length increase regularly, the graph will be a straight line, so that any deviation from the straight line indicates a change in form of the animal.

The effect of food

Experiment: To examine the effect of feeding on the rate of development, and length of the gut of tadpoles.
Apparatus: Frog spawn, iodine in potassium iodide solution, four dishes for keeping frog spawn.
Method: Keep equal amounts of frog spawn in the four dishes—large beakers will do if suitably aerated with water-weed. To two of the aquaria add a few drops of iodine solution. Later, when the tadpoles are free-swimming and feeding, place small pieces of meat in one of the 'iodine' aquaria, and in one of the 'water only' aquaria. This means that two groups will be solely herbivores and two groups herbivores. Note the time taken for the various groups of tadpoles to reach the various stages of development. When the tadpoles are fairly large select a few from each group, kill and dissect them. Measure the length of the intestine.

What conclusions can be drawn as to the effect of diet on development and gut length?

Hormones
Experiment: To show the effect of thyroid extract on the development of tadpoles.
Apparatus: Frog spawn, 7 small aquaria, thyroid extract tablets.
Method: Make up a dilution series of thyroid extract, containing $\frac{1}{2}$, $\frac{1}{4}$, $\frac{1}{8}$, $\frac{1}{16}$, $\frac{1}{32}$, $\frac{1}{64}$, $\frac{1}{128}$ grains per litre. Set up 7 aquaria with water-weed, using these solutions. Put equal amounts of frog spawn in each aquarium and record the rates of development, the growth rate of limbs and shrinkage of the tail. Are any of these solutions too concentrated? What happens if they are?

Environment
Experiment: To show the effect of the environment on the development of tadpoles.
Apparatus: Frog spawn, 8 small aquaria.
Method: Set up 8 aquaria with water-weed and an equal amount of frog spawn. Arrange them so that they have the following treatments: LHB, LHW, LCB, LCW, DHB, DHW, DCB, DCW. Where L = good light, D = dark, H = warm place, C = cold place, W = white background, B = black background.

Keep a record of the various stages of the development of the tadpoles. What conclusions do you draw regarding the effect of light, temperature, and background colour? Notice that half the aquaria have any one treatment. Are any effects combinations of treatments?

REPRODUCTION

Examine and draw prepared slides of sections through the mammalian ovary and testis and spermatozoa. Under a student microscope you will see little of the structural detail of the spermatozoa. Compare your slide with photographs of a higher magnification (see Figure 72 and 73).

Regeneration

Experiment: To show regeneration in planarians.
Apparatus: Live planarians, slides, sharp scalpel, brush, Petri dishes.
Method: Label three dishes and partly fill them with water. Place a planarian on the slide using the brush. When it is fully

T.S. OF A SMALL PORTION OF THE OVARY

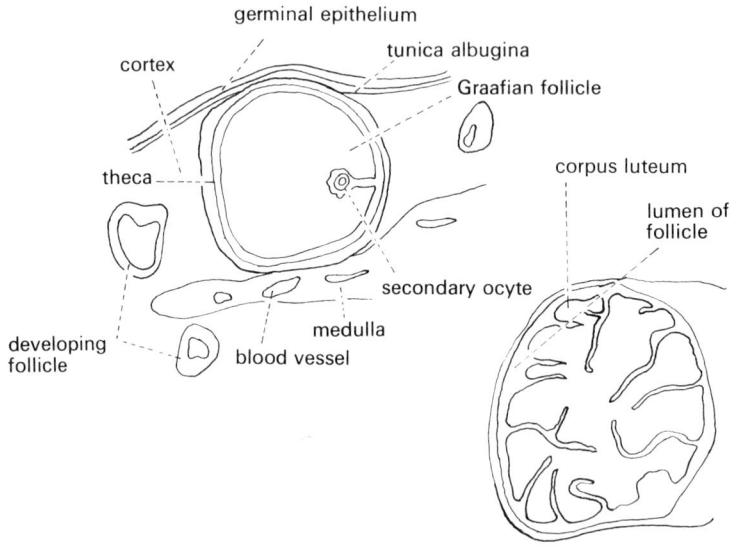

Fig. 72.

T.S. OF CORPUS LUTEUM

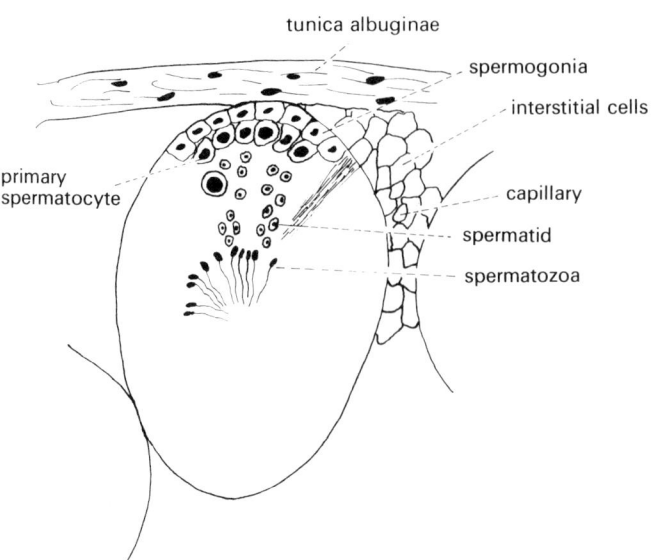

Fig. 73. Diagram of a transverse section through a seminiferous tubule.

extended cut off the head and posterior end. Place the pieces in separate dishes. Repeat until there are 6 pieces in each dish.

Repeat the experiment, cutting the planarians down the middle line in the longitudinal plane; cutting them in the same plane, but only from the head to the pharynx; and cutting them similarly from the posterior end to the pharynx.

To reduce the chance of failure, (1) change the water every three days, (2) remove any dead pieces and change the water immediately, (3) keep the culture cool in dim light, (4) do not attempt to feed.

Keep a record of observations daily.

8 Variety of Animals

PROTOZOA

Class: Rhizopoda

Amoeba

Examine a prepared slide of *Amoeba* under the microscope, and draw it (see Figure 74). This organism is in three dimensions, but has been pressed during the preparation of the slide. You will possibly have to focus at different levels to see all the structures.

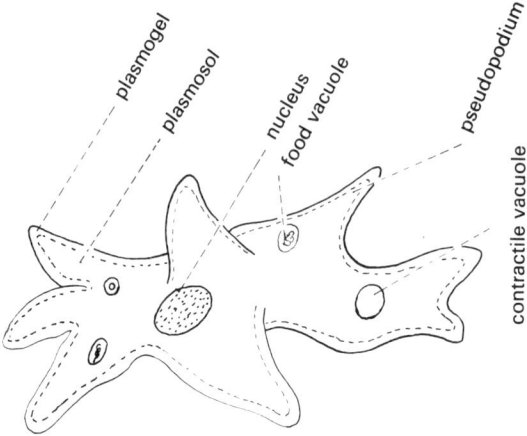

Fig. 74. Amoeba.

Now look at a culture of living Amoebae under low power. Notice the continual movement of the protoplasm. Select one animal and make a number of outline sketches to show the change in shape during the formation of *pseudopodia*. Similarly make a series of drawings at timed intervals to show the formation and disappearance of the *contractile vacuole*. Notice that this has a constant position.

You may be fortunate in seeing a specimen undergoing binary fission.

Irrigate the slide with 0·1% acetic acid, and observe the avoiding reaction of the Amoebae.

Monocystis

Examine prepared slides from the seminal vesicles of an earthworm. The parasite is usually stained a deeper purple than the surrounding sperm morulae.

Plasmodium

Examine prepared slides of blood from a person infected with malaria. This will contain the various stages of the parasite within the erythrocytes.

Class: Flagellata

Look at electron microscope photographs of sections through flagella and cilia. Notice their complex structure. You are dealing with unicellular organisms showing a high degree of complexity.

Euglena

Look at a culture of *Euglena* under low and high power of the microscope. Notice the method of movement. Allow the culture to dry out. As this occurs the movement of the flagella becomes clearer. To see this you will have to reduce the light intensity. With small organisms and thin sections, you do not always get the best definitions with the strongest light.

Examine prepared slides of *Euglena* and using this and the living material, draw the specimens (see Figure 75).

Experiment: To show phototaxis in *Euglena*.
Apparatus: Culture of *Euglena*, large test-tube or beaker.
Method: Cover the test-tube with paper, to leave only a small slit on one side for the light to enter. Use a colony of *Euglena* which is sufficiently numerous to cloud the water green. Place the tube in the light for a day. Remove the paper. What position has the *Euglena* taken up? Account for your result. Repeat the experiment using lights of different intensities.

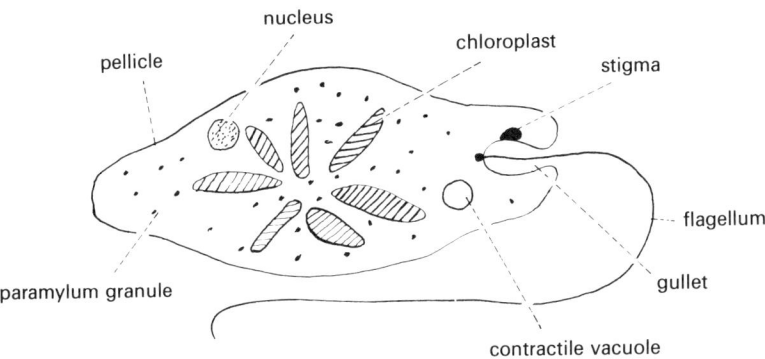

Fig. 75. Euglena viridis.

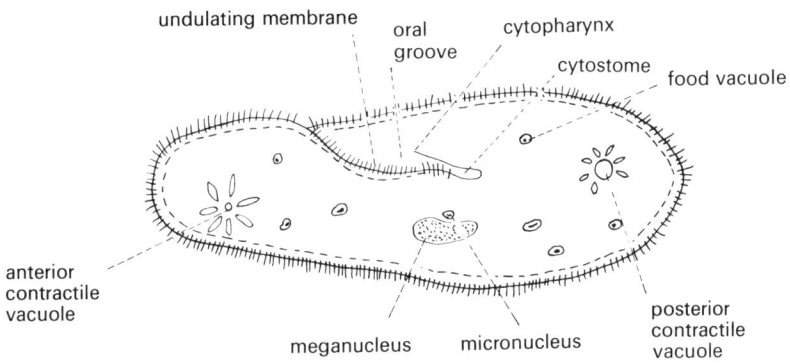

Fig. 76. Paramecium.

Class: Ciliata

Paramecium

Examine prepared slides of *Paramecium*, and draw them (Figure 76).

Make a culture of *Paramecium* in the following way: Boil some hay or dry grass in water to make an infusion. Leave it for a few days to allow bacteria to grow in it. Add some paramecia to this and leave them to multiply. The culture can be renewed by adding a few ml. of the old culture to some new hay infusion. A few ml. of pond water can be used instead of a pure culture of *Paramecium* in making the original culture. This will give a mixed culture of small pond organisms, which will usually include *Paramecium*.

Mount a drop of the culture on a slide and examine under the microscope. Notice the movement. The animals can be slowed down by placing the culture on a few strands of cotton-wool before covering it with a cover-slip, or by placing one drop of nickel sulphate solution on the slide, or a weak gelatine solution.

When cotton-wool fibres are added to the culture, notice the 'avoiding action' taken when the animals hit against the fibres.

Observe the formation of and the position of the contractile vacuoles. Look for stages of asexual reproduction.

Add a few drops of carmine solution to the slide, and observe the flow of the particles over the surface of the *Paramecium*. This should give an idea of the movement pattern.

Experiment: To observe the ingestion and digestion of food by *Paramecium*.
Apparatus: Microscope, coverslip and slide, pipette, 0.4% Congo Red solution, yeast, *Paramecium* culture, beakers, burner.
Method: Prepare a culture of yeast, and add

an equal volume of Congo Red solution. Boil this mixture for about 10 minutes, stirring constantly. This stains the yeast cells. Allow the mixture to cool. Add a little of this prepared material to a drop of the *Paramecium* culture on a slide. Within a few seconds the yeast cells are ingested, and will show up red in the food vacuoles. Follow the course of ingestion carefully with the subsequent formation of the food vacuoles. The Congo Red also acts as an indicator. Observe any changes in colour which will indicate changes in the pH of the vacuolar contents.

Look at a prepared slide of *Paramecium*, showing the various stages in conjugation.

Vorticella

Examine prepared slides of *Vorticella* and draw it (see Figure 77). Notice the basic similarity between its structure and that of *Paramecium*, namely, the two nuclei, cilia, contractile vacuoles, gullet, and mega- and micro-nuclei.

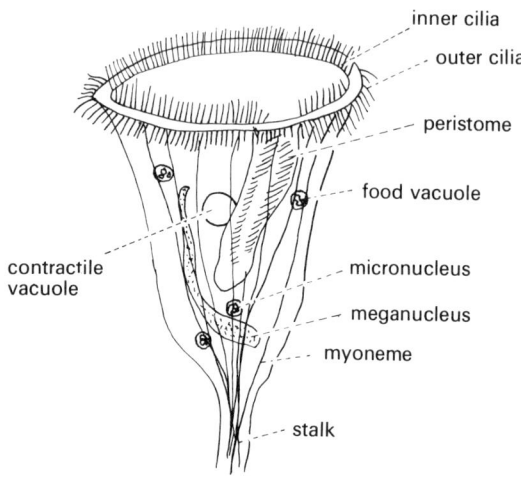

Fig. 77. Vorticella.

COELENTERATA

In examining members of this phylum, notice that the basic body pattern is similar in all classes. They are all radially symmetrical, with a ring of tentacles, and no anus. The body wall is diploblastic (two-layered). The division

into classes depends on the presence or absence and the dominance of the hydroid and medusoid stages in the life-cycle.

Class: Hydrozoa

Obelia

If available, examine living specimens of *Obelia*. It is found in the sea, usually attached to the underside of seaweed. The colonies are fine and branching, usually about one inch high.

Examine prepared slides of the colony and medusa (Fig. 78).

Hydra

There are three species of *Hydra* which may be found in fresh water, attached to stones on water-weed; *H. viridis* is green, due to the presence of the unicellular green alga, *Zoochlorella*, *H. fusca* is brown, and *H. vulgaris* is practically colourless.

Place a living *Hydra* in a small dish of fresh water, so that it is free to move, but can be examined with a hand lens. Notice the expansion and contraction of the body and tentacles. When the animal has settled, introduce some 'water fleas' (*Daphnia*) or similar small crustacean, and observe the *Hydra* capture them with the tentacles and ingest them.

Your specimen may have 'buds' or sex organs on it.

Examine and draw (see Figure 79) prepared slides of a section through *Hydra* and similar ones through the ovary, and testis.

Class: Scyphozoa

Draw a fresh, or preserved specimen of *Aurelia* (Figure 80) or other jelly-fish. Notice the basic similarity to the structure of the medusa of *Obelia*.

Class: Anthozoa

Draw a fresh or preserved specimen of *Actinia* (Figure 81), or other sea anemone. Notice the basic similarity to the structure of *Hydra*.

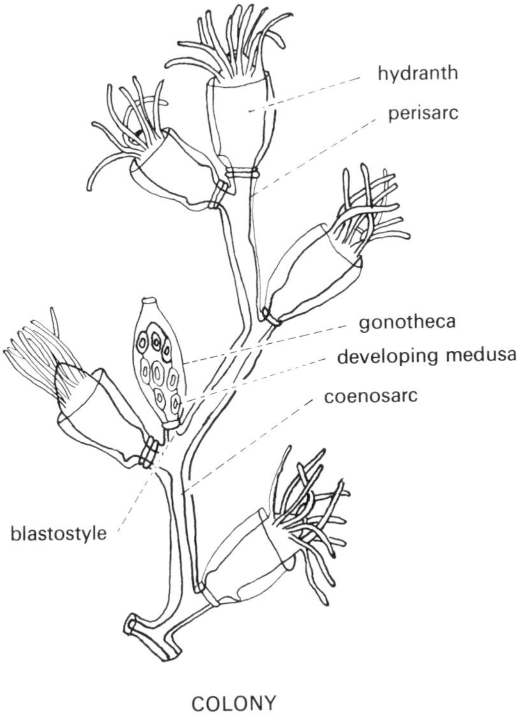

hydranth

perisarc

gonotheca

developing medusa

coenosarc

blastostyle

COLONY

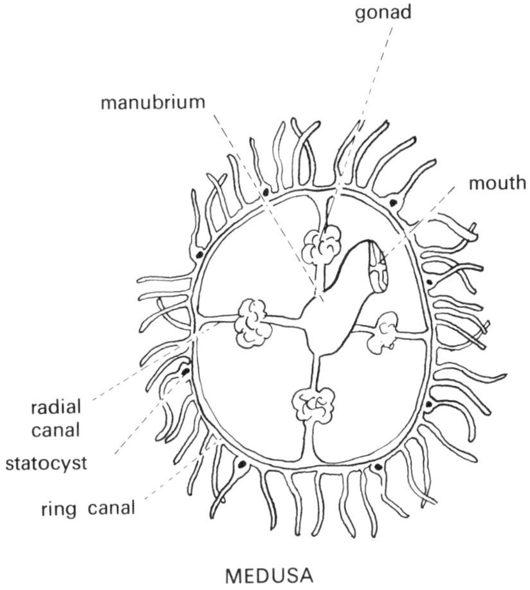

gonad

manubrium

mouth

radial canal

statocyst

ring canal

MEDUSA

Fig. 78. Obelia.

interstitial cell   nematocyst

musculo-epithelial cell

sensory cell

mesoglea

muscle tail of endoderm cell

endoderm cell

glandular cell

BODY WALL

Fig. 79. Sections through the wall of Hydra.

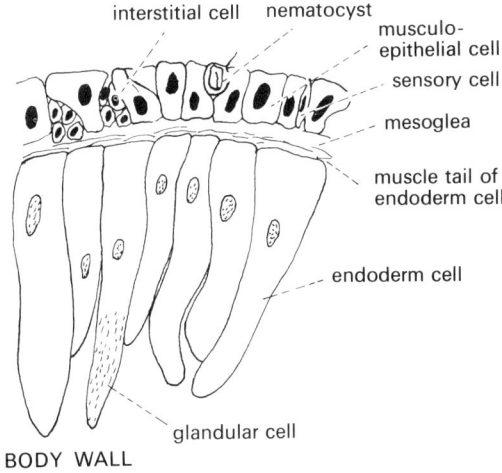

spermatogonia   spermatozoids

oocyte

T.S. TESTIS

T.S. OVARY

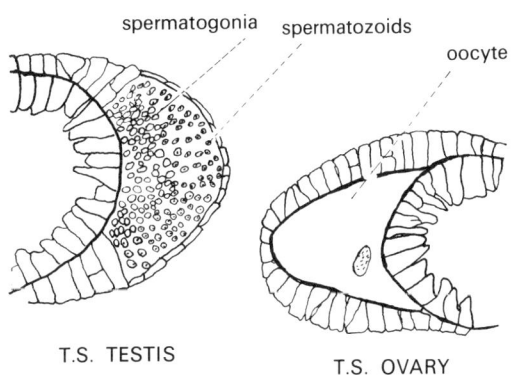

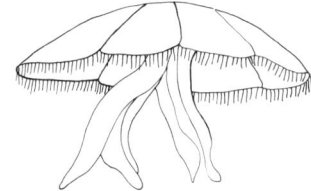

Fig. 80. Aurelia (Jellyfish).

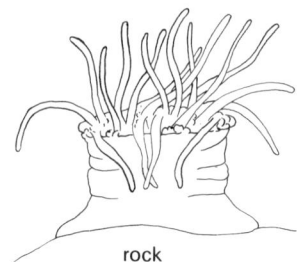

rock

Fig. 81. Actinia (Sea anemone).

Examine a piece of coral.

PLATYHELMINTHES

Class: Turbellaria

Planarians

These are free-living in fresh or sea water. They are relatively easy to find under stones or mixed with water weed. They can be trapped by placing a jam-jar with raw meat or a dead worm in the stream. They can be cultured in aquaria, but care must be taken to avoid fouling the water with the meat on which they are fed. The genera most likely to be caught in fresh water are *Planaria* or *Dendrocoelum*.

On the living specimen notice the tentacles, eye spot, and protrusible pharnyx (see Figure 82). Notice that there is definite

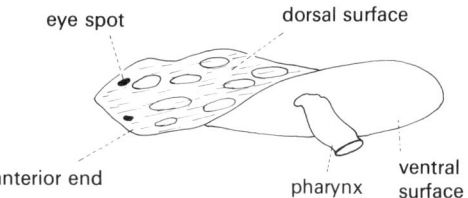

eye spot   dorsal surface

anterior end   ventral
              pharynx   surface

Fig. 82. Living planarian.

cephalization in this group, and that the dorsal and ventral surfaces are different in colour.

Once the animal is freely moving, touch it with a needle and notice the response. Now cut off the 'head' and allow the animal to recover from the shock. Touch the decapitated animal with a needle again. Does it respond? What indication does this give of the control of the body-movement by the ganglia and nerve net?

Examine a prepared slide of a whole mount of a planarian and a transverse section through the body (Figure 83). In the former, you may well have difficulty in recognising the different systems. The transverse section (Figure 84) shows the triploblastic, acoelomate characteristics of the phylum.

Regeneration in planarians has been discussed earlier.

Class: Trematoda

Liver fluke *(Fasciola hepatica)*

Examine prepared slides of a whole liverfluke, and transverse sections through the body (see Figures 85a and b). Notice the similarity of the basic body pattern between the liverfluke and the planarian.

Examine slide of the miracidium, sporocyst, and redia.

Class: Cestoda

Tapeworms *(Taenia* spp. or *Dipylidium caninum)*

Examine whole mounts of a tapeworm. Draw a representitive portion of it (see Figure 86). Examine mounts of the cysts in meat (beef or or pork).

Draw a mount of the scolex and of a proglottid, showing the structure (Figure 86). Examine a transverse section of a proglottid. Notice the basic similarity between this and similar sections of members of the other platyhelminthes.

Look at a slide of a flame cell (Figure 87). These are found terminating the branches of the excretory canal in all these platyhelminths.

INTESTINAL AND EXCRETORY SYSTEM

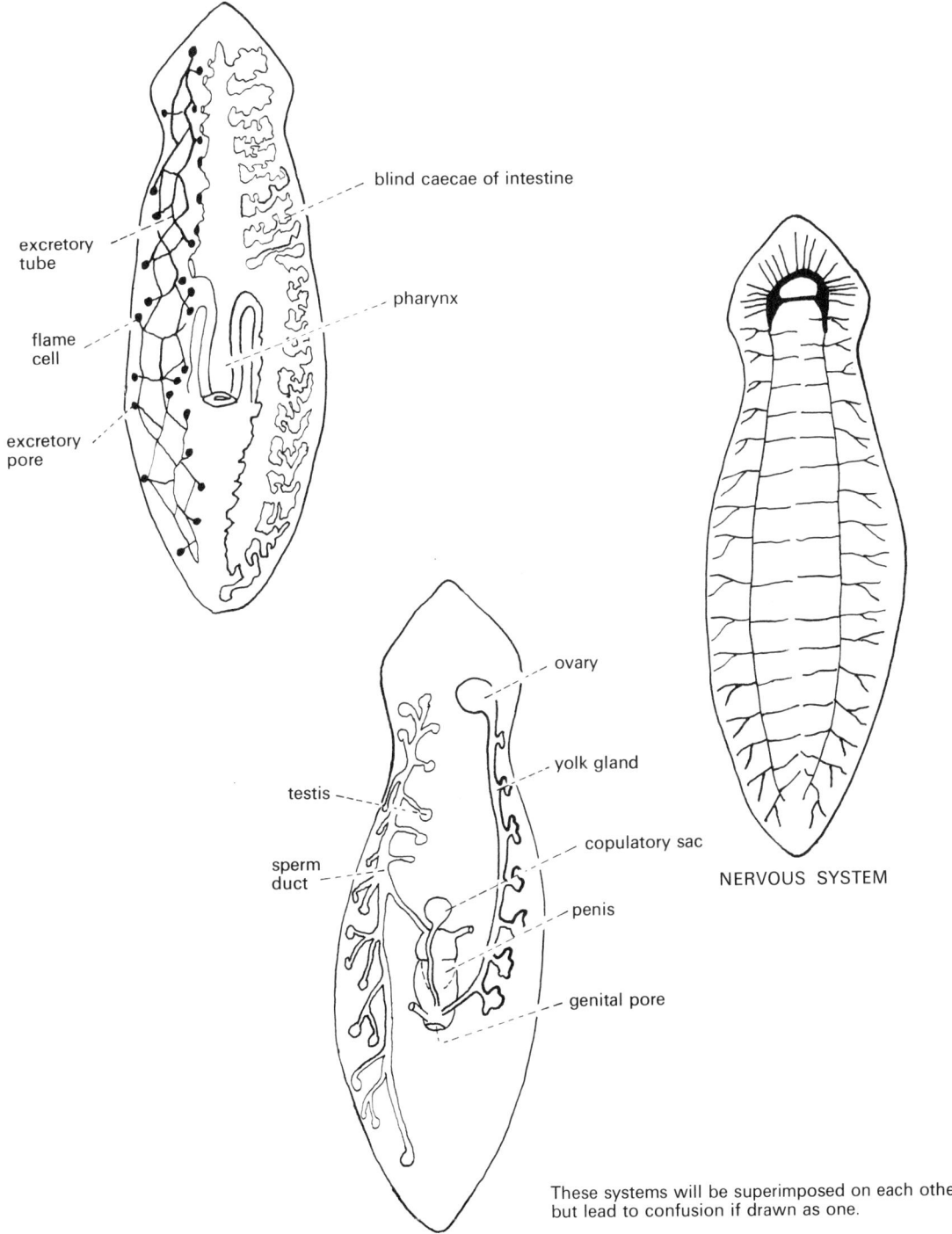

blind caecae of intestine

excretory
tube

pharynx

flame
cell

excretory
pore

ovary

yolk gland

testis

copulatory sac

sperm
duct

penis

genital pore

NERVOUS SYSTEM

These systems will be superimposed on each other
but lead to confusion if drawn as one.

REPRODUCTIVE SYSTEM

Fig. 83. Systems seen in whole mount of a planarian.

cilia
rhabdite
nucleus
circular and
oblique muscle
longitudinal
muscle
reproductive
organs
intestine
parenchyma

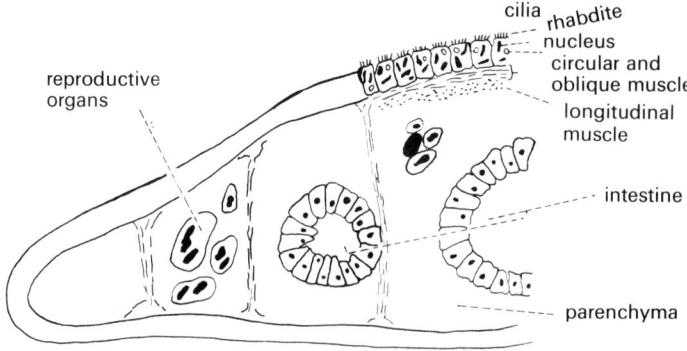

Fig. 84. Transverse section through planarian in anterior region reproductive organs.

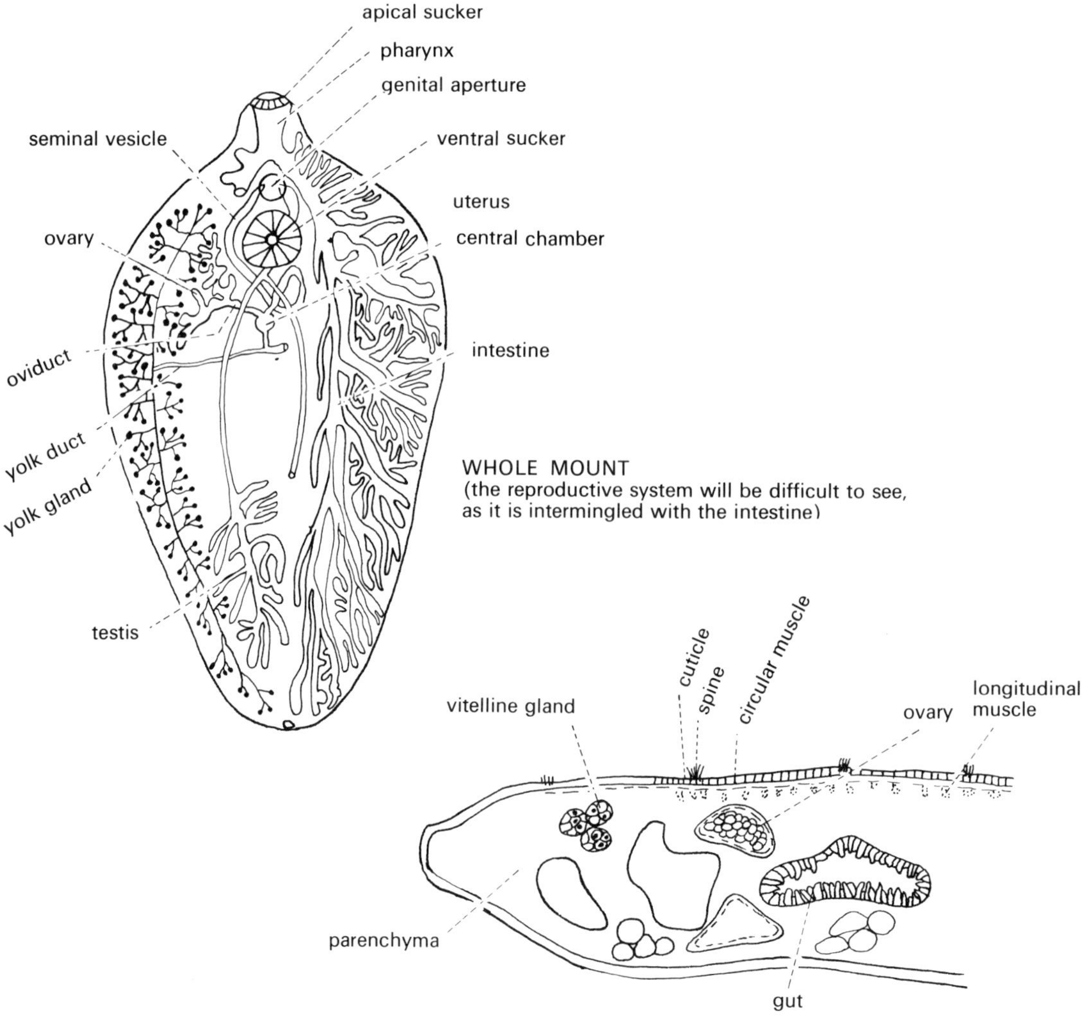

apical sucker
pharynx
genital aperture
seminal vesicle
ventral sucker
uterus
ovary
central chamber
oviduct
intestine
yolk duct
yolk gland
testis
WHOLE MOUNT
(the reproductive system will be difficult to see, as it is intermingled with the intestine)
cuticle
spine
circular muscle
longitudinal
muscle
vitelline gland
ovary
parenchyma
gut
TRANSVERSE SECTION
through the region of the ovary

Fig. 85a. Fasciola Hepatica.

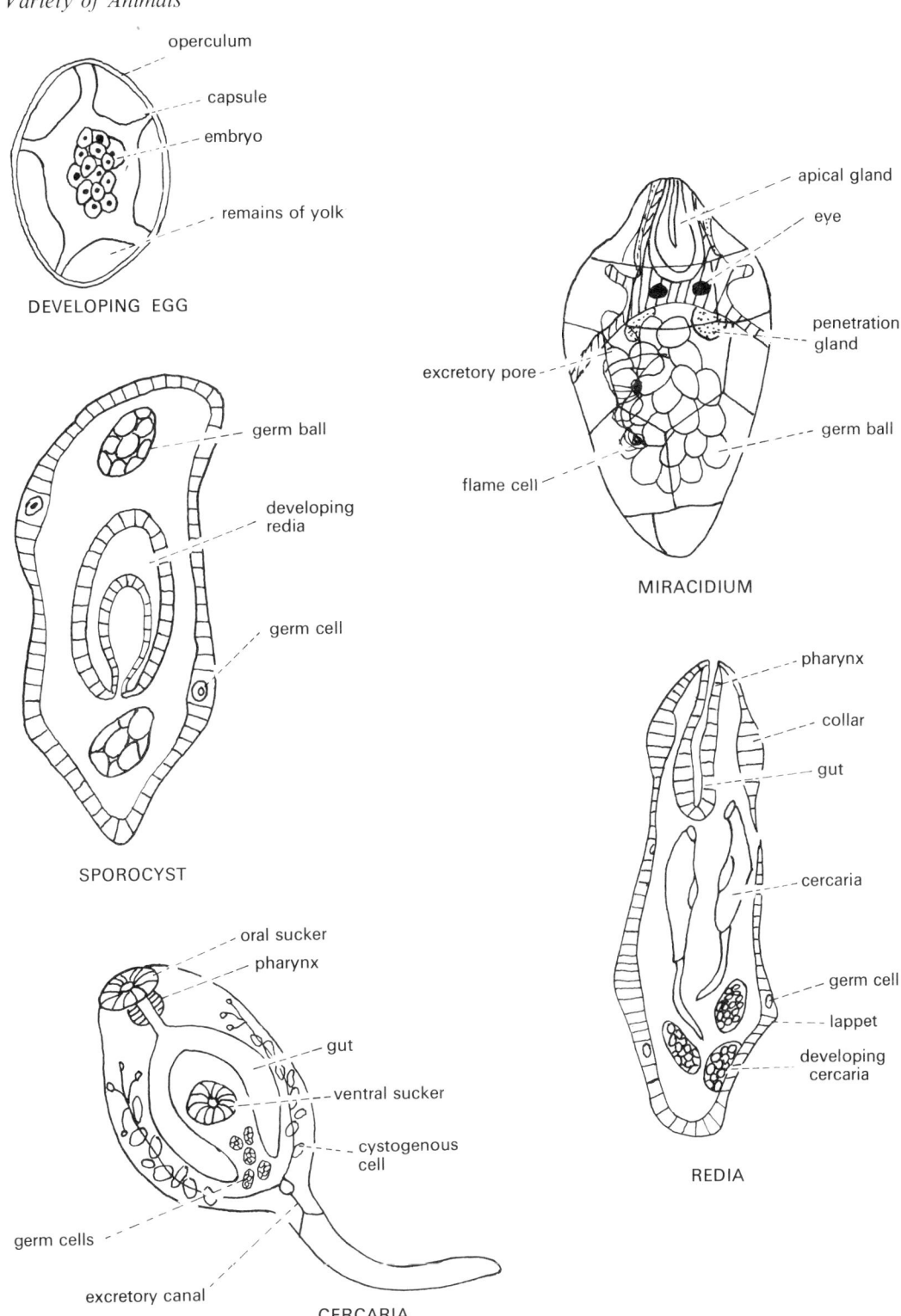

Fig. 85b. Fasciola Hepatica.

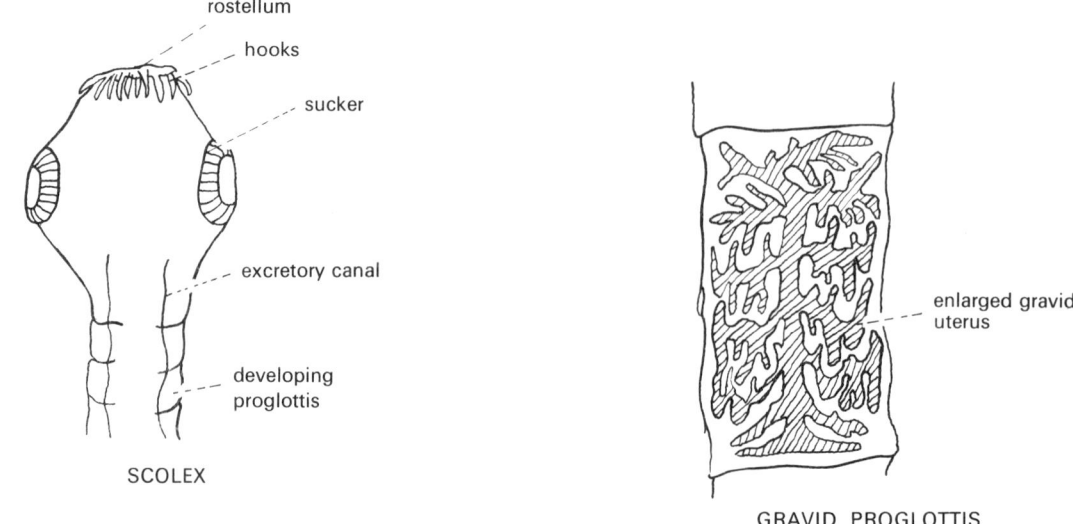

SCOLEX

GRAVID PROGLOTTIS

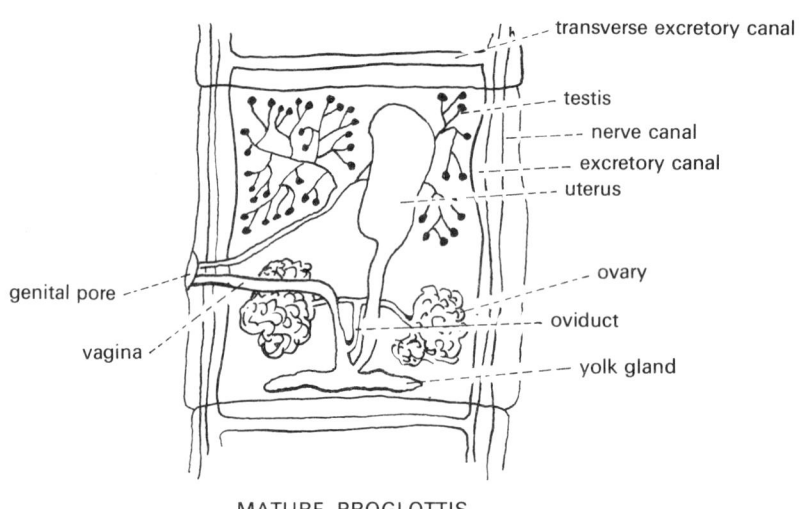

MATURE PROGLOTTIS

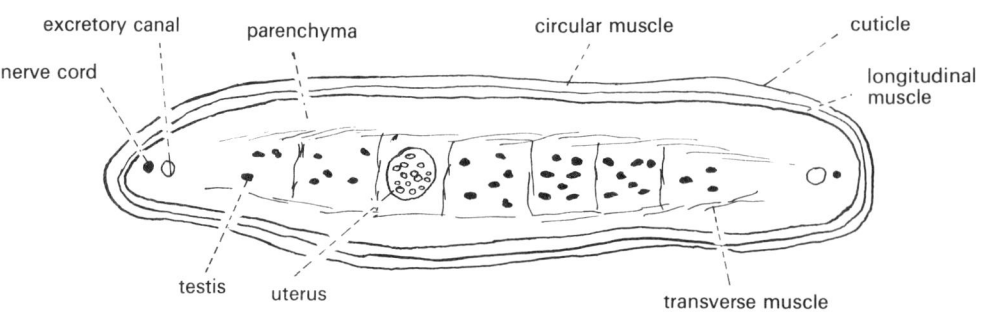

TRANSVERSE SECTION OF MATURE PROGLOTTIS

Fig. 86. Taenia (Tapeworm).

NEMATODA

This is a widely distributed phylum, its members living in various habitats.

Examine preserved specimens of *Ascaris lumbricoides* or *A. megalocephala,* slides of soil eelworms, *Heterodera* (one of the plant eel worms), and the free-living *Turbatrix aceti* from vinegar. Notice the great variety in size, but the basic similarity in structure.

Draw preserved male and female specimens of *Ascaris* sp. (Figure 88). There is no segmentation, and complete lack of sense organs or appendages.

Draw a transverse section through *Ascaris* sp. (Figure 89); remember that the body cavity is *not* a coelom. Notice the thick cuticle, the organized nerve cord, and the absence of muscle around the intestine.

ANNELIDA

Class: Polychaeta

Nereis

This is the common ragworm, or clamworm. It is found crawling under rocks and in the mud of the inter-tidal zone. The related *Arenicola marina* (lug worm) is found buried in the sand of the sea-shore. Its presence is easily recognized by the worm-casts produced in the inter-tidal zone at low tide.

Examine whole specimens of *Nereis* making detailed drawings of the head, terminal segment, and a single body segment (Figure 90). The body is *metamerically segmented,* and bilaterally symmetrical, with definite *cephalization.*

Draw a slide of a transverse section through a segment and a parapodium (see Figure 91). This animal has a *coelom,* which is shown on the slide by the investing of the intestine with muscle. Notice the complex muscles of the parapodium.

Examine and draw a slide of a *trochophore* larva. This larva is of interest as it is similar to that of the *Mollusca,* the adults of which differ considerably from those of the polychaetes. Is there any evolutionary significance here?

Class: Oligochaetae

Earthworm *(Lumbricus terrestis)*

The other commonly found genus, which is very like *Lumbricus* is *Allolobophora (Eisenia).* This differs from *Lumbricus* mainly in the position of the clitellum and the shape of the prostomium.

Examine a freshly-killed earthworm, using a hand-lens, and draw the external features (Figure 92).
Notice the clear evidence for metameric segmentation. External sense-organs are absent and the posterior part of the body is flattened. How does this show adaptation to the habitat? The parts are shown in the diagram. The *anus* opens on the *pygidium* which is not a true segment.

Draw your fingers gently over the lower surface from tail to head and notice the roughness caused by the *chaetae.*

Place a living worm on a rough surface (damp filter-paper) and observe the peristaltic waves of contraction passing over the body as it moves. Touch various parts of the body and observe the contractions.

Keep the worm in the dark for several hours. Then, in a darkened room, shine a narrow beam of light on it. Notice its reaction. Remember that there are no obvious external light-sensitive organs.

Examine a prepared slide of a transverse section of a worm (Figure 93).

Apart from the structures labelled in the diagram, notice, (1) that the body cavity is a true coelom (it is lined by the muscle of the body wall and by the muscular wall of the intestine), (2) the relatively simple musculature of the chaetae (compare with *Nereis*), (3) the complex structure of the longitudinal muscle, (4) the three giant fibres in the ventral nerve cord, and (5) the structure of the chaetae.

Dissection

Use a freshly-killed worm, and dissect under water in a dissecting dish. Pin the specimen down, with the dorsal surface uppermost. Place a pin through the head as far forward as possible (you may damage the cerebral

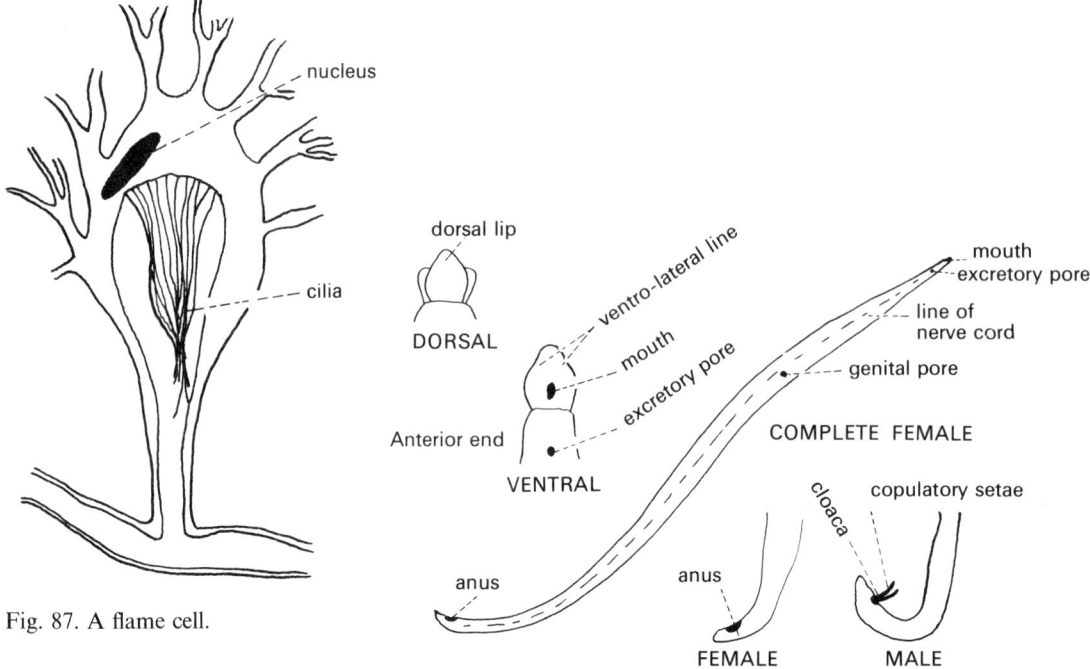

Fig. 87. A flame cell.

Fig. 88. Ascaris (Roundworm).

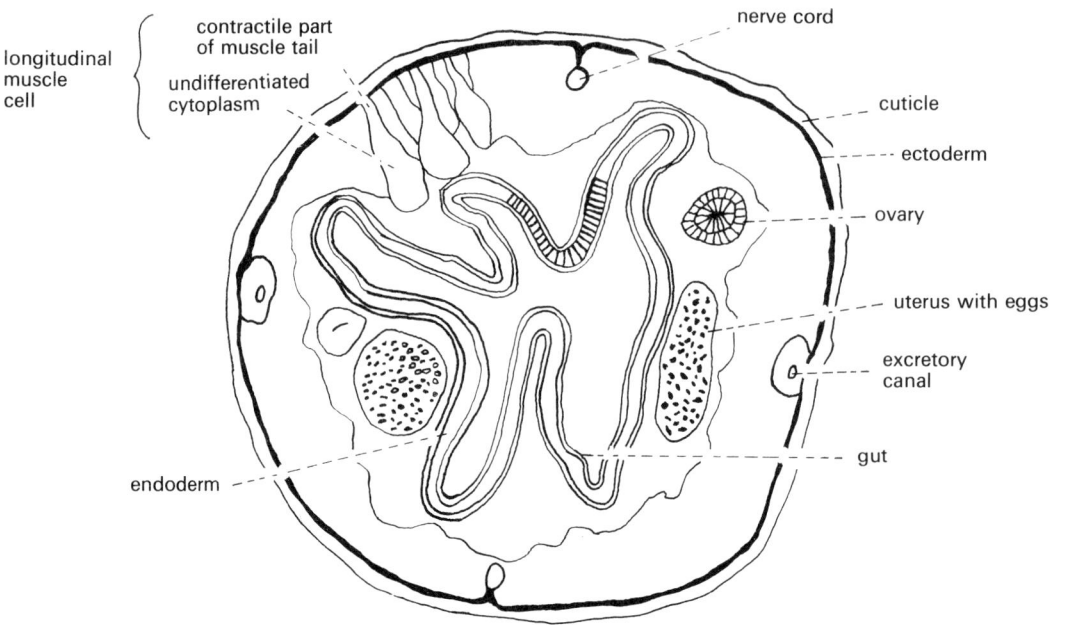

Fig. 89. Transverse section through female Ascaris.

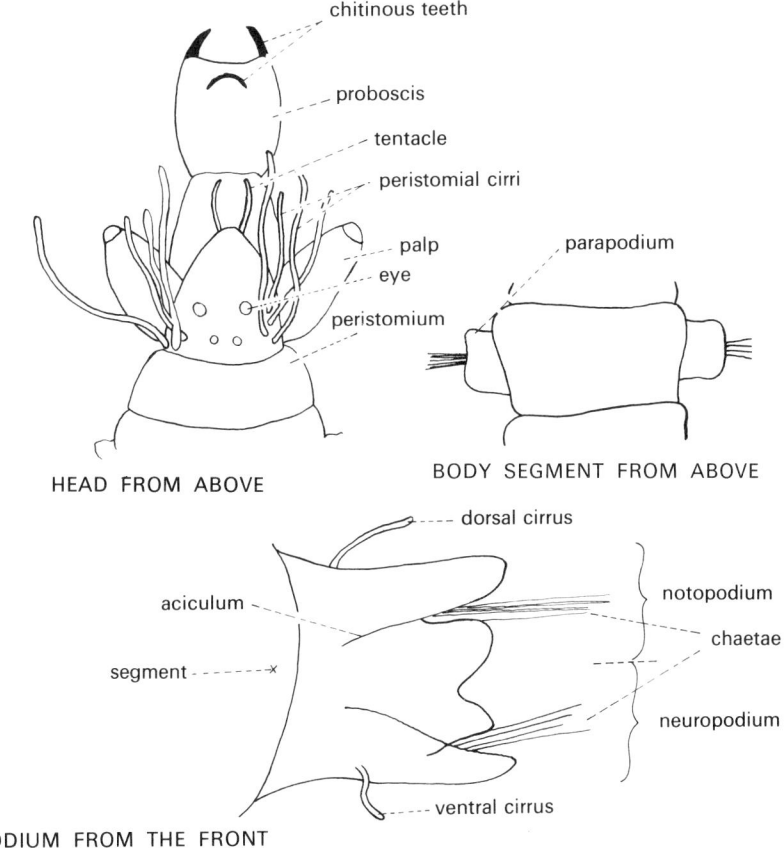

Fig. 90. External features of Nereis.

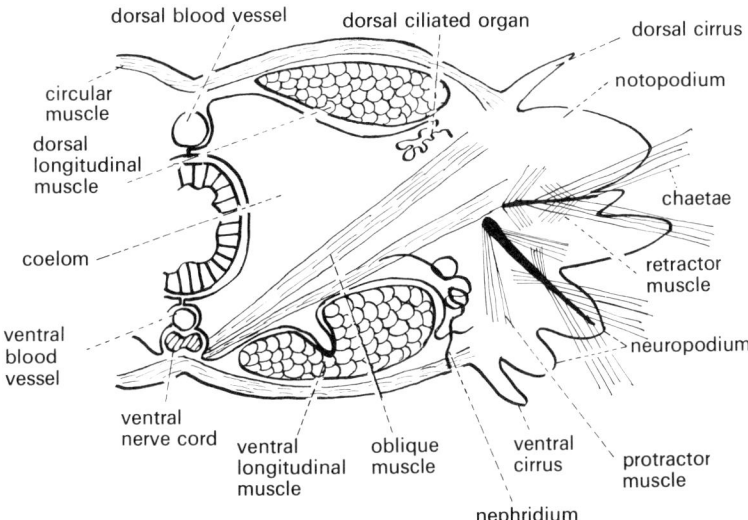

Fig. 91. Transverse section through segment of Nereis.

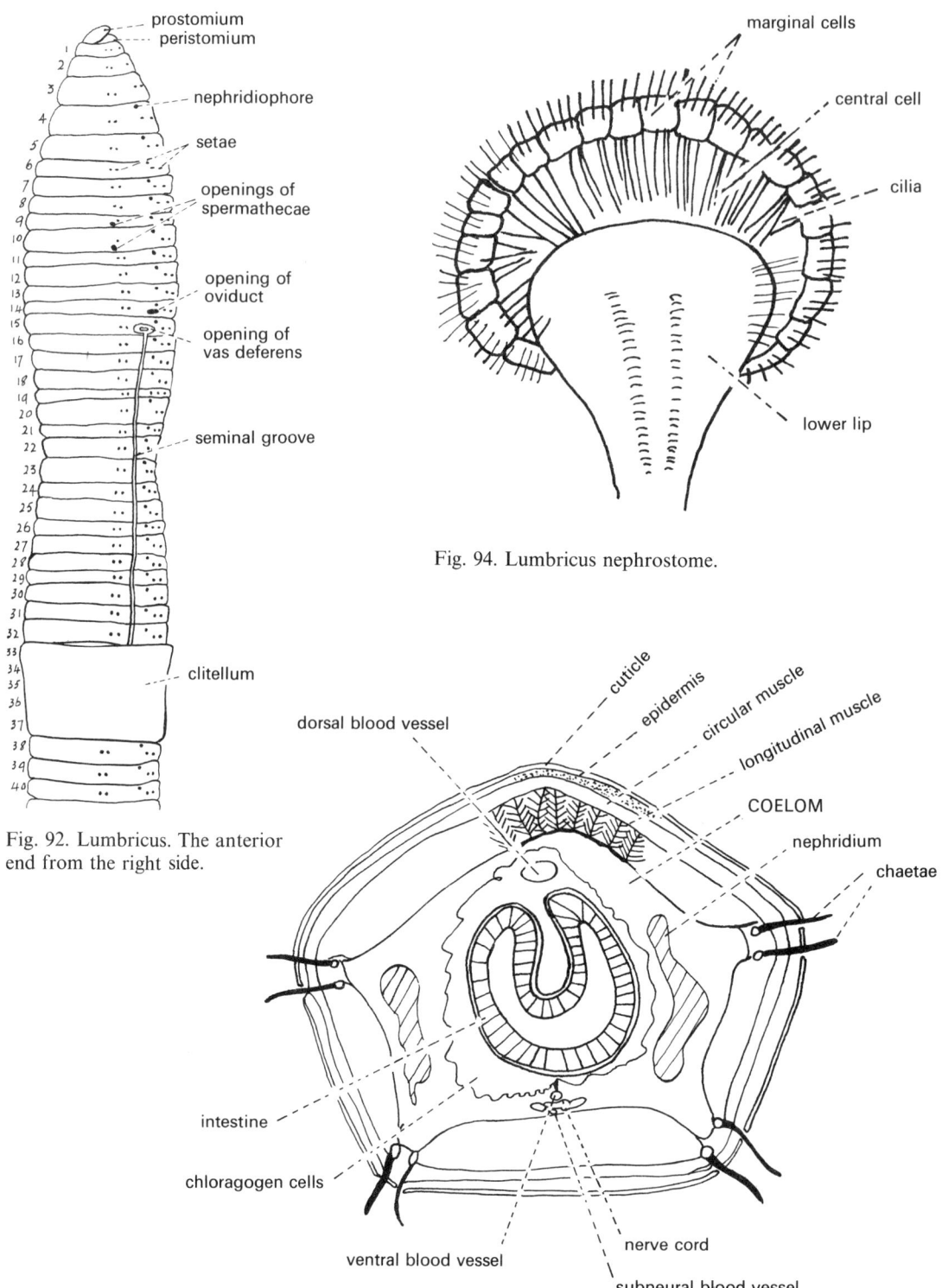

prostomium
peristomium
1
2
3
4
5
6
7
8
9
10
11
12
13
14
15
16
17
18
19
20
21
22
23
24
25
26
27
28
29
30
31
32
33
34
35
36
37
38
39
40

nephridiophore

setae

openings of
spermathecae

opening of
oviduct

opening of
vas deferens

seminal groove

clitellum

Fig. 92. Lumbricus. The anterior
end from the right side.

marginal cells

central cell

cilia

lower lip

Fig. 94. Lumbricus nephrostome.

cuticle
epidermis
circular muscle
longitudinal muscle

dorsal blood vessel

COELOM

nephridium

chaetae

intestine

chloragogen cells

ventral blood vessel

nerve cord

subneural blood vessel

Fig. 93. Transverse section of the Earthworm.

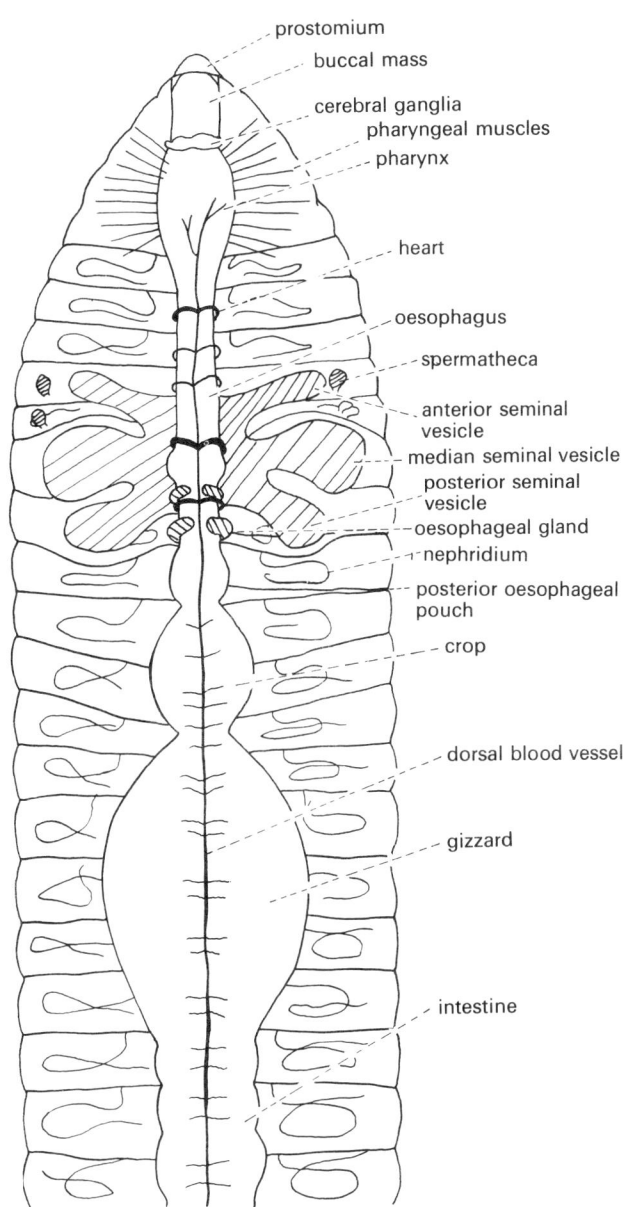

Fig. 95. General dissection of Lumbricus.

ganglia if it is pinned too far back), and another through the extreme posterior end. Cut through the body wall from about segment 30 to segment 2 (see Figure 92), along the mid-dorsal line. The body wall is quite tough and slippery, and this incision is best begun by holding a piece of the body-wall in a fine forceps at the point of insertion and making a small transverse slit with a fine scissors. Using this hole as a starting point, proceed forward. Gently pull the body-wall to either side, cutting through the septa as required, and pin it out. It is convenient to put pins in segments 5, 10, 15, 20, etc., to aid in numbering the segments for identifying the various organs.

As soon as you have opened a few segments, pick out one of the *nephridia* and mount it on a slide in physiological saline solution. Examine it under the microscope. If the nephridium has been removed quickly enough, you should see the beating of the cilia (Figure 94).

Once the body wall is pinned out, the dissection will appear as shown in Figure 95.

The structures in segments 9–12 may be obscured by the seminal vesicles which will then have to be displaced. Notice the distension of the oesophagus in segments 10 and 13 to form the *oesophageal pouches* and in segments 11 and 12 to form the *oesophageal glands*. The *crop* is thin-walled, and the *gizzard* thick-walled. Posterior to the gizzard, the surface of the intestine is obscured by the loose, yellow-brown *chloragogen cells*.

The nephridia are loose, light-coloured bodies lying in pairs, one on either side of the intestine, in each segment, except the first three and the last one (evidence of metameric segmentation).

The blood system will have been greatly damaged by opening the body-wall, but the dorsal blood vessel and five 'hearts' should be visible.

You will probably have difficulty in distinguishing the female reproductive system. First locate segment 13. Displace the seminal vesicle and oesophagus to one side, and the *ovaries* will appear as small triangular-shaped bodies against the anterior wall of the segment. Use a hand lens to locate them. Penetrating the posterior wall of the same segment is the small *ovarian funnel*, with the *oviduct* leading from it into segment 14. With the seminal vesicles still displaced towards the mid-line of the body, locate the *spermathecae* against the posterior wall of the septa of segments 9 and 10. They will lie roughly mid-way between the nerve cord and the edge of your dissection.

Of the male reproductive system, the *seminal vesicles* are obvious as large yellow-white bodies with three pairs of lobes in segments 9, 11, and 12. Open one of the vesicles and make a smear of its contents on a slide. This can be fixed with 70% alcohol, stained with haematoxylin, and mounted in glycerine if required. Examine your mount under the microscope and see the stages in the development of the spermatozoa, from the masses of spermatogonia and spermocytes to the individual spermatozoa. Look for the stages in the life-history of the parasite *Monocystis*.

Continue the dissection of the male system by removing the oesophagus to expose the median region of the seminal vesicle, then gently rake the top wall off it and wash out the contents with a pipette. You should now see the small testes hanging from the front septa of segments 10 and 11, near the central line. The sperm funnels open near the rear septa of the same segments, so that the anterior funnels lie near the posterior testes. From the funnels, the slightly coiled *vasa efferentia* run to fuse on either side of segments 12. These then join to form the paired *vas deferens* which open by pores on segment 15 (Figure 96).

Examine the nervous system by cutting through the oesophagus near the *cerebral ganglia* and removing the seminal vesicles and part of the intestine (see Figure 97). From the cerebral ganglia run a pair of *circumoeso-phageal commissures (connectives)* around the oesophagus to the *sub-oesophageal*

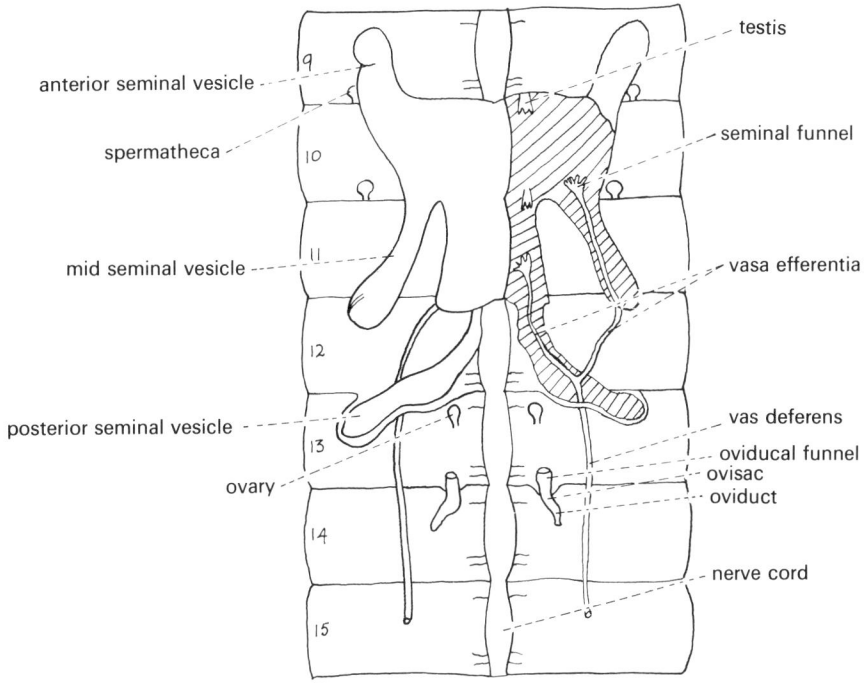

Fig. 96. Reproductive system of Earthworm. (The wall of the left seminal vesicle is removed.)

ganglion below. Each segment has a ganglion with three pairs of nerves running from it, (evidence of metameric segmentation), the paired nerve cord, and the *prostomial nerves* running forward from the cerebral ganglion.

ARTHROPODA

This is a very large phylum including all the bilaterally symmetrical animals with an exoskeleton and paired jointed limbs. Some classes and many orders you will not deal with, and what follows is a survey of the more important groups, concluding with the dissection of the locust.

Class: Crustaceae

The Crayfish *(Astacus fluviatilis)*

You might have to work with a similar crustacean, as the crayfish is becoming scarce in Great Britain. The labelling and structures will be essentially similar.

Examine and draw the dorsal and ventral aspects of the animal (see Figure 98).

Although the body is divided into head, thorax and abdomen, the head and thorax are almost fused, and covered by the common *carapace.* The segments of the abdomen are covered individually by plates, called *sclerites.* The one above is the *tergum,* and the one

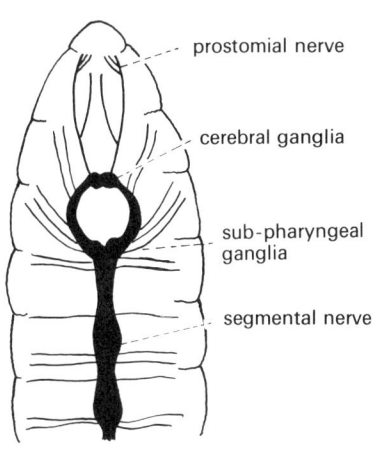

Fig. 97. Nervous system of Lumbricus.

below the *sternum*, while the side-plates are *pleura*.

Notice the paired antennae and the compound eyes. Cut away the side of the carapace and see the gills under it. The flattened tail fan *(uropods* and *telson)* is used for swimming backwards when the animal is disturbed.

Each segment, except for the first, bears a pair of appendages each consisting of a basal part *(protopodite)* with two apical branches *(endopodite* and *exopodite)* (see Figure 99). These are highly modified, or may be absent from some limbs. Remove one of each pair of limbs and draw it. Notice particularly, that although they have the same basic pattern, the appendages are modified for various functions.

The Water Flea (*Daphnia* spp.)

Examine living material under the microscope. Notice the beating of the heart. The animal

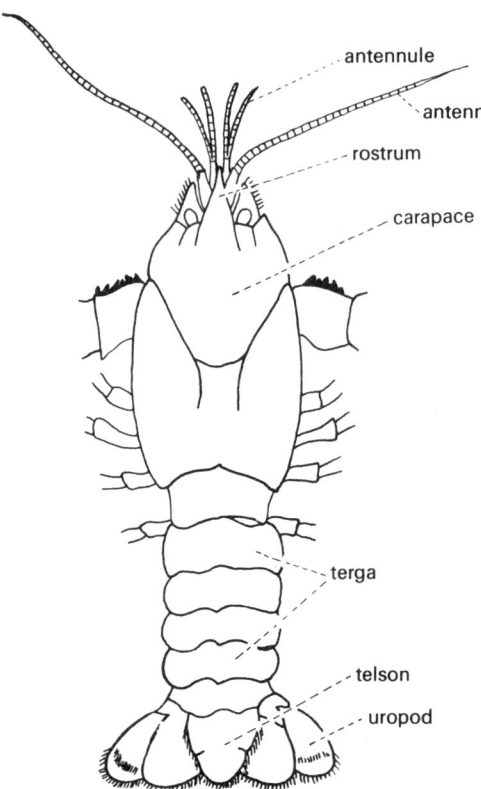

Fig. 98. Dorsal view of the Crayfish, showing the leg bases only.

bears a superficial resemblance to the crayfish. The paired antennae are the diagnostic feature which place this animal in the Crustaceae (Figure 100).

The Woodlouse (*Oniscus* spp.)

The woodlouse (Figure 101), has no gills and breathes by trachea. It is placed in the Crustaceae for the following reasons (1) there are two pairs of antennae, (2) the first thoracic segment is fused with the head, forming a rudimentary cephalothorax, (3) the body is dorso-ventrally flattened.

Look for these features on your specimen, and draw it from above.

Subphylum: Myriapoda; Class: Chilopoda

The Centipede (*Scolopendra* spp.)

The sub-phylum is characterized on all the segments of the body, except the first three, by the presence of trachea, and the single pair of antennae (see Figure 102).

Examine the centipede and notice these points.

Class: Insecta

This is an extremely large class, and at best you can only examine a few representative members. All have one pair of antennae, a body divided into head, thorax and abdomen, and three pairs of walking legs on the thorax. Breathing is by trachea.

Examine a prepared slide of the trachea, and draw them.

Order: Orthoptera (Cockroach, grasshopper, locust, stick insect and leaf insect).

The Cockroach (*Periplaneta americana*)

Pin the specimen out with the wings on one side extended, taking care not to rip the wing. Notice the long antenna (Figure 103), and the complex venation of the hind wing.

The head lies at right angles to the line of the body. Examine and draw the head from the front (see Figure 104). Notice the large *compound eyes*, the *maxillary* and *labial palps*, and the large *mandibles*.

Draw a prepared slide of the mouthparts

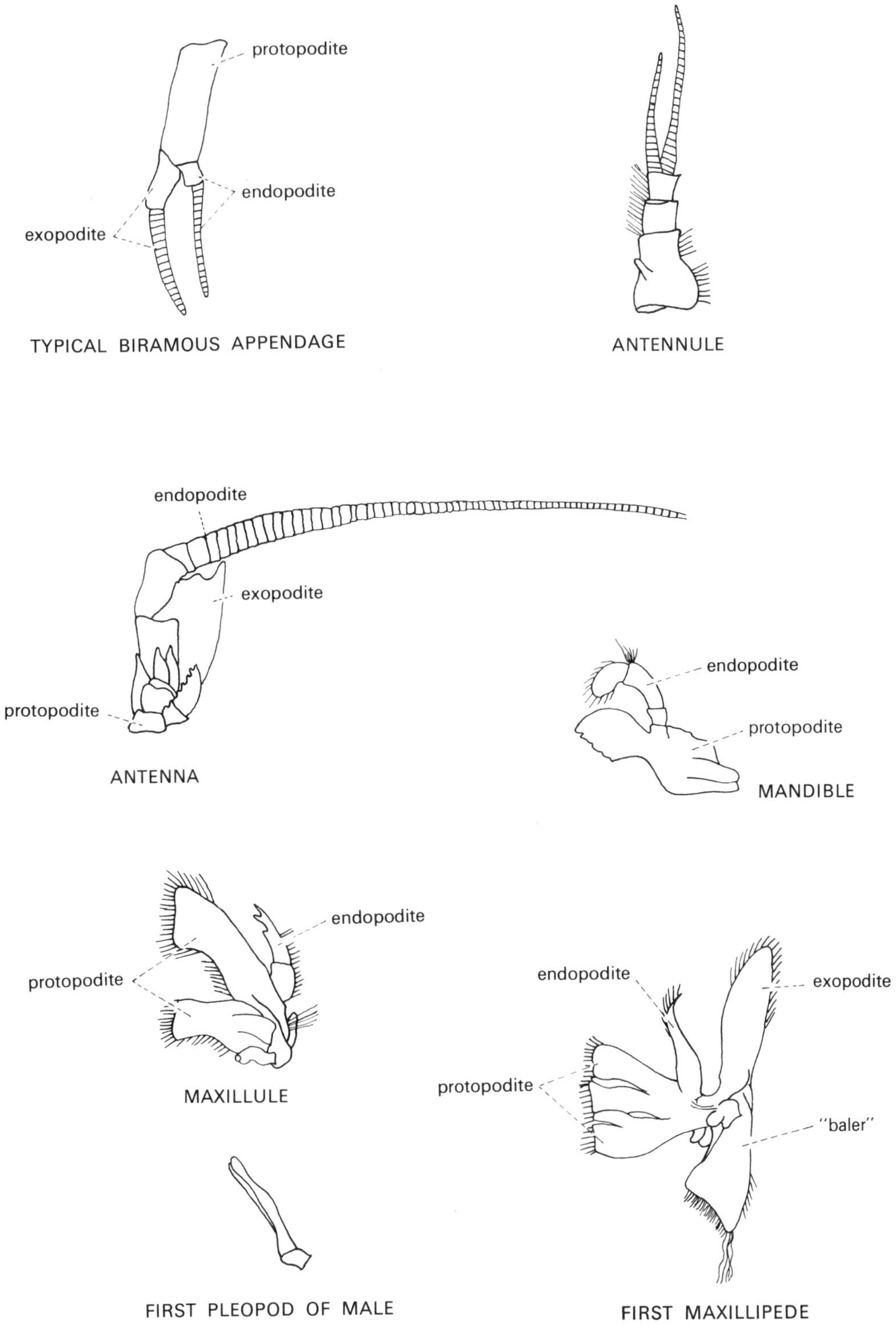

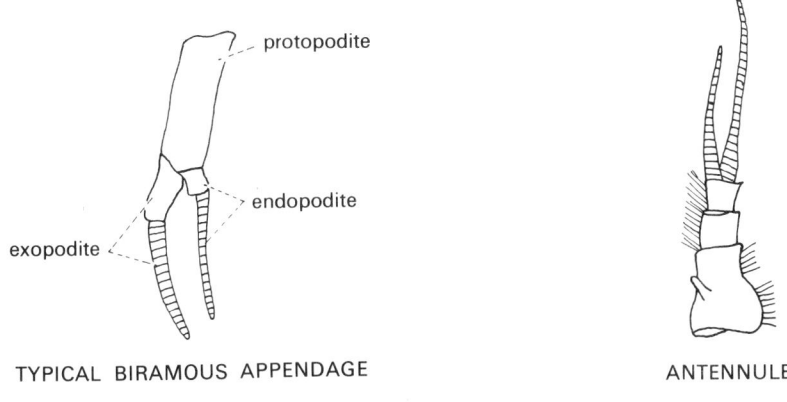

protopodite

endopodite

exopodite

TYPICAL BIRAMOUS APPENDAGE

ANTENNULE

endopodite

exopodite

protopodite

ANTENNA

endopodite

protopodite

MANDIBLE

endopodite

protopodite

MAXILLULE

endopodite

exopodite

protopodite

"baler"

FIRST PLEOPOD OF MALE

FIRST MAXILLIPEDE

Fig. 99a. Appendages of the Crayfish.

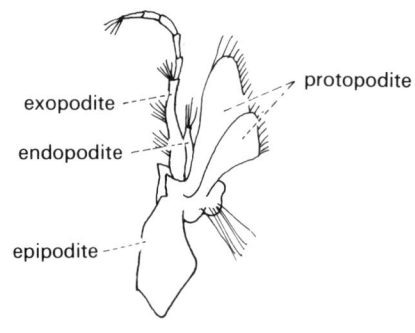

exopodite
endopodite
protopodite
epipodite

FIRST MAXILLIPEDE

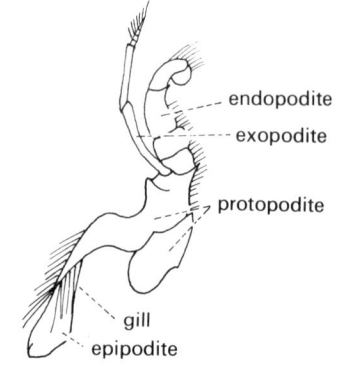

endopodite
exopodite
protopodite
gill
epipodite

SECOND MAXILLIPEDE

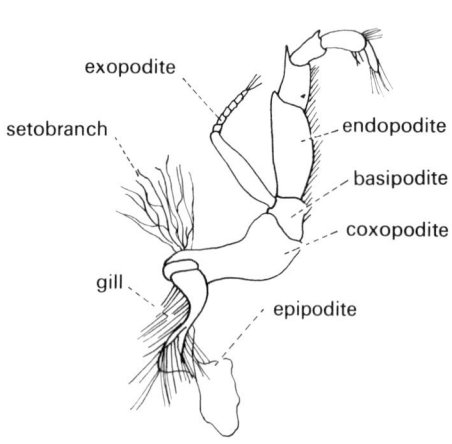

exopodite
setobranch
endopodite
basipodite
coxopodite
gill
epipodite

THIRD MAXILLIPEDE

endopodite

CHELA

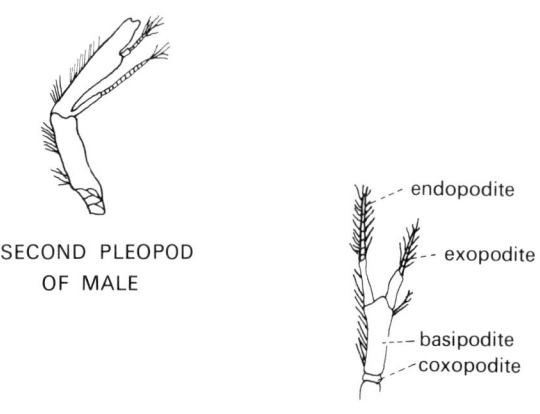

SECOND PLEOPOD
OF MALE

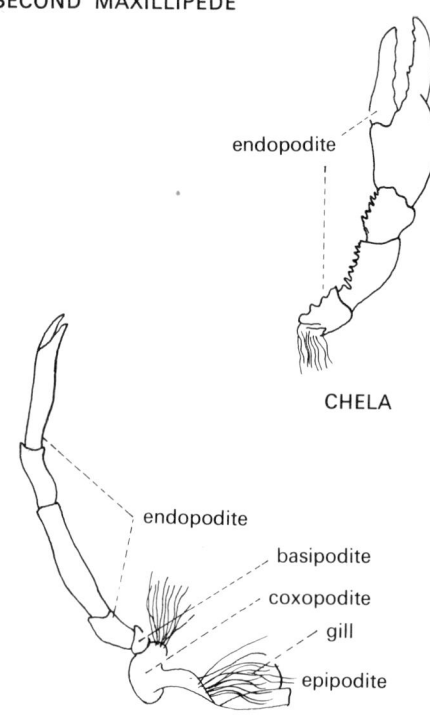

endopodite
basipodite
coxopodite
gill
epipodite

FIRST WALKING LEG

endopodite
exopodite
basipodite
coxopodite

TYPICAL PLEOPOD

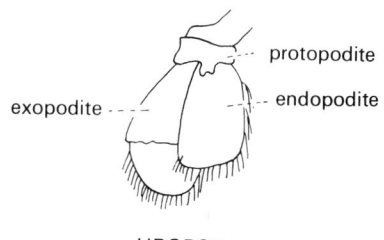

exopodite
protopodite
endopodite

UROPOD

Fig. 99b. Appendages of the Crayfish.

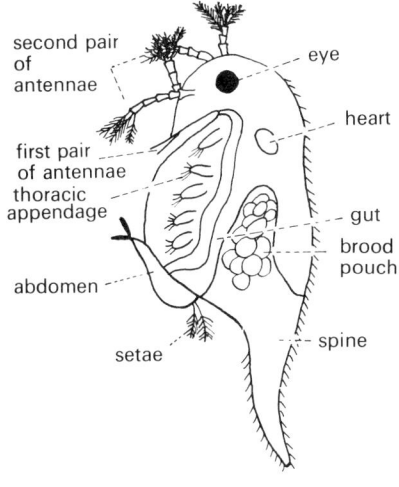

Fig. 100. Daphnia.

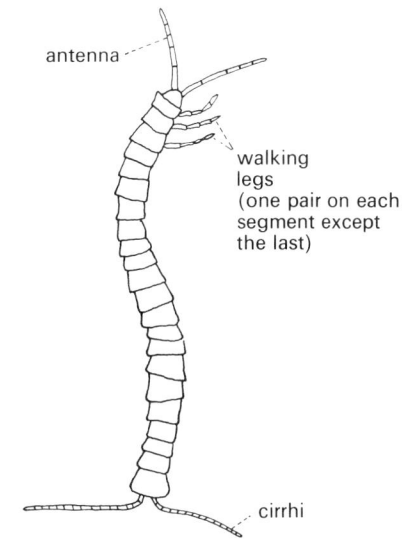

Fig. 102. A Centipede.

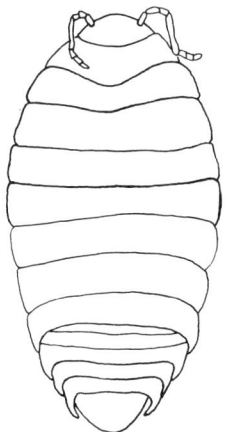

Fig. 101. Woodlouse from above.

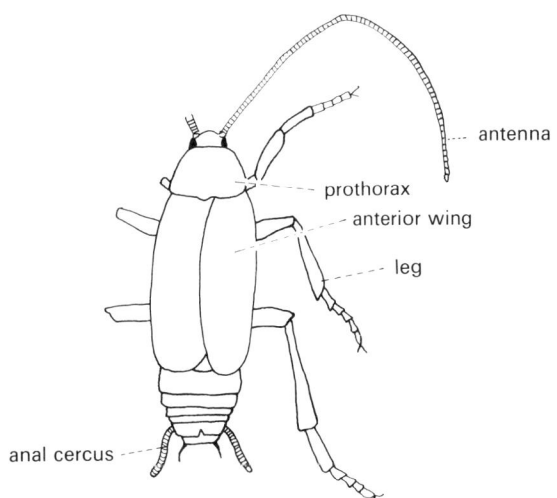

Fig. 103. Dorsal view of male Cockroach.

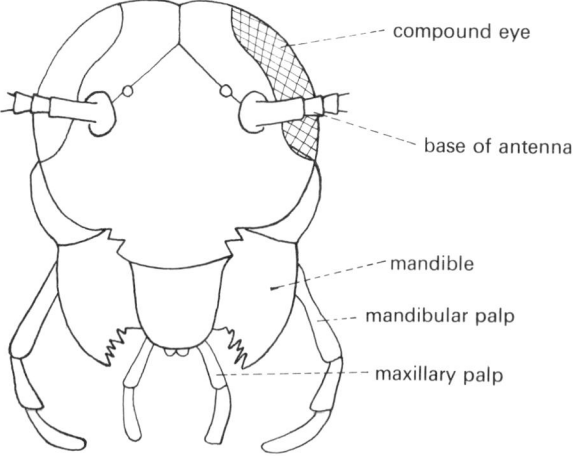

Fig. 104. Front of the head of the Cockroach.

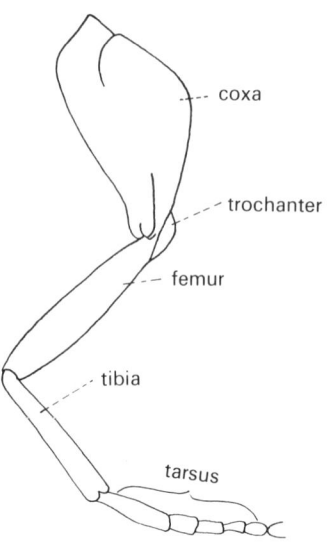

Fig. 106. Leg of the Cockroach.

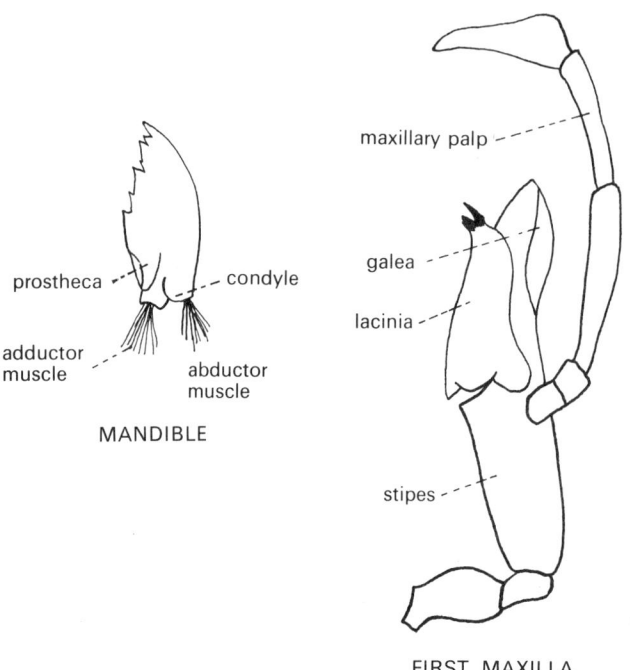

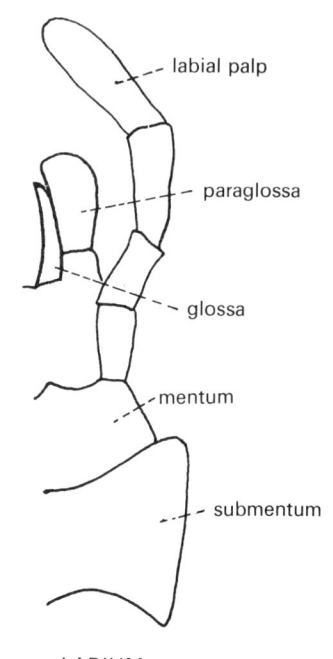

Fig. 105. The mouthparts of the Cockroach.

(Figure 105), which are typical of the biting type found in insects. Remove one of the legs and draw it (see Figure 106). Notice that the structure of the mouthparts and leg are basically similar to those of the crayfish.

The anatomy of the locust will be described at the end of the section on insects.

Order: Hemiptera (Aphids, leaf hoppers, assassin bugs, stink bugs, bugs.)

The Greenflies and Blackflies *(Aphids)*

There are several genera and species of aphids which are pests on a wide variety of plants. They have piercing mouth-parts and are easily identified by the *cornicles*. The males are rarely seen, and the females are usually wingless, but may have wings.

Draw a mounted specimen (see Figure 107).

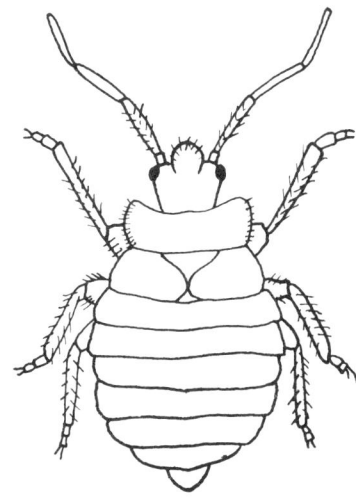

Fig. 108. Bedbug.

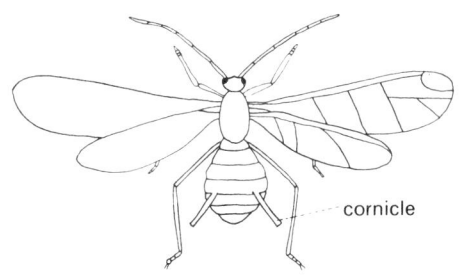

Fig. 107. A winged female Aphid.

The Bed Bug *(Cimex* spp.)

The body is hairy and dorso-ventrally flattened. The wings are vestigial and the head is set back into a notch in the thorax (Figure 108). The mouth-parts are modified for piercing and sucking. Are the vestigial wings an adaptation to the habitat? Draw a specimen, and a slide of the mouth-parts.

Order: Lepidoptera (Moths and Butterflies).

The members of this order have scaly wings, and the mouth-parts formed into a long *proboscis* for sucking.

The Cabbage White Butterfly *(Pieris brassicae)*

Study the various stages in the life-history (see Figure 109).

Examine the eggs with a hand lens, or under the low-power of the microscope. Notice the shape, colouring and sculpturing.

Look at the larva carefully to see the spiracles, noting the segments on which they occur. The mouthparts are of the simple biting type with the *spinneret* between them, and the eyes simple ocelli.

Notice that the thoracic legs are jointed, i.e. are true legs, while the prolegs and clasper are extensions of the body and are not jointed.

The pupa shows the developing adult structures inside it. The *dorsal* and *lateral projections* are for the attachment of the silk strand which holds the pupa to its support.

The male and female adults differ in the wing markings. The tips of the fore-wings are dark, and there is a spot on the leading edge of the hind-wing in both sexes, but the female has in addition two spots on the fore-wing. Notice the clubbed antennae, which distinguish the butterflies from the moths. Examine one under the microscope and see the structure, noticing the distribution of the hairs.

Mount a few wing scales and examine them.

Remove the head and examine under the microscope. Notice the long, coiled *proboscis*

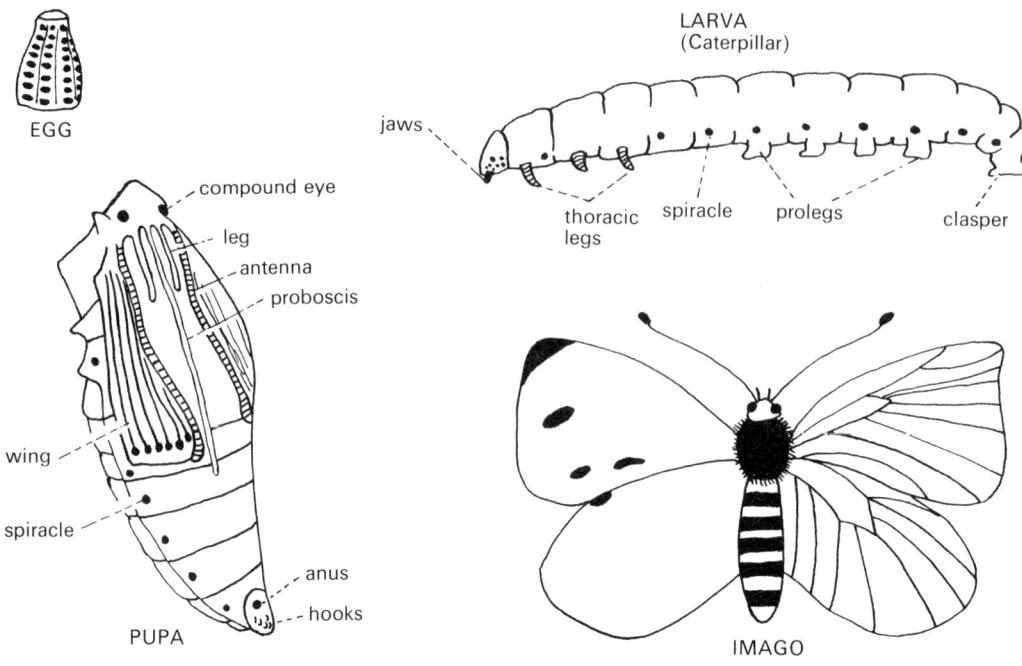

Fig. 109. Stages in the life history of the Cabbage White Butterfly.

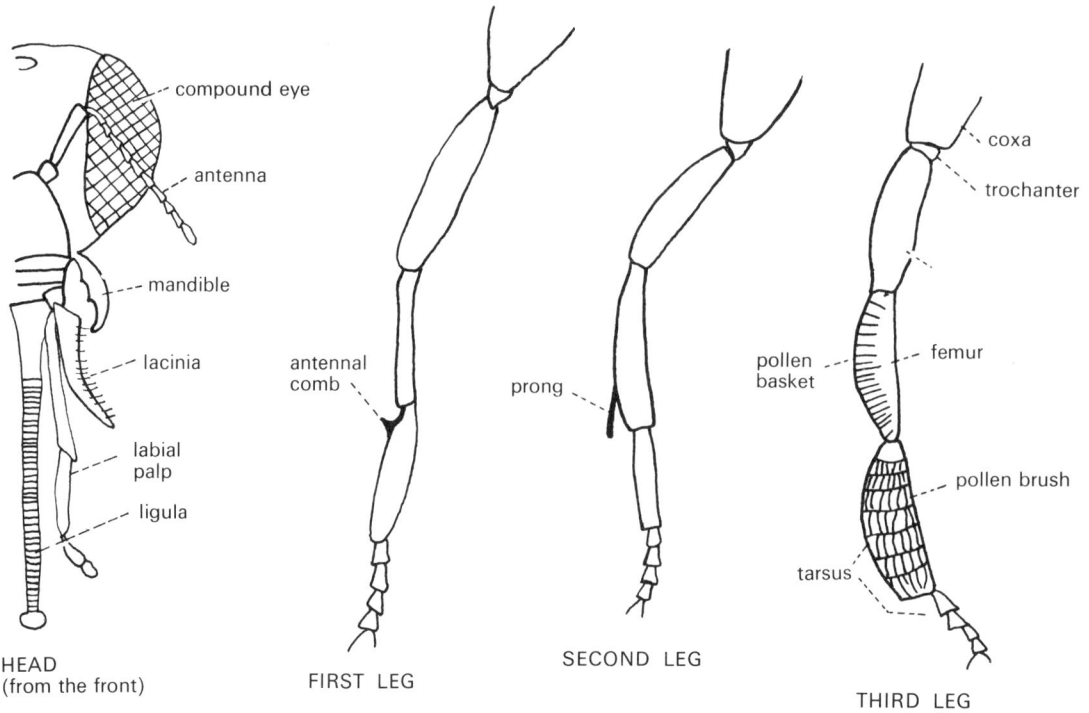

Fig. 110. Head and legs of Honey Bee.

with strengthening ridges, and the structure
of the compound eye.

Order: Coleoptera (Beetles)

This is a very large order, with a great
diversity of larval form. Examine a few of
the adults, e.g. Ladybird, Ground Beetle.
The mouthparts are of the normal biting
type, and the fore-wing is hardened to form
the protective *elytra*.

*Order: Hymenoptera (Bees, wasps, ants,
ichneumon flies,* etc.)

The Honey Bee *(Apis mellifica)*

Examine a bee larva. There are no legs.

In the pupa the structures of the developing
adult can be seen clearly.

Compare the external features of the queen,
drone, and worker. They differ considerably
in size and proportion.

The head of the worker bee (Figure 110),
bears compound eyes, three single simple
eyes (ocelli) and a pair of antennae. See these
with a hand lens. Examine a prepared slide
of the mouthparts of a worker; they are
modified for biting and lapping.

Remove the legs from one side of the
worker, and examine them individually.

Order: Diptera (Flies, gnats, mosquitoes)

All the adults in this order have one pair of
wings, the second pair being represented by
the *halteres*.

The Housefly *(Musca domestica)*, **Blowfly**
(Calliphora erythrocephala)

Examine the egg, larva and pupa of either
of these types (Figure 111). Notice the
extreme simplicity of the external structure.
The larva lacks legs and the mouthparts are
reduced to hooks. Are these adaptations to
the habitat?

Draw the adult from above. Lift the pair
of scaly appendages behind the wings to
expose the *halteres*. The antennae are
extremely short, with their function partly
assumed by the stiff bristles *(aristae)*
attached to them.

Look at a slide of the mouthparts which
are adapted for sucking up liquids.

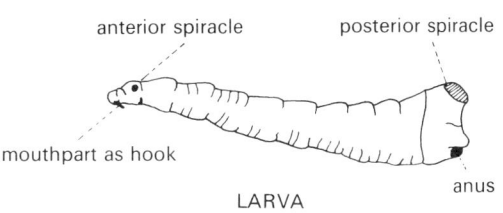

LARVA

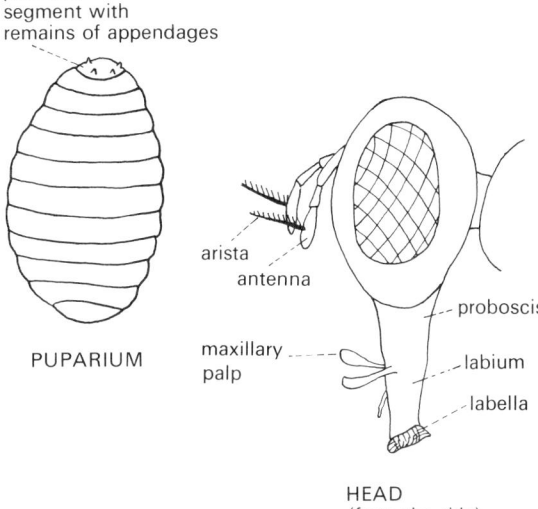

PUPARIUM

HEAD
(from the side)

Fig. 111. The Housefly.

The Mosquito *(Culex spp.)*

Examine and draw an egg-raft (see Figure
112).

On the larva look especially for the spiracle
at the end of the *siphon*. The siphon is on the
same segment as the spiracle of the housefly
larva, and may be an adaptation to an aquatic
habitat. The *mouth-brushes* around the mouth,
trap food particles from the water. The last
abdominal segment terminates in a number
of plates which contain trachea, and may act
as gills.

The pupa has similar adaptations to the
aquatic environment. Draw one, looking
especially for signs of the developing adult
organs.

Look at a slide of the adult female. Note
the feathered antennae and the halteres.

Examine a slide of the mouthparts of the
female, which are modified for piercing the
skin and sucking blood.

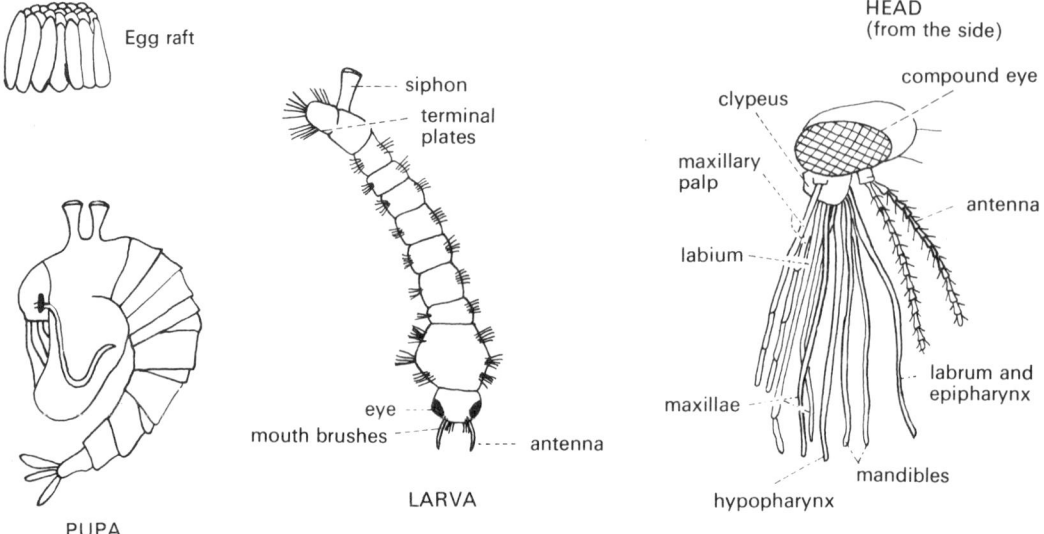

Fig. 112. The Mosquito (Culex).

Order: Aphaniptera (Fleas)

These are all wingless, and parasitic on birds and mammals.

The Flea *(Pulex irritans)*

Examine the slide of the larva (see Figure 113). Notice the extreme simplicity—the absence of wings and eyes, the simple antennae, and the biting mouthparts.

The pupa is enclosed in a cocoon, which will have to be removed before preparing the slide.

The body of the adult is laterally compressed, with the legs well modified for jumping. The hairs are long and point backwards. The mouthparts are for sucking. Are these adaptations to the habitat?

Dissection of the Locust

Cut the wings off your specimen, and pin it dorsal side up in a dissecting dish. Push the pins through the tops of the legs and the extreme posterior of the body (through the external genitalia). Continue with the animal covered in water.

Make an incision through the cuticle to the right side of the body, and continue it up the length of the body. Gently lift the

cuticle off to the left. Make a similar incision through the muscles beneath the cuticle. At the anterior and posterior ends of the incision, cut across laterally to the left as far as possible. Turn this flap of muscle over to the left and pin it down. In doing this you will have damaged the blood vessels, especially in the thorax, but you should now see the longitudinal heart lying in the muscles to the left of the dissection. Notice how extremely muscular the thorax is. The *tympanum* will show as a fine membrane on the first abdominal segment on the left side.

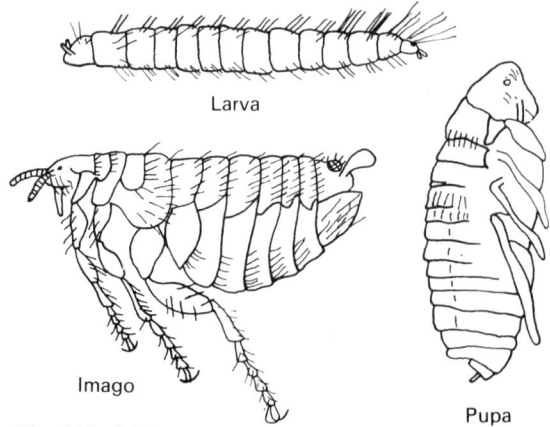

Fig. 113. A Flea.

The long thread-like bodies lying in the body cavity and connected to the posterior end of the mesenteron (mid-gut) are the *Malpighian tubules*. Dorsal to the heart are a series of *trachea* which are also found running through the muscles, while generally distributed around the internal organs is a yellowish-white *fat body*. Many of the trachea, and the fat body will be removed during the dissection.

Displace the intestine to the right, and pin it down. Pay particular attention to the posterior end, so that the reproductive system is fully exposed. This will expose a pair of *salivary glands* anteriorly (each of which is white and like a bunch of grapes), the nerve cord, and the reproductive system (see Figure 114).

In both male and female reproductive systems, note especially, the position of the gonads on the dorsal surface of the intestine, and the course of the ducts.

Draw your dissection.

Although the nervous system is partially exposed, the dissection can be improved by making lateral incisions in the head capsule to just above the eyes. Lift off the top of the capsule, cutting through muscles as necessary. Pin the two sides of the head so as to stretch them slightly. Pick off any muscles left obscuring the nerves in the head, using a fine forceps. If the insect is now placed in 70% alcohol overnight, the nerves become whitened, and more distinct. Starting in the abdomen now remove the ventral diaphragm which overlies the nerve cord, and the fatty sheath surrounding it. This requires great care. Follow the nerve cord forward, removing any muscles that obscure it.

Notice that there are ganglia formed by the fusion of the ganglia of individual segments, but that the basic pattern of the nervous system is the same as that of the Annelids.

MOLLUSCA

Class: Lammellibranchiata

The freshwater Mussel *(Anodonta cygnea)*
Examine a living specimen in an aquarium.

Notice the position in which the mussel lies in the sand, and the external *foot*. If a few grains of dye (e.g. carmine) are placed near the opening of the shell the movement of the water can be observed. There is a definite direction of flow from the anterior to the posterior through the *inhalent siphon* and out through the *exhalent siphon*.

Look at a killed specimen. The blunt end is anterior, and the two *valves* of the shell show its bilateral symmetry. The shell is characteristic of most molluscs. Prise open the shell and cut through the muscles holding it as close to the shell as possible. Cut through the *ligament* of the *hinge*. Lift off the shell and turn it over to see the areas of muscle attachment (Figure 115).

Now that the shell has been removed, you will see a lobe of the *mantle*. This is a characteristic of the Mollusca.

Lift the mantle lobe to see the side of the animal. This will expose the gills, and the muscular *foot* which is characteristic of the Mollusca (see Figure 116).

The Snail *(Helix pomatia)*

Examine a killed specimen with the body extended (Figure 117). The *tentacles* are characteristic of the Gasteropoda, the larger, posterior ones bearing the eyes. The flattened *foot* is used in locomotion. On the right side you will see the *genital aperture* with a groove leading from it. Under the shell, near the posterior end of this groove is a small opening into the *mantle cavity*. You can pass a fine seeker through this opening *(pneumostome)* into the mantle cavity.

Remove the shell by cutting from the upper anterior edge, follow the coils around, being careful not to cut downwards into any organs, The parts enclosed in the snail's shell, correspond roughly to those parts permanently inside the shell of a mussel, and the complex of organs is called the *visceral hump*. Notice again the fine membraneous mantle, which is suffused with blood vessels. This is the breathing organ.

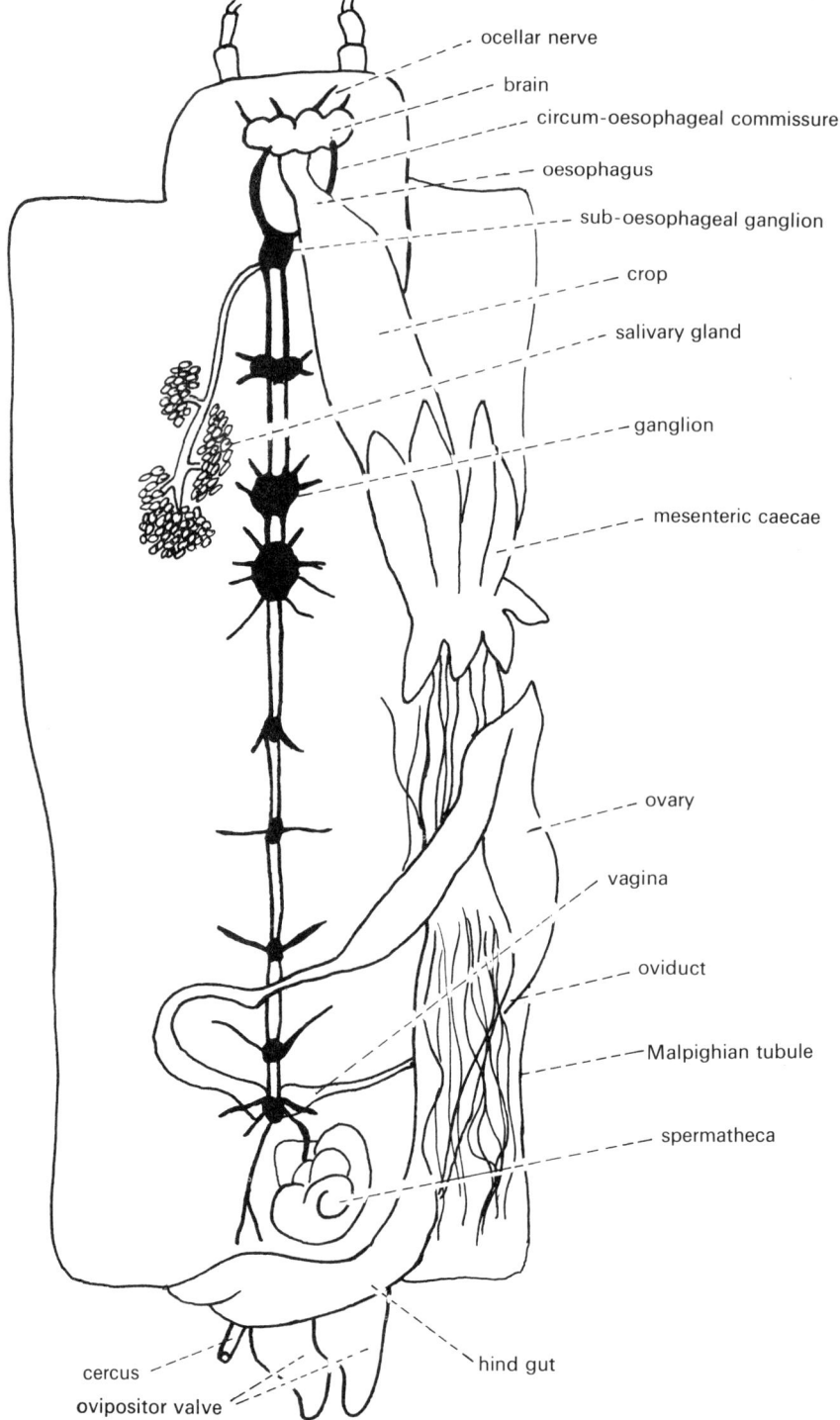

ocellar nerve

brain

circum-oesophageal commissure

oesophagus

sub-oesophageal ganglion

crop

salivary gland

ganglion

mesenteric caecae

ovary

vagina

oviduct

Malpighian tubule

spermatheca

cercus

ovipositor valve

hind gut

Fig. 114. General dissection of the Locust.

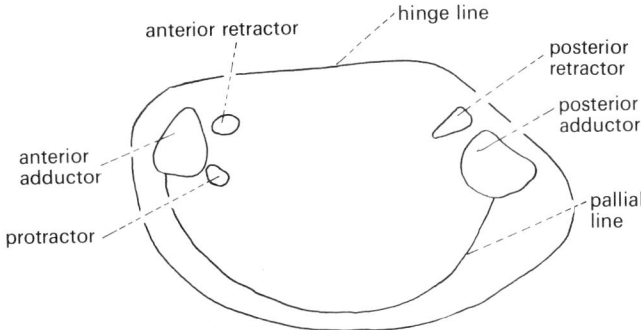

Fig. 115. Muscle impressions on inside of Mussel shell.

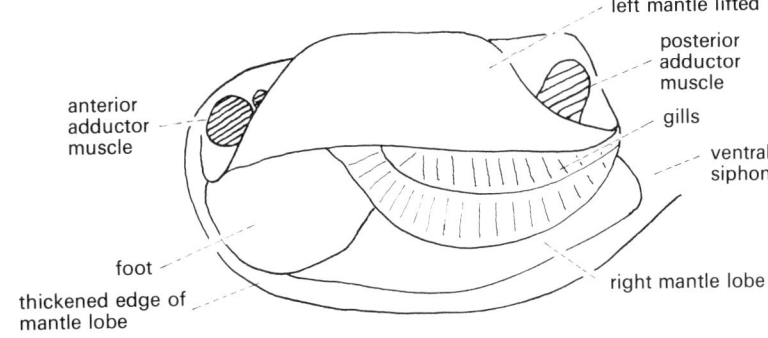

Fig. 116. The Mussel with left valve removed and left mantle lifted.

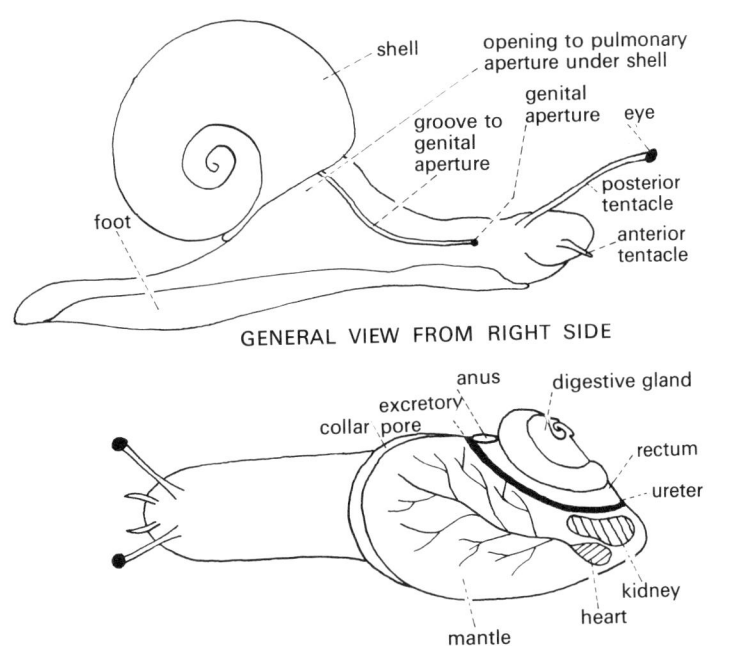

Fig. 117. The Snail.

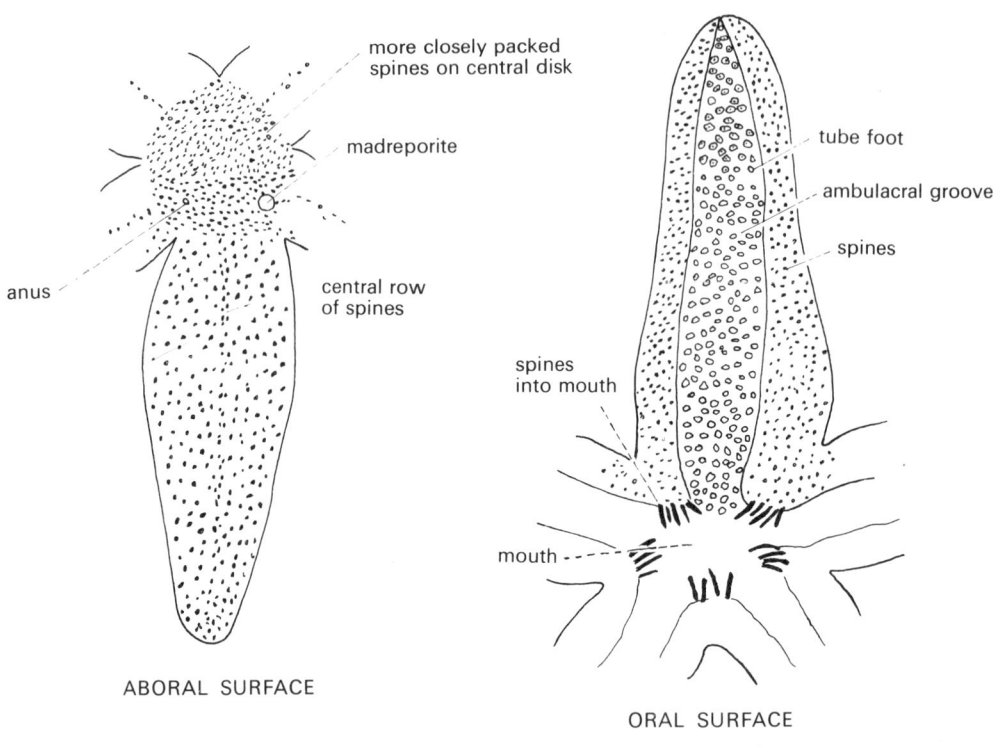

more closely packed
spines on central disk

madreporite

anus

central row
of spines

ABORAL SURFACE

tube foot

ambulacral groove

spines

spines
into mouth

mouth

ORAL SURFACE

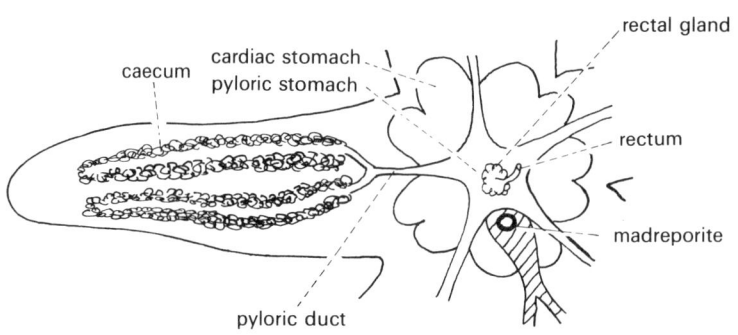

caecum

cardiac stomach
pyloric stomach

rectal gland

rectum

madreporite

pyloric duct

GENERAL DISSECTION FROM ABORAL SURFACE

Fig. 118. Starfish.

ECHINODERMATA

This phylum includes the Starfish, Sea Urchins, Brittle Stars, and Sea Cucumber. Its most striking characteristic is the radial symmetry.

The Star Fish *(Asterias rubens)*

Examine a specimen from the top *(aboral)* side and bottom *(oral)* side. Notice the mouth and anus, and the arrangement of the calcareous plates. Examine the parts of the arm in detail (Figure 118).

Pin your specimen in a dissecting dish, aboral side upwards. From the disk, make incisions down the centres of each arm. Pin back the skin and see the radial symmetry of the internal organs.

CHORDATA

Sub-phylum: CRANIATA

Class: Pisces

The Dogfish *(Scyliorhinus canicula)*

External features

Draw the animal from the side (Figure 119). Observe the position of the various parts, especially the length of the tail, i.e. the muscular region behind the pelvic fin and anus. This may be removed in your specimen or supplied separately.

From the ventral side, make drawings of the mouth and cloacal regions of both the male and female (Figure 120).

Draw a prepared slide of a longitudinal section through the skin, to show the structure of the *denticles* (Figure 121). Notice the similarity to the structure of the mammalian tooth.

The skeleton

The skeleton is cartilaginous, and it is impossible to prepare one in class. Use prepared mounts.

Draw the chondrocranium (see Figure 122), from the dorsal, ventral and lateral aspects. Make sure that you have labelled all the foramina.

Draw the visceral skeleton from dorsal, ventral, and lateral aspects (Figure 123).

Draw the pectoral and pelvic girdles (Figure 124), and fins from above.

Make the following drawings of the vertebrae (see Figure 125): 1. Side views of trunk and caudal vertebrae; 2. Transverse sections of the same vertebrae; 3. Saggital sections through the trunk vertebrae.

We now proceed to the dissection proper.

If possible, soak the preserved specimen overnight in water. This will soften the tissues and remove some of the formalin.

Fix the animal, ventral side up on the dissecting board, by pushing the spikes through the base of the fins.

Make a median incision through the skin only, from just in front of the pelvic girdle, to just behind the pectoral girdle. At the ends of this incision, make lateral incisions. Peel back the two flaps of skin thus formed to the sides. This will reveal the musculature of the body wall. The thick zig-zag of the muscular *myotomes* with the fine *myocommata* between them are clearly visible.

The abdominal viscera and associated blood vessels

Make a median incision and lateral incisions in the muscles of the body wall to give two

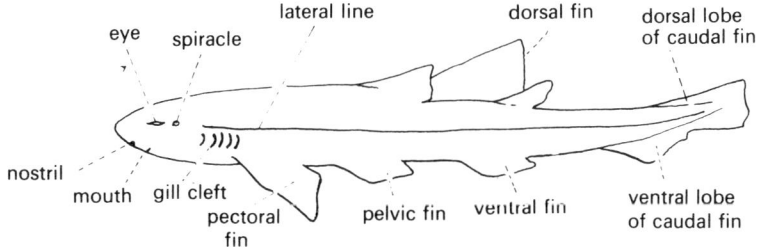

Fig. 119. The Dogfish.

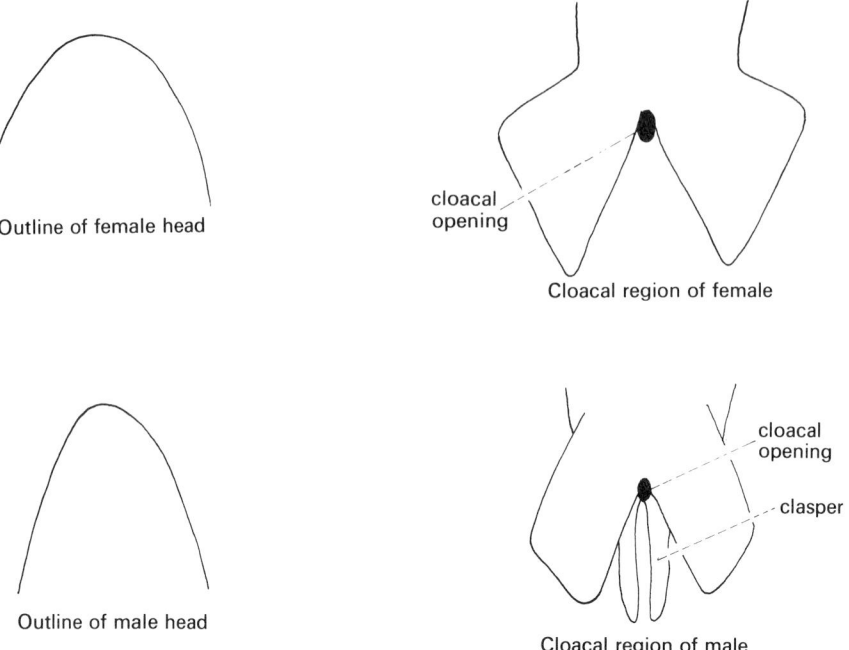

Fig. 120. External features of the Dogfish.

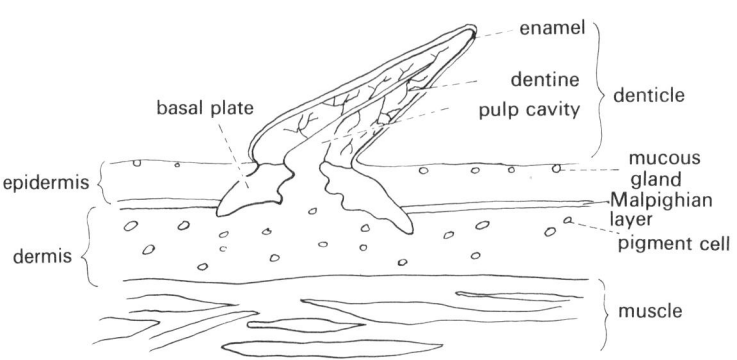

Fig. 121. Vertical section through the skin of the Dogfish.

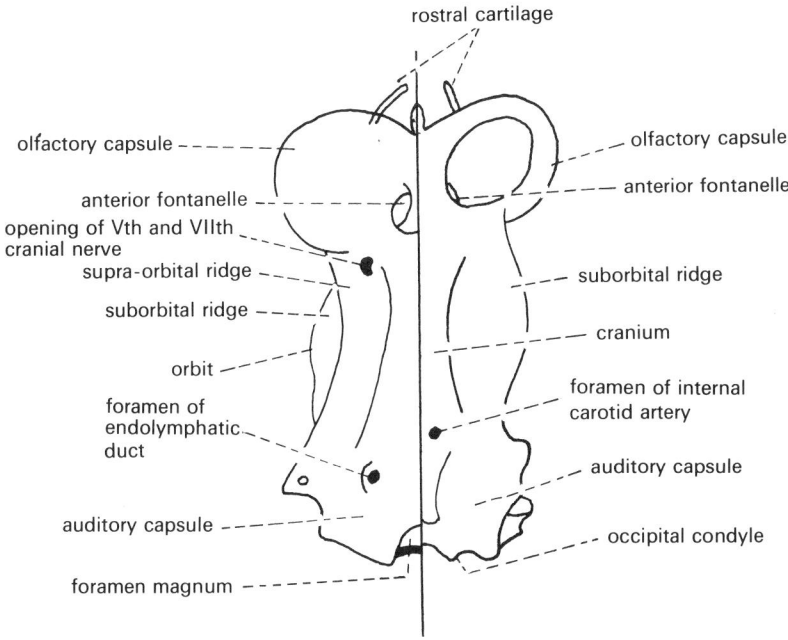

Fig. 122. The chondrocranium of the Dogfish.

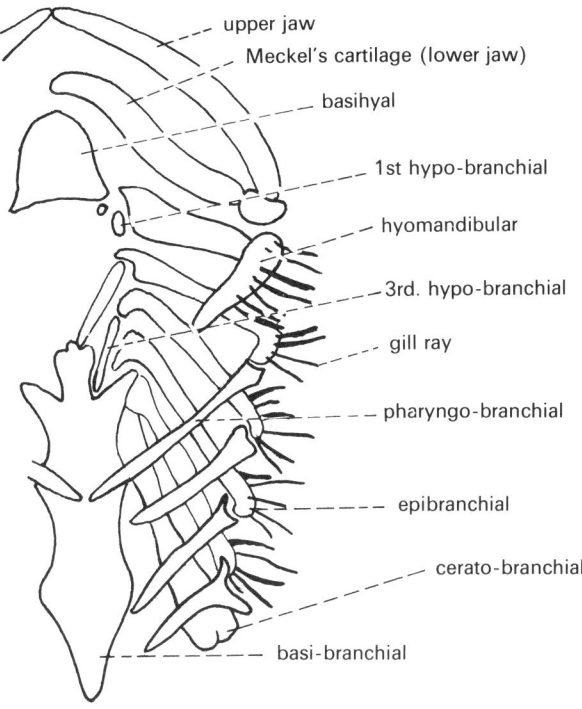

Fig. 123. The visceral skeleton of the Dogfish from above.

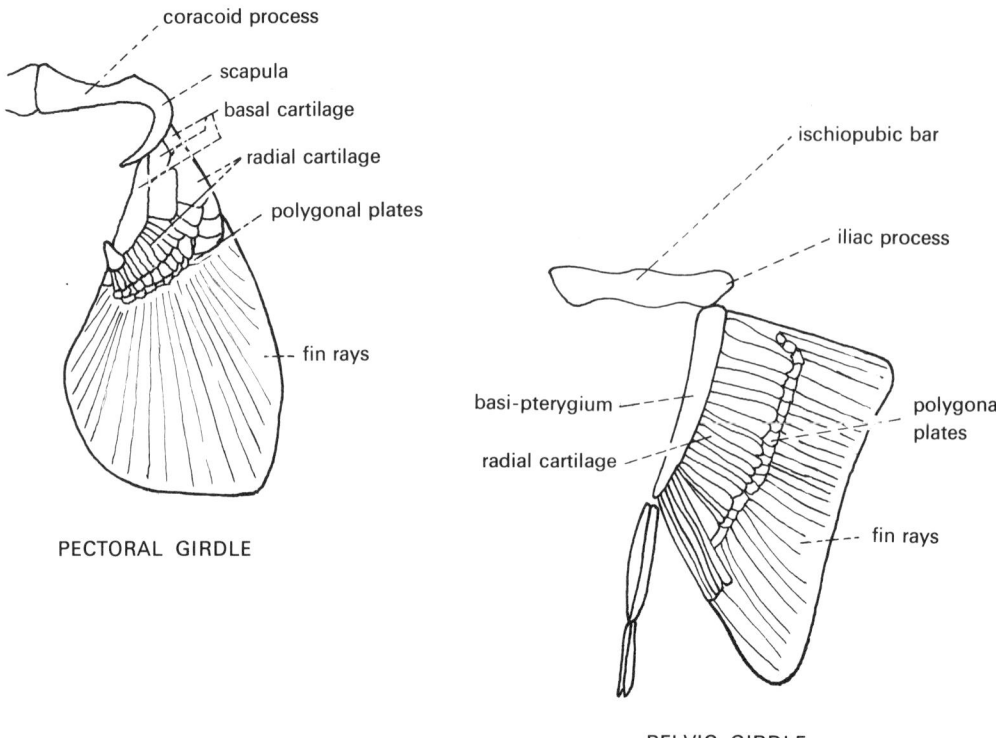

Fig. 124. The limb girdles of the Dogfish.

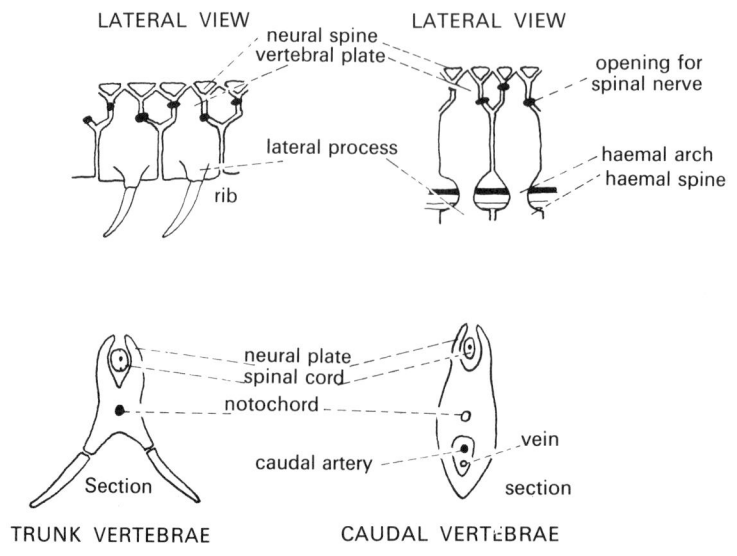

Fig. 125. Vertebrae of the Dogfish.

flaps of tissue as above. Pin back to expose the viscera, and draw them *in situ* (Figure 126).

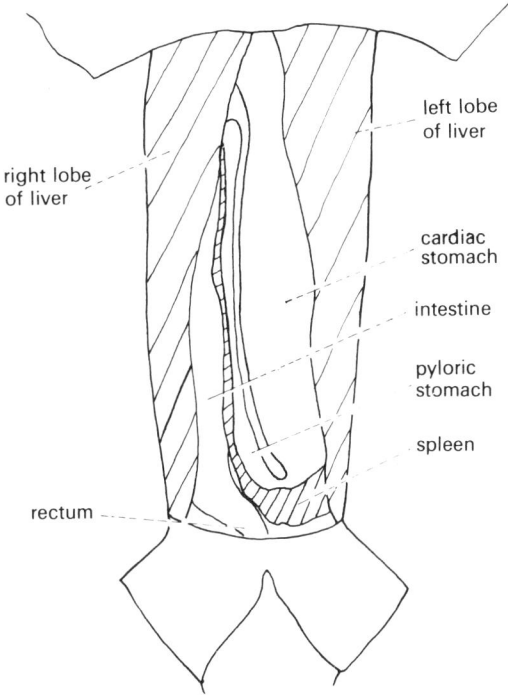

right lobe
of liver

left lobe
of liver

cardiac
stomach

intestine

pyloric
stomach

spleen

rectum

Fig. 126. The viscera of the Dogfish *in situ.*

Dissection of the anterior part of the alimentary canal cannot be attempted without destroying important dissections of the vascular system. The pharynx and buccal cavity will either be shown to you in demonstration, or you will see them later when you dissect out the blood supply to the gills.

The dominant structure in the abdomen is the liver, which is lobed, and grey-yellow in colour. Displace the larger lobe over the area of the pectoral fin, and the right-hand lobe over to the animal's right (Figure 127). The spleen lies on the pyloric stomach near the median line. Displace this to the left, revealing the pancreas. Displace the intestine to the right to reveal the structure below it and draw.

The arteries to the gut may be difficult to distinguish in a preserved specimen, and are probably best discovered by finding the dorsal aorta, and following their various branches out into their respective organs.

Displace the whole alimentary canal over to the right, taking care not to cut the connective mesentary. Deflect the two lobes of the liver forward (Figure 128). This will expose the *dorsal aorta* running medianly and the arteries running to the various organs. From the posterior of the animal, the main arteries to locate are the *posterior mesenteric* to the rectal gland, *lienogastric* to the stomach, spleen and pancreas, *anterior mesenteric* to the posterior end of the intestine and gonads, and the *coeliac* to the anterior end of the stomach, the pyloric intestine, liver and pancreas.

The next step is to expose the hepatic portal system (see Figure 129). The main branches of the *hepatic portal vein* arise from the junction of the two liver lobes, and run underneath the intestine, giving off fine branches to the pancreas, and finally branching into the *posterior intestinal vein* to the posterior part of the intestine, and the *posterior lienogastric vein* to the spleen and stomach. These can be seen by deflecting the liver forward, the stomach and spleen over to the left, and the intestine over to the right.

Having examined the abdominal viscera, the internal structure of the intestine can now be examined. Open the ileum just above the rectal gland, by a small slit. Through the external surface of the ileum you will see a spiral blood vessel; cut up spirally from the slit along the line of this vessel, revealing the spiral valve of the ileum. At the anterior end of the spiral valve the ileum ends, and is joined by the duodenum, and shortly by the stomach. Make a straight incision along the length of both these organs. On the inner wall of the duodenum, you should see the opening of the *bile duct* and the *pancreatic duct*. Examine the stomach content and try to identify the food the animal eats. You are very likely to find white thread-like nematode parasites in the digestive system, and sometimes in the actual body cavity.

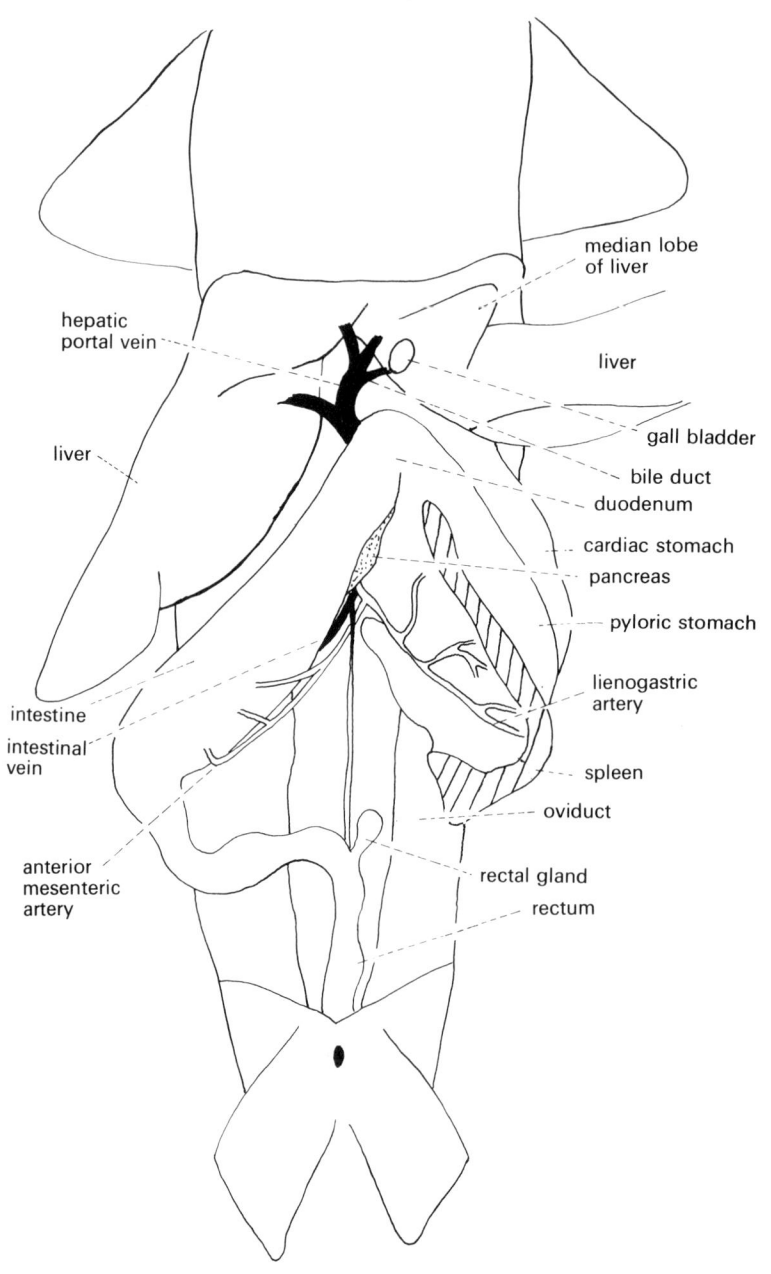

Fig. 127. The viscera of the Dogfish displaced.

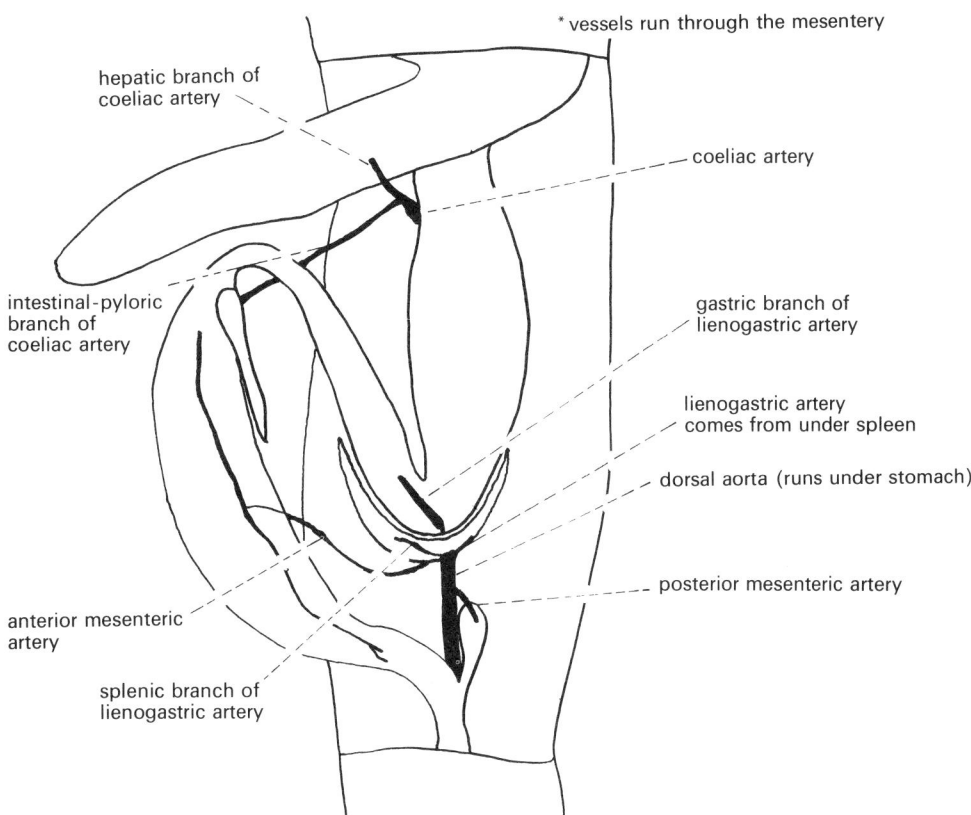

Fig. 128. Arteries arising from the dorsal aorta.

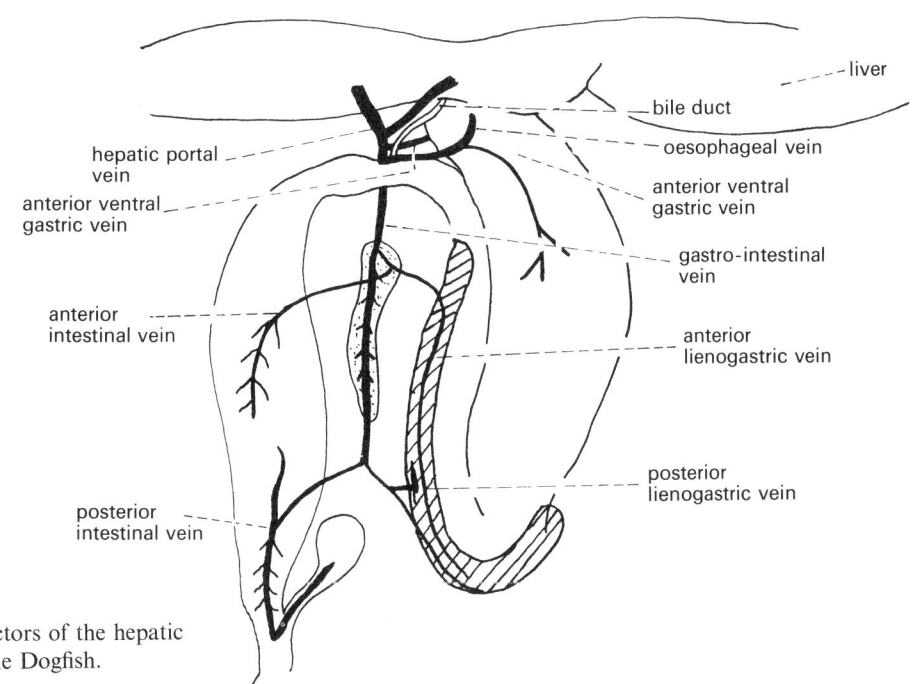

Fig. 129. The factors of the hepatic portal vein of the Dogfish.

Urinogenital system

Remove the intestine by cutting through the oesophagus close to the stomach, and through the rectum. Lift out the intestine by cutting through the mesenteries. Remove the lobes of the liver by cutting them off close to their origin.

The male (Figure 130)

The testes which are already visible on removal of the intestine are extremely delicate and should be handled carefully. Deflect them over to the right, exposing and slightly stretching the *vasa efferentia* and the

mesorchium. Remove the tough *peritoneum* from the rest of the system which lies on the dorsal wall of the body. The mesorchium leads into the *vas deferens* (Wolffian duct) which lies on top of the *mesonephros* (non-functional kidney). Unravel the *vas deferens* as far as possible, and pin it over to the right.

Remove the body wall in the pelvic region, cutting through and removing the pelvic girdle if necessary. Pull the halves of the wall apart, and pin them. This will reveal the *seminal vesicles*, *sperm sac* and the five urinary ducts from the *metanephros* (functional kidney). Separate these and display them. Draw your dissection.

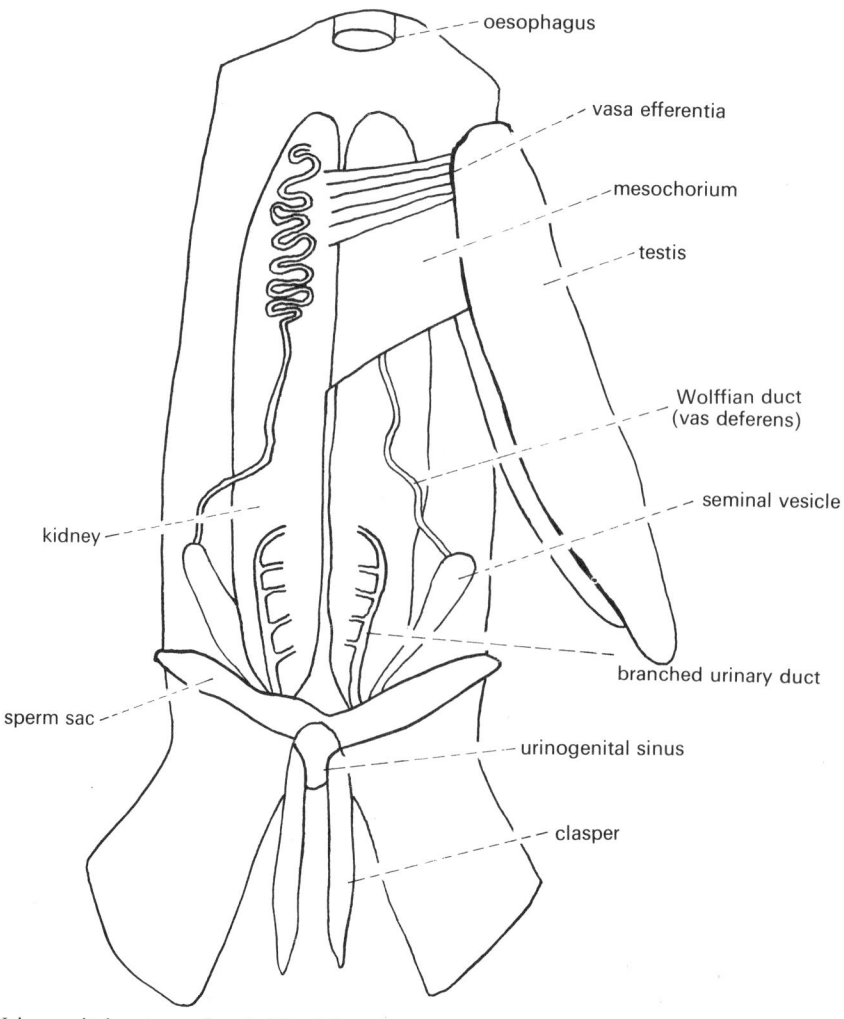

Fig. 130. Urinogenital system of male Dogfish.

The female (Figure 131)

Deflect the single ovary over to the right so as to cover the right oviduct, and to expose the *mesovarium*. Notice the eggs in various stages of development showing through the tissues of the ovary. Remove the peritoneum to expose the non-functional mesonephros. Remove the body wall in the pelvic region, as for the male. Draw your dissection.

The vascular system

The more obvious blood vessels in the abdomen have already been investigated, including the hepatic portal system, but it now remains to demonstrate the *renal portal system* (Figure 132). This can be done, by cutting off the tail near the pelvis (if this has not already been done) and pushing a fine wire into the *caudal vein* which runs through the *heamal arch* of the exposed vertebra. The wire will emerge in one of the kidneys, showing the branching of the caudal vein to form the renal portal system.

Afferent branchial arteries

The part of the body behind the pectoral fin

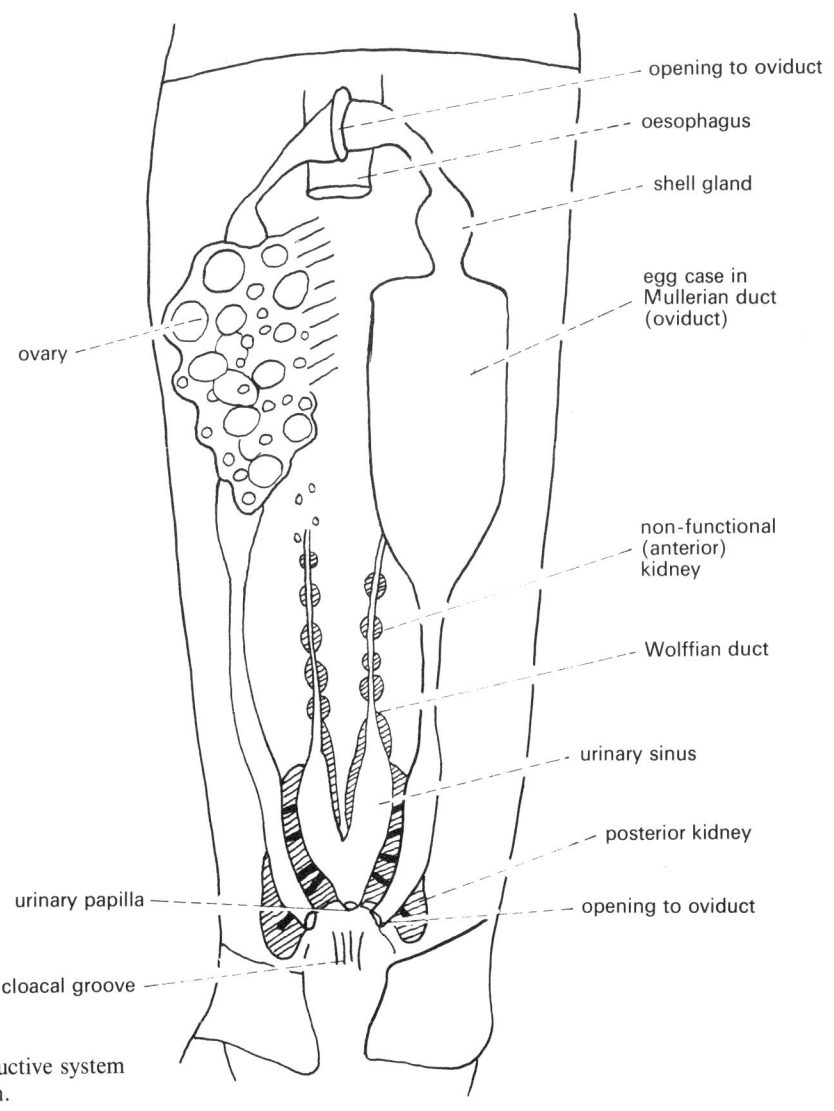

Fig. 131. Reproductive system of female Dogfish.

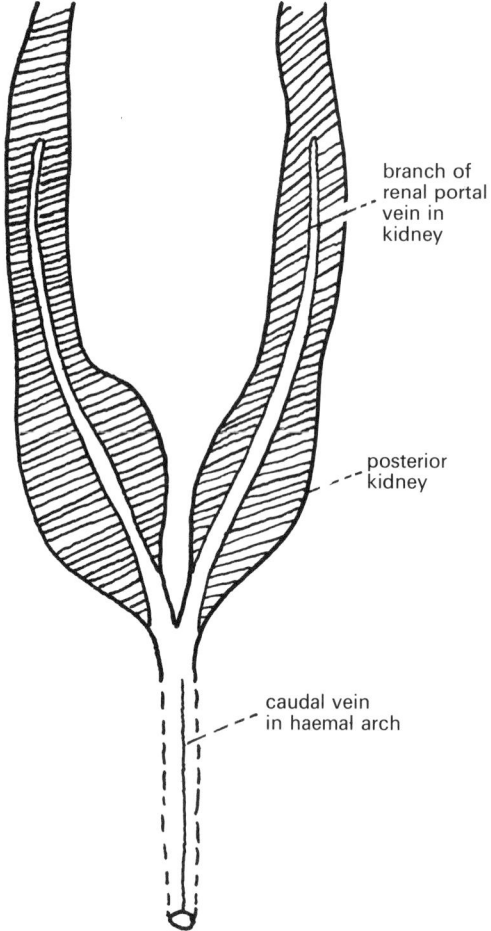

branch of
renal portal
vein in
kidney

posterior
kidney

caudal vein
in haemal arch

Fig. 132. Renal portal system of the Dogfish.

can now be cut off and discarded, but leave
the pectoral fins attached to the part retained,
as they are useful to pin the animal down by.

Re-fix the animal on the board, ventral side
upwards, and push a test-tube into the mouth
and down into the pharnyx. This stretches the
area to be dissected.

1. Remove the skin and connective tissue
from the pectoral girdle area to the lower jaw,
and laterally to the extreme edges of the
ventral surface. This will reveal the *transverse
branchial muscle* and the *mandibular muscle*
lying attached to the lower jaw.

2. Make a *shallow* median longitudinal
incision through the transverse branchial
muscles and lift them individually to either

side of the dissection; then remove them
completely and cleanly. This will reveal a
large muscle running medianly from the
pectoral girdle to the lower jaw *(coraco-
mandibular muscle)* and the three pairs of
extra-branchial cartilages on either side of it.

3. Cut through the muscle near the jaw, and
near the girdle. Lift it off to reveal a pair of
muscles running similarly *(coracohyoid
muscles)*.

4. Remove this muscle as before. This will
reveal the *thyroid gland* just posterior to the
insertion of the coracohyoid muscle (about
one-third from the front of the animal in the
median line).

5. Lift the thyroid gland and the *ventral
aorta* and the origins of the two *innominate
arteries* will be exposed. Trace the *first* and
second afferent branchial arteries laterally from
the *innominate* artery on either side (6).

7. The thin-walled *inferior jugular sinus* runs
on either side, from about half way along the
innominate artery, and immediately posterior
to it, to the pectoral girdle. You may well
have damaged its ventral wall, but in any
case, lift this wall by inserting a seeker into
the cavity through the opening behind the
innominate artery, and remove the wall.[8]
This will reveal the *3rd, 4th* and *5th afferent
branchial arteries* which you will see through
the dorsal wall of the sinus.

Locate the *ventral aorta* lying in the mid
line between the five pairs of *coracobranchial
muscles*. To display the aorta it will be
necessary to cut through a small amount of
muscle. Now cut through the origin of these
five muscles on the pectoral girdle. Each
muscle can now be lifted separately, with their
bases between the arteries running to the gills.

Cut through the insertions of these muscles
separately and remove them, leaving only the
stumps between the branchial arteries. The
bases of all the afferent branchial arteries
should now be visible and have to be traced
out to the gills (Figure 134).

Working from the outside, open the gill
clefts to where the gill lamellae end. In all
these operations, great care must be exercised

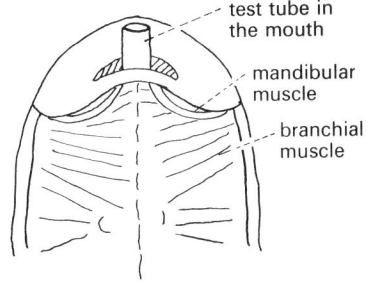

1. SKIN REMOVED

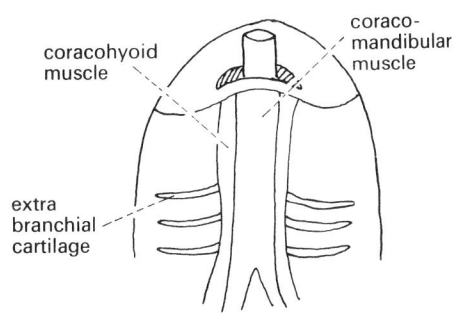

2. BRANCHIAL MUSCLES REMOVED

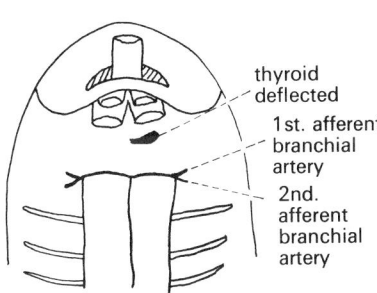

3. AFTER REMOVAL OF CORACO-MANDIBULAR MUSCLE

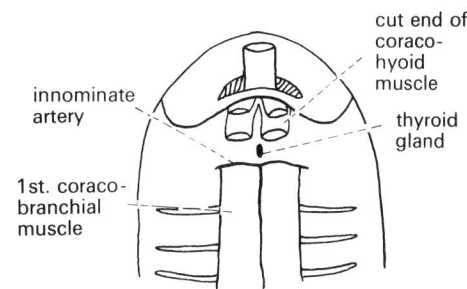

4. AFTER REMOVAL OF CORACO-HYOID MUSCLE

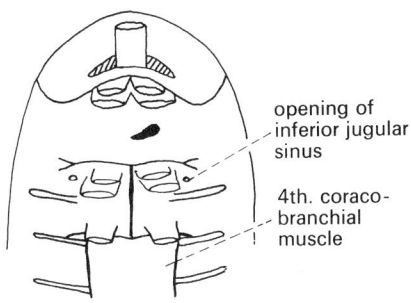

5. AFTER DEFLECTION OF THYROID AND REMOVAL OF TISSUE

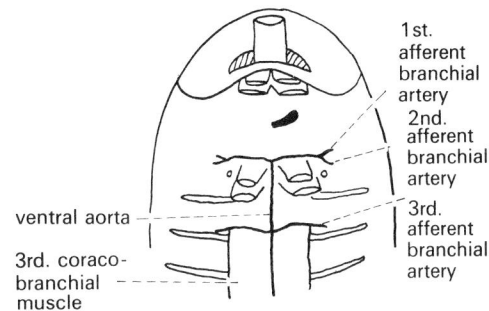

6. AFTER REMOVAL OF 1st AND 2nd CORACO-BRANCHIAL MUSCLE

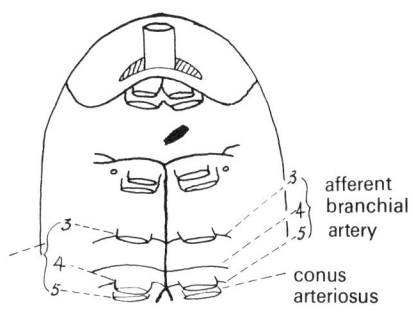

7. AFTER REMOVAL OF 3rd. CORACO-BRANCHIAL MUSCLE

8. AFTER REMOVAL OF 4th AND 5TH CORACO-BRANCHIAL MUSCLES

Fig. 133.

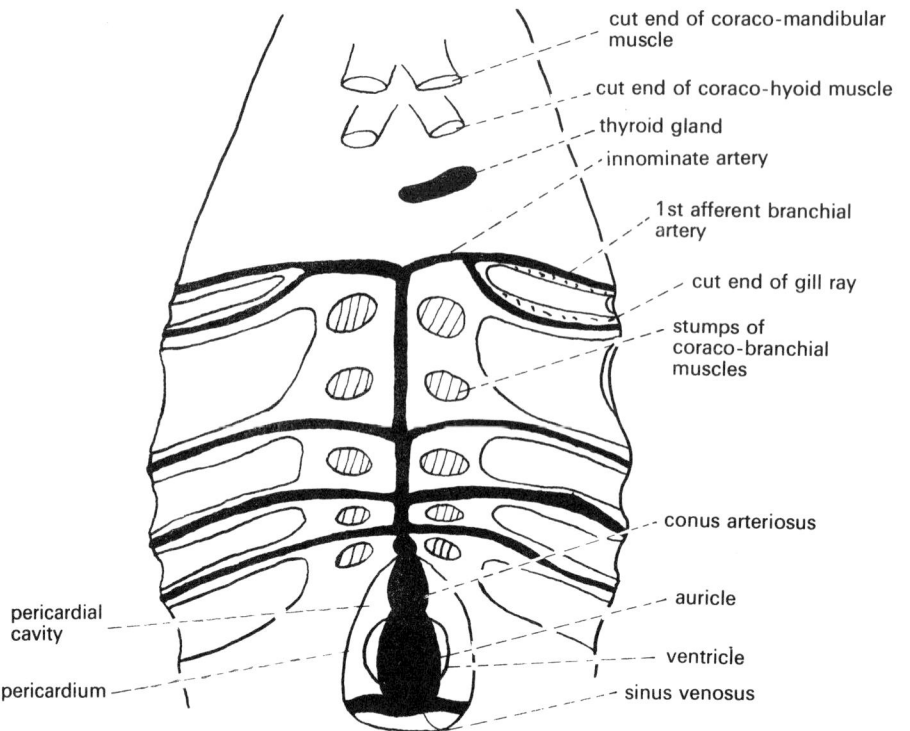

Fig. 134. Completed dissection of afferent branchial arteries.

so as not to cut through the arteries. Remove any tissue overlying each gill arch; cut through the posterior hemibranch over each arch, and lift them out. This will expose the gill rays and the afferent branchial arteries. Once the arteries are visible, the gill rays can be removed, and the anterior hemi-branch cut down to each afferent branchial artery. The dissection can now be cleaned up to expose the afferent branchial system completely.

The heart

Remove the cartilage and other tissue from the base of the ventral aorta to expose the heart. Draw it. Open the *sinus venosus* by making a transverse cut in the ventral wall. Clean the blood from it. This will expose the opening of the various sinuses. Draw it (see Figure 135).

Remove the heart from the animal, and cut it in half longitudinally. Draw to show the internal structure.

The efferent branchial system (Figure 136)

Cut through the right (your right) hand side of the pharynx, along through the gill slits. This will give the ventral part of the animal as a loose flap, which should be opened over to the left and pinned down. This exposes the roof of the buccal cavity and pharynx. Wash this to remove the mucus. The *1st, 2nd* and *3rd epibranchial arteries* are now visible as faint lines through the mucous membrane. Cut through the oesophagus, and deflect the remains of the viscera to the left. Cut through the *posterior cardinal sinus* exposing the paired *subclavian arteries* (running to the pectoral fins) and the *coeliac artery*, all arising from the *dorsal aorta*.

Lift off and pull away the mucous membrane from the surface of your dissection. This will expose the dorsal aorta and the four epibranchial arteries, close to which lie the *pharyngo-branchial cartilages*. These latter now have to be removed.

The first pair are fused by a small V-shaped piece of cartilage in the mid-line. Remove this, and then lift up each of the pharyngo-branchial cartilages, removing them by cutting through the attaching muscles.

Now remove the *epibranchial cartilages* which lie immediately to the outside of the pharyngo-branchial cartilages in your dissection. This will show the loops of the *efferent branchial arteries* around the gill pouches.

Removal of the *hyoid arch* (the cartilage in front of the first gill pouch) will reveal the *hyoidian artery*.

The other vessels forming a network, and arising from the anterior part of the dorsal aorta, can now be cleaned up and exposed, but some of them are difficult to trace and expose fully.

The Cranial nerves

This dissection should be carried out on the side that has not been cut in investigating the efferent branchial system.

Dissection of the orbit (Figure 137)

The eyeball has to be removed to reveal the cranial nerves passing through the orbit. Cut through the skin joining the eyeball to the outer skin. This will free the eyeball which can now be pulled outwards from the socket in which it is held by the *4 rectus muscles* and the *2 oblique muscles*. It may be necessary to deflate the eyeball, by cutting into it and removing the lens. Pull the eyeball as far as possible out of the socket, and cut through the six muscles and the *optic nerve* close to the eyeball which can now be removed and discarded.

The lower eyelid should be pulled down and pinned, and the cartilage over the top of the orbit (supraorbital ridge) cut away, but do not damage any nerves here. The following cranial nerves will now be visible.

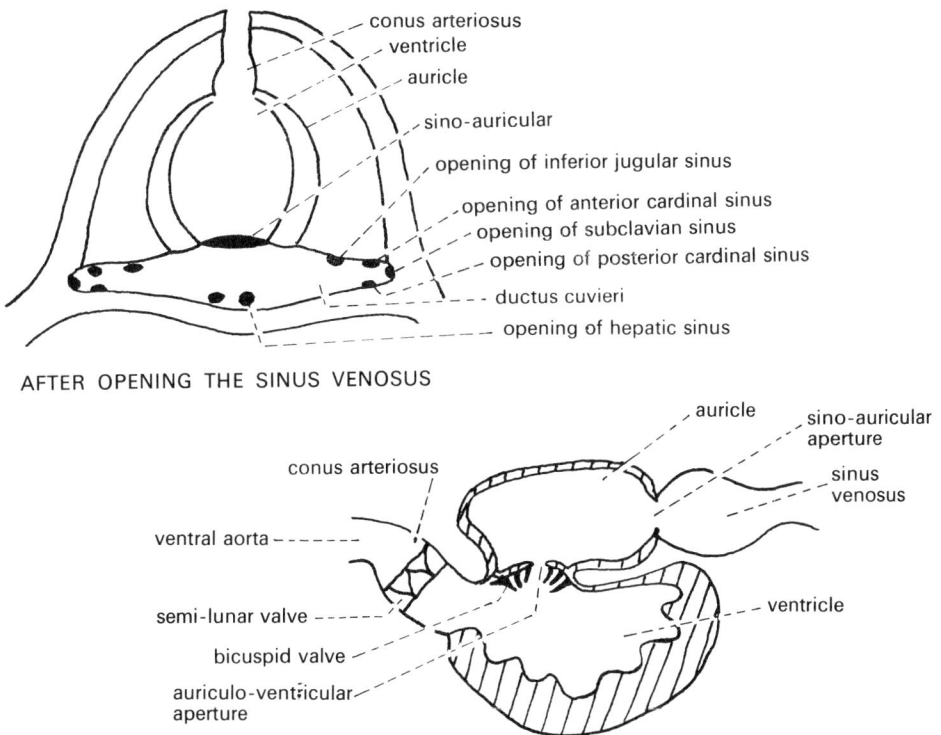

Fig. 135. Dissection of the heart of the Dogfish.

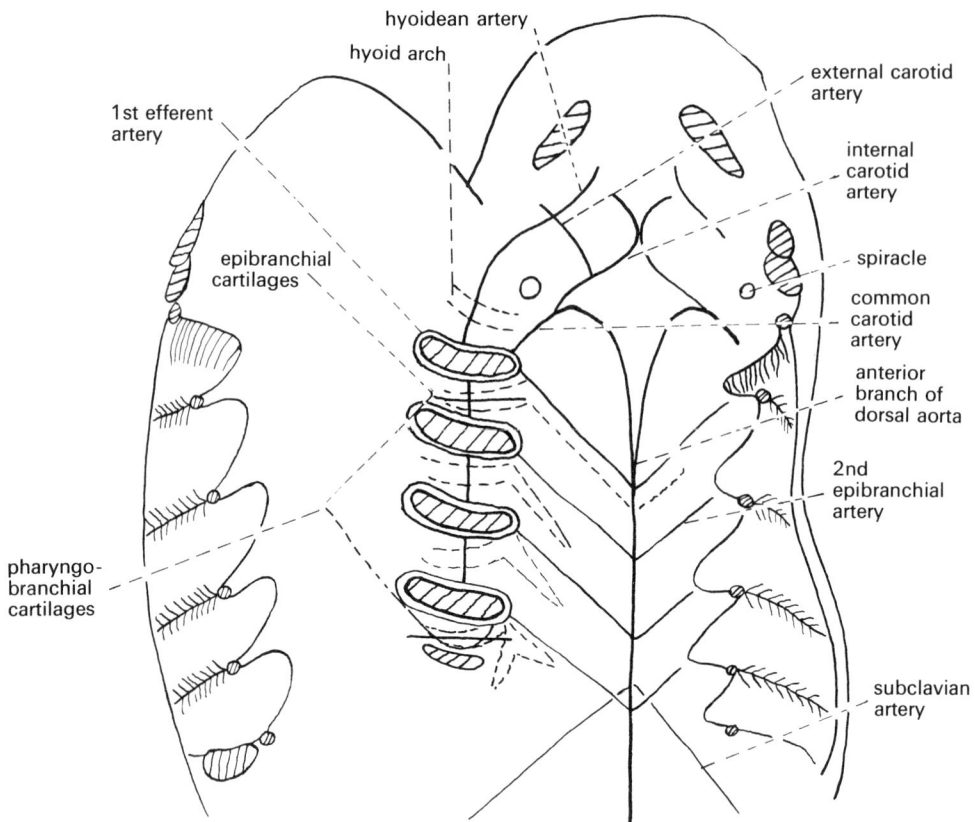

Fig. 136. Efferent branchial system of the Dogfish.

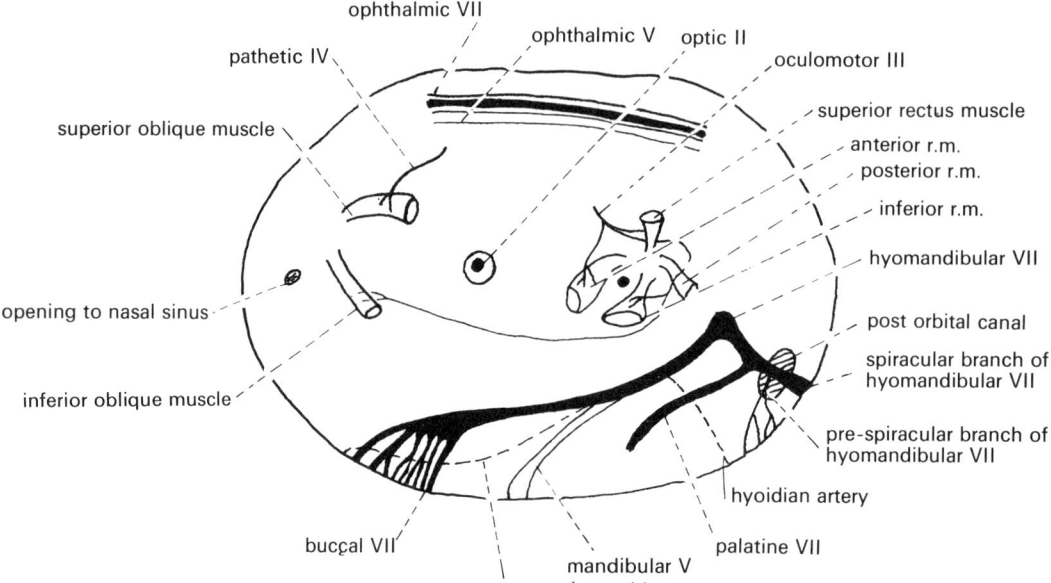

Fig. 137. Dissection of the orbit of the eye of the Dogfish.

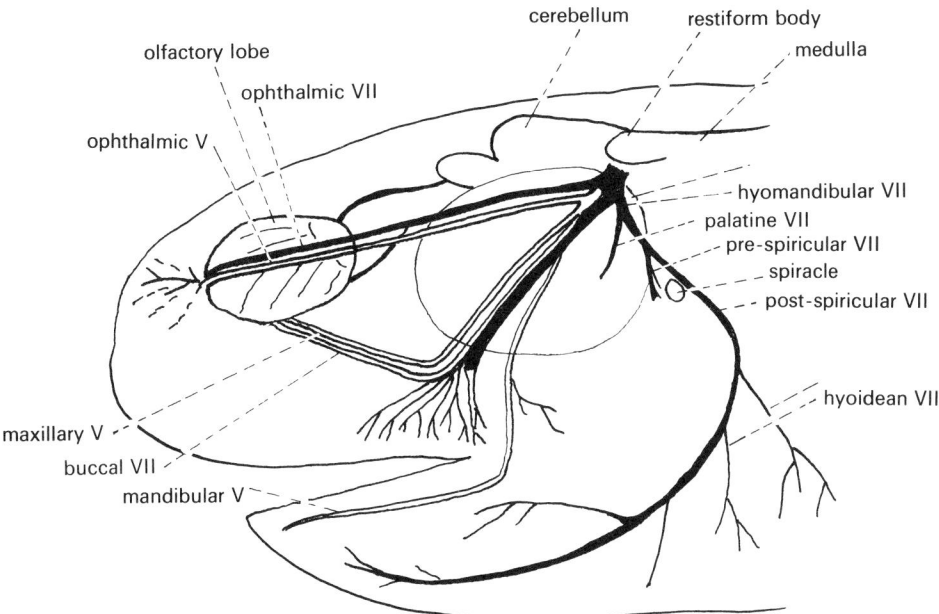

Fig. 138. The Vth and VIIth cranial nerves of the Dogfish.

II—*Optic*—entering the centre of the orbit, and already cut through in removing the eyeball.

III—*Oculomotor*—enters the orbit near to, and a little to the front of the insertion of the four rectus muscles. A long branch runs to the inferior oblique muscle, and can be seen by lifting the muscle. Other branches run to the rectus muscles and are made visible by moving these muscles.

IV *Trochlear* (Pathetic)—runs to the superior oblique muscle. It is very short and can be seen by deflecting the muscle outwards.

V—*Trigeminal*—this is a very large nerve entering the orbit just behind the rectus muscles, and running forward along the bottom of the orbit, closely associated with the VII. This branch divides into two, the *maxillary* running forward, and the *mandibular* running downwards. Another branch of the Vth runs across the top of the orbit—the *ophthalmic* branch—with a branch of the VIIth.

VI—*Abducens*—this is a very fine small nerve that can be seen by displacing the posterior rectus muscle forwards.

VII—*facial*—this enters the orbit with the Vth, but branches immediately to form the *buccal* which runs alongside the maxillary V. A second branch—*hyomandibular*—passes out through the posterior side of the orbit, and gives rise to a branch—*palatine* which runs parallel to the buccal branch. The third branch—*ophthalmic*—of VII runs across the top of the orbit with the *ophthalmic* V.

The Vth and VIIth Cranial nerves (Figure 138)
These two nerves are clearly visible in the orbit, and this should be the starting point for the dissection. Start with the ophthalmic branches, and trace them forward into the snout, where they pass over the olfactory organ. Remove the skin and muscle over the snout, and the nerves are easily discovered.

Next trace the maxillary branches over the cartilage of the upper jaw, and the mandibular V along the cartilage of the lower jaw. The mandibular V runs backwards slightly after leaving the orbit to pass behind the angle of the jaw.

The hyomandibular VII should now be traced. After giving off the palatine branch, it branches again near the spiracle to give the short *prespiracular VII* immediately anterior to the spiracle, and the much larger *post spiracular VII*. Trace the latter carefully downwards when it branches again to form the *external mandibular* branch to the ampullary canals of the lower jaw, and the *hyoidian branches* to the hyoid arch.

These nerves should now be traced into the brain. To do this the rectus and oblique muscles will have to be removed, together with their associated nerves. Then carefully chip away the cartilage of the cranium until the origin of the nerves is revealed in the brain.

The IXth and Xth cranial nerves (Figure 139)

Continue this dissection on the same side as you have used for the Vth and VIIth cranial nerves. Find the *anterior cardinal sinus*. This lies immediately behind the spiracle (or the postspiracular VII which you have exposed), and above the gill slits. It is easy to find as it is softer to the touch than the surrounding muscles. Open the sinus by a longitudinal incision and remove the roof of it. This will reveal the *IXth cranial nerve* at the anterior

end of the sinus running to the first gill slit and the *branchial branches of the Xth* running to the other gill slits. Remove the thin lower wall of the sinus to fully expose these nerves which lie immediately below it. The IXth and the branchial branches of the Xth, further branch each into a *pretrematic* and *post-trematic* branch, running to the anterior and posterior edges of the gill clefts respectively. Carefully dissect out these branches.

Trace these branches of the Xth to their origin which is in the *visceral Xth*. This gives branches to the stomach, heart and other organs, but it cannot be traced to its ultimate destination as the organs have been removed. Trace the visceral Xth as far as possible anteriorly and posteriorly.

Slightly more dorsally lies the *lateral Xth*, a branch which innervates the lateral line. It lies deeper in the muscle than the others, but its origin near the brain should now have been uncovered. Trace it backwards for a short distance.

Finally, follow both the IXth and Xth to their origins in the brain. In doing this you will destroy the auditory capsule, and the VIIth (auditory) cranial nerve.

The Brain

Remove the rest of the cartilage of the cranium, over the dorsal surface of the brain,

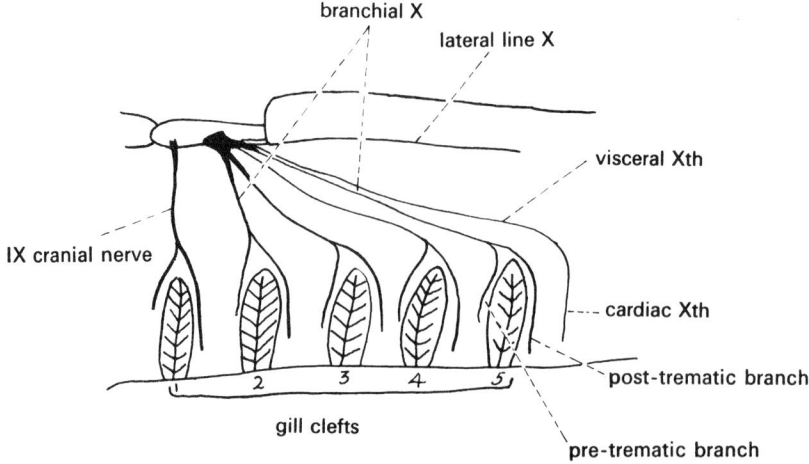

Fig. 139. The IX and Xth cranial nerves of the Dogfish.

working forward to expose the olfactory organ, and backwards removing some of the neural arches of the spine to expose the spinal cord. This is best done by removing the rest of the skin and underlying tissues from the skull and then making a small hole in the centre of the cranium, using the tip of a small scalpel. Be careful not to push the point into the underlying brain. Using this hole as a starting point, chip the cartilage away from the brain until it is completely exposed. During this operation, you will almost certainly remove the *pineal body,* and the *pineal stalk.* Before trying to handle the brain, it is advisable, though not essential, to soak the whole dissection in 70% ethanol for 2–3 days to harden the brain tissue.

Starting from the spinal cord, lift the brain out of the cavity, working forward and cutting the roots of the cranial nerves as near to the cartilage as possible. Lift the complete brain out and draw it in dorsal, ventral and lateral aspects (Figures 140, 141, 142).

Now cut and draw a median longitidinal section through the brain to show the ventricles (Figure 143). Notice the relative thicknesses of the nerve tissue forming the walls of the ventricles.

Sections through the body

Examination of body sections is important to help to realise the relationships between the different systems which have been dissected separately. A new specimen will be necessary if you are to cut the sections yourself, but you will probably be provided with prepared mounts. The sections should be about one inch thick, and taken from the tail *(caudal),* between the pectoral and pelvic fins *(visceral* or *abdominal)* and across the gill *(branchial* or *pharyngeal)* region (Figure 144). The exact

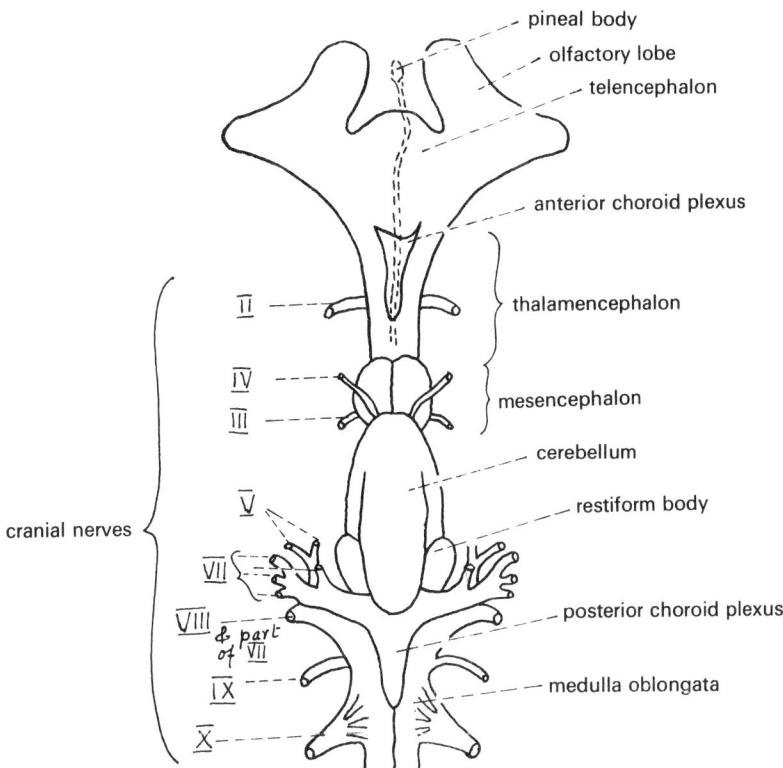

Fig. 140. The dorsal view of the brain of the Dogfish.

A Textbook of Practical Biology

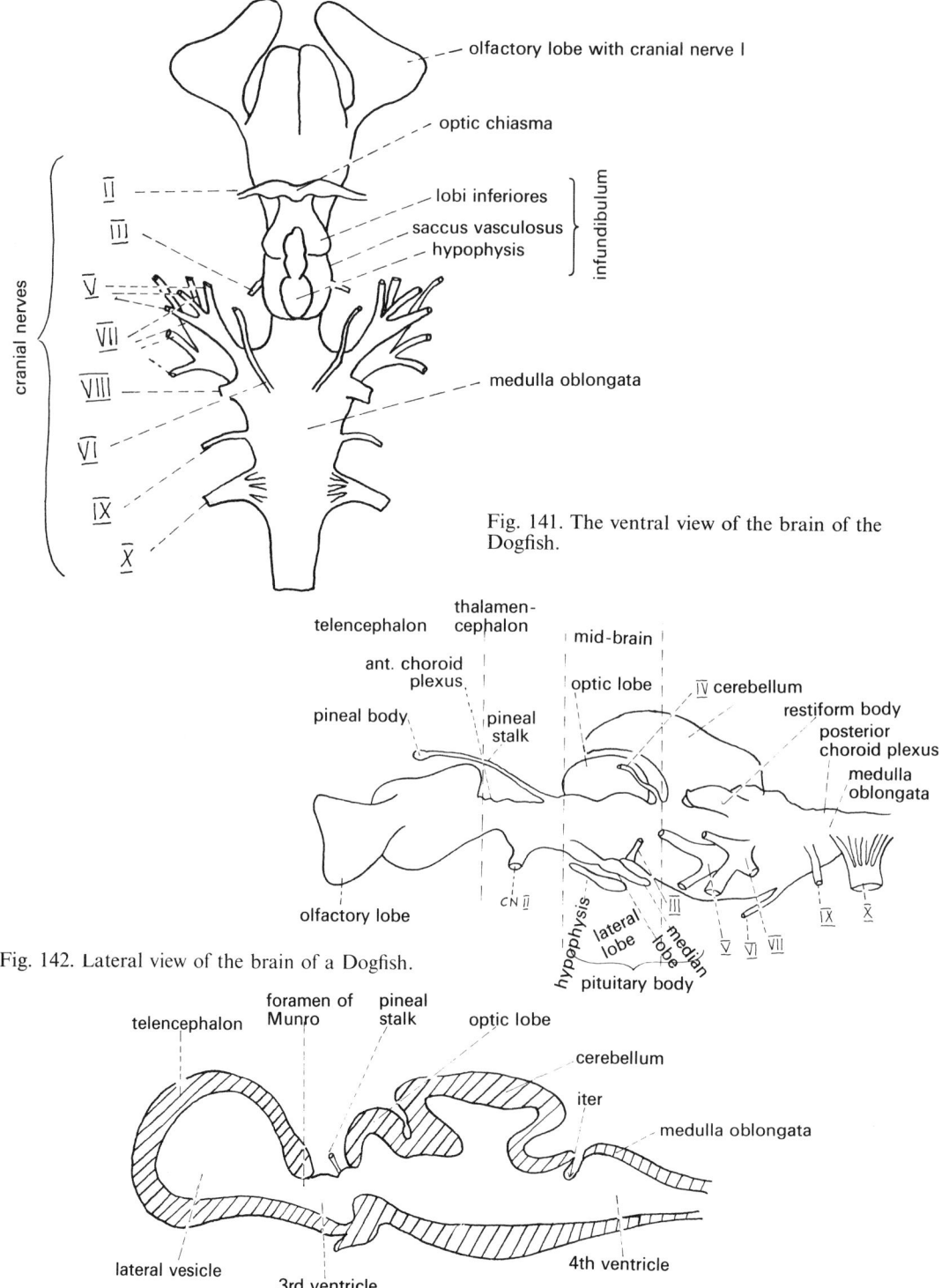

Fig. 141. The ventral view of the brain of the Dogfish.

Fig. 142. Lateral view of the brain of a Dogfish.

Fig. 143. Median longitudinal section through the brain of the Dogfish.

structures seen will depend on the precise position of the sections.

The Bony or Teleost Fish e.g. Minnow *(Phoxinus)*

Apart from the skeleton being of bone rather than cartilage (as in the dogfish, an *elasmobranch* fish), the bony fish is distinguished externally by the absence of the spiracle, the covering of the four gill slits by an *operculum*, and the body covering of bony scales (Figure 145).

Draw your specimen from the side. Lift the operculum to see the gill slits. Remove a scale and examine it under the microscope.

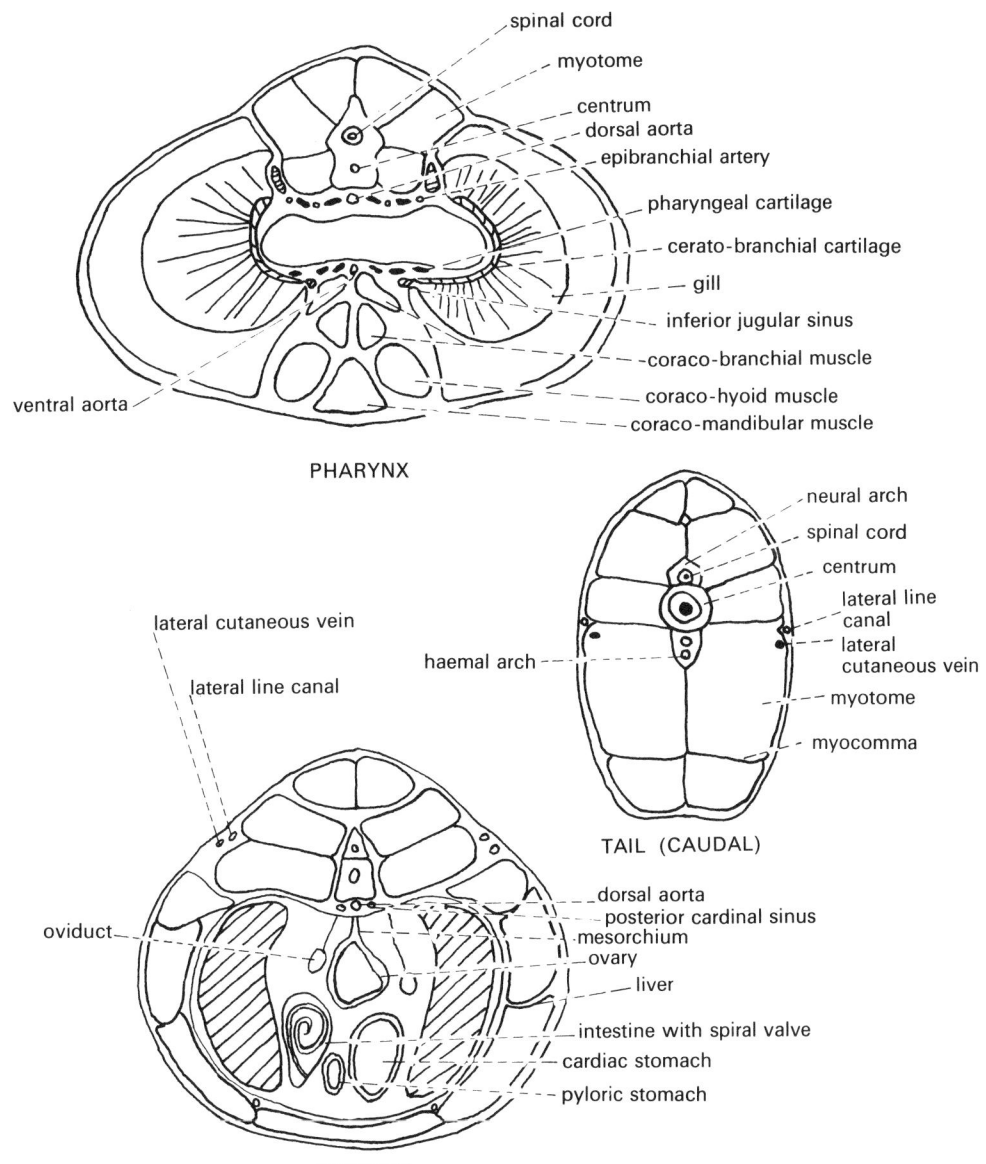

Fig. 144. Transverse sections through the body of the Dogfish.

Class: Amphibia

The Frog *(Rana temporaria)*

Examine a living specimen to note the following:

1. The jumping movement: Notice the predominant use of the muscular back legs in leaping.

2. Breathing movements: The air is pumped in through the nostrils by the movement of the floor of the mouth (the 'buccal pump' mechanism). This is brought about by the movement of the *hyoid*. After the pumping has proceeded for sometime, the nostrils are closed and the air forced into the

Repeat with another frog in a box painted black, using moss or horticultural peat to keep it moist. After 24 hours, place the two frogs side by side and notice the difference in colour. This adaptation is brought about by the contraction of the pigment-containing *melanophores* (see Figure 146), giving the light colour, and their relaxing giving the dark colour. In conjunction with this, examine prepared slides of the skin.

External features

Draw a side view of the head, noticing the position of the *nostril* and *eye* both of which

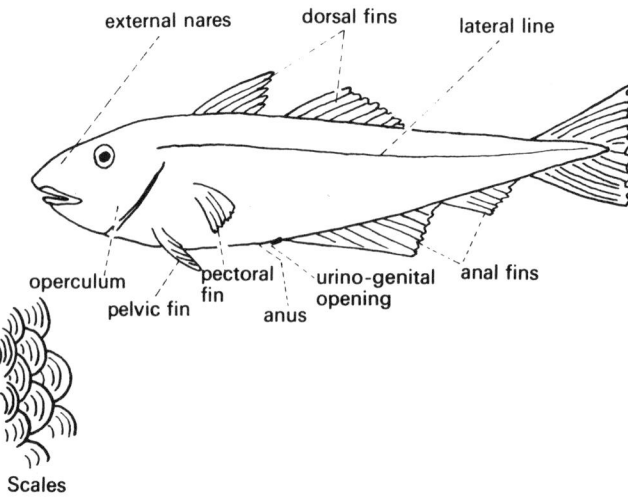

Fig. 145. A typical Teleost fish.

lungs. How often are the nostrils closed in relation to the number of pumping actions of the hyoid? Is there any difference before and after activity?

3. Feel the skin. What do you note? Has this any connection with its function as a respiratory surface?

4. Under the skin are large lymph spaces. These are clearly demonstrated in the hip region, where the lymph 'hearts' can be seen pulsating.

5. Place a frog in a box, the inside of which has been painted white. Keep it moist by putting damp white cotton-wool in with it.

could be high on the head when it is in a certain position. Why is this? The mouth is extremely wide, open it and see the position of the tongue and its attachment. Do not damage any structure in doing this. Examine the *eye* noticing the *nictitating membrane*. Just behind the eye, and a little lower is the *tympanic membrane*.

Draw the fore-limb and the hind limb. Notice the difference in size. The fore-limb has four digits, and in the male a dark-brown or black *nuptial pad*; the digits are not webbed. There are five webbed digits to the hind limb.

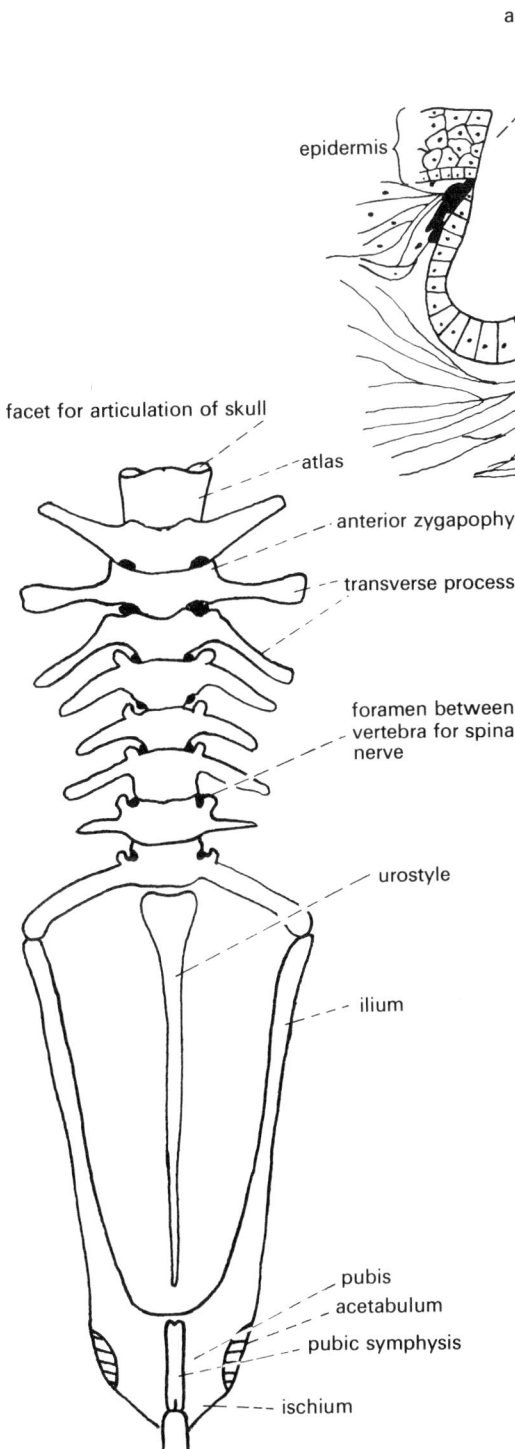

aperture to mucous gland

mucous gland

epidermis

Malpighian layer

melanophore

blood vessel

connective tissue

Fig. 146. A section through the skin of the Frog.

facet for articulation of skull

atlas

anterior zygapophysis

transverse processes

foramen between vertebra for spinal nerve

urostyle

ilium

pubis

acetabulum

pubic symphysis

ischium

Fig. 147. Ventral view of the axial skeleton of the Frog.

The skeleton

Use a mounted articulated skeleton, and separate bones.

The axial skeleton

Make a drawing of the complete vertebral column from above. (Figure 147). Notice the absence of *transverse processes* on the *atlas* and the large transverse processes on vertebrae 2–4, for the attachment of powerful back-muscles, while those of vertebrae 5–8 are smaller, having the less powerful abdominal muscles attached to them. The processes of the 9th vertebra articulate with the iliac bones of the pelvic girdle. They are large as they take part in the thrust when the animal is jumping.

Draw the *atlas,* and any of vertebrae 2–7 (which may be considered typical) from the front and above (Figure 148). The centrum of the atlas is reduced, and there are two large anterior *condyles* for the articulation of the skull. The typical vertebra has *post zygopophyses* and *anterior zygopophyses*, the former facing downwards and the latter facing upwards. This will help in orientating the bone. These vertebrae are structurally similar to those of the rat.

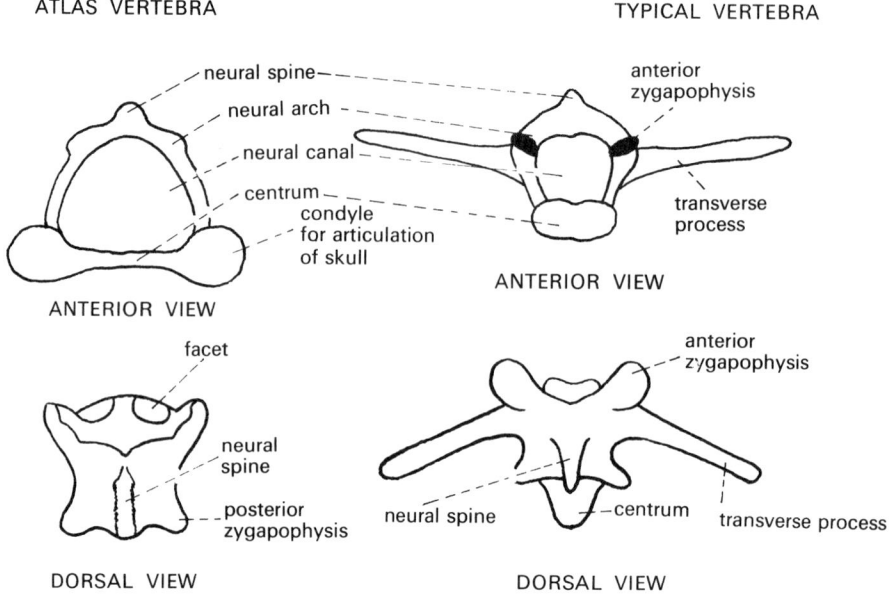

Fig. 148. Vertebrae of the Frog.

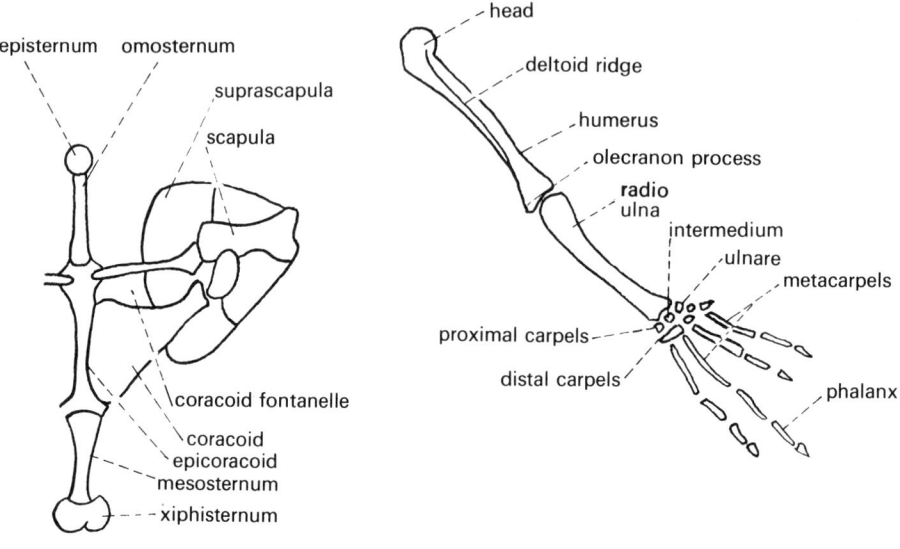

VENTRAL VIEW OF RIGHT SIDE
OF PECTORAL GIRDLE

RIGHT FORE-LIMB FROM ABOVE

Fig. 149. Anterior appendicular skeleton of the Frog.

The pectoral girdle and fore-limb

Draw the girdle from below, and the limb (Figure 149). The limb has the typical pentadactyl pattern, but the typically separate radius and ulna are represented by the single *radio-ulna*, with a smaller number of wrist bones than is typical.

The pelvic girdle and hind limb (Figure 150)

The girdle has already been drawn in dorsal view, with the axial skeleton, but now draw it from the side. In the hind limb, the tibia and fibula are united as the *tibio-fibula*, and the two 'ankle' bones—*calcaneum* and *astragalus*—are much larger than is typical,

The points to notice are:

1. The lower jaw made of two fused *Meckel's cartilages*.

2. The *hyoid apparatus* lying beneath the lower jaw. Recall its function during breathing.

3. The simple *columella*.

4. The extremely large *orbit*, which is open below (except for a membrane) into the buccal cavity.

5. The opening for the cranial nerves.

6. The very fine teeth, especially those on the *vomers*, the latter forming the floor of the nasal cavity.

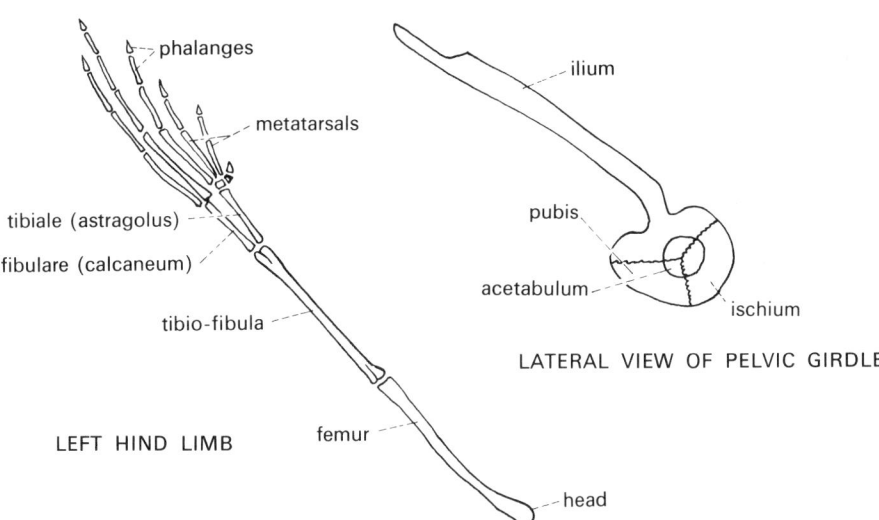

Fig. 150. Posterior appendicular skeleton of the Frog.

with a reduction in the number of the other 'ankle' bones.

The Skull

Draw the skull in dorsal, lateral and ventral aspects (Figure 151). The skull is essentially of cartilage, and some of this remains in the mature skull. Some of the cartilage has become ossified during development to form *cartilage* (or *replacement*) *bone*. Other, overlying membranes have also become ossified to form *membrane* (or *dermal*) *bones*.

General dissection

It is preferable to use a freshly killed specimen, especially for the dissection of the vascular system.

Pin the animal in a dissecting dish, but do not fill it with water until the ovaries (of the female) have been uncovered, and removed if too large. If they are moistened, the albumen covering the eggs will swell, and they will obscure the rest of the organs. Stretch the limbs and pin through the feet. Lift the

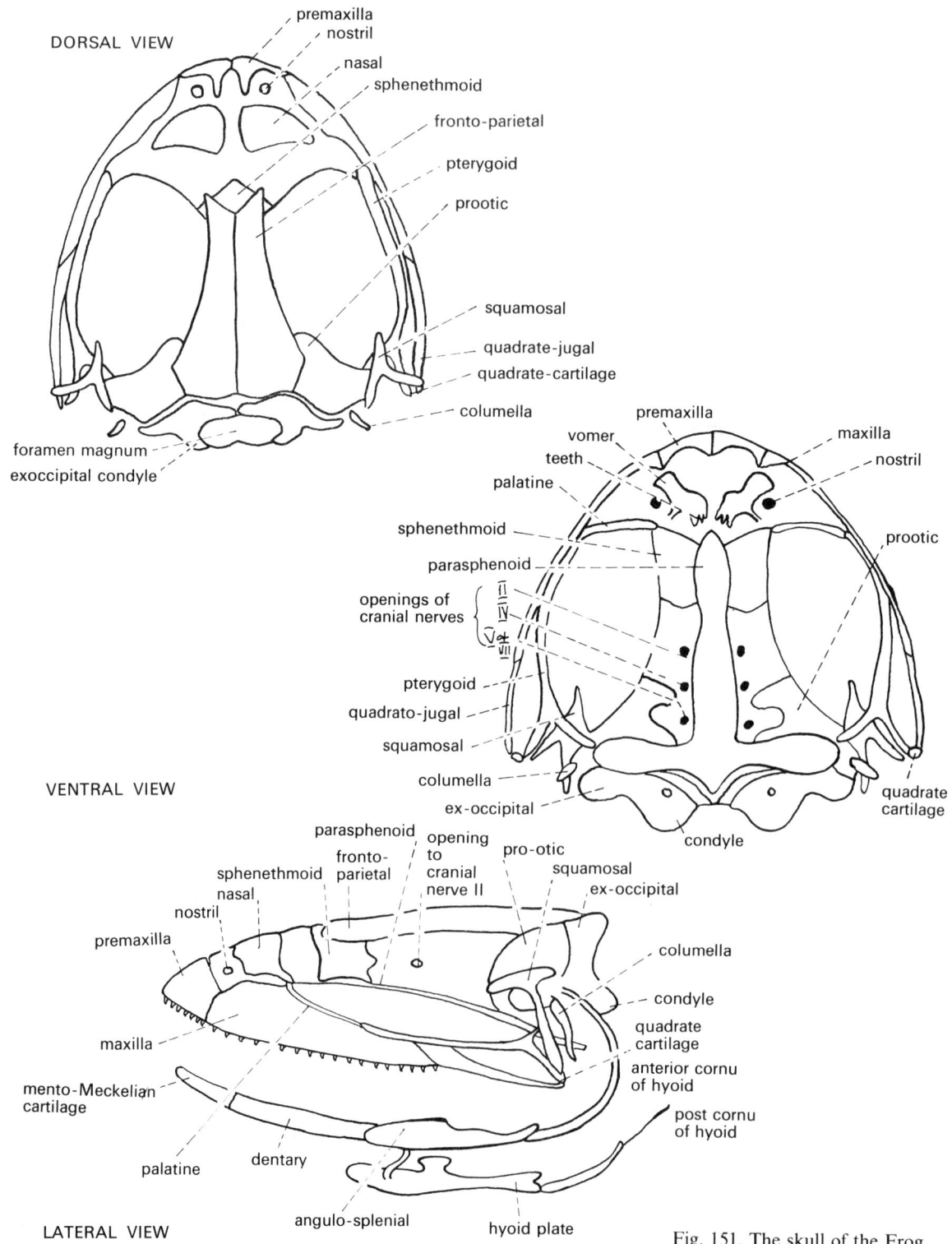

Fig. 151. The skull of the Frog.

skin, which is quite loose due to the presence of lymph cavities, and make a median incision from the cloaca region, up to the jaw (Figure 152). From this incision, cut outwards to the limbs, taking care not to cut through any of the *cutaneous blood vessels*. Pin back the skin.

Cut through the mid-line of the *pelvic girdle* and the associated muscles, with scissors. Care should be taken not to cut the *pelvic veins*.

Un-pin the dissection, and re-pin it, stretching the limbs apart. This will help in

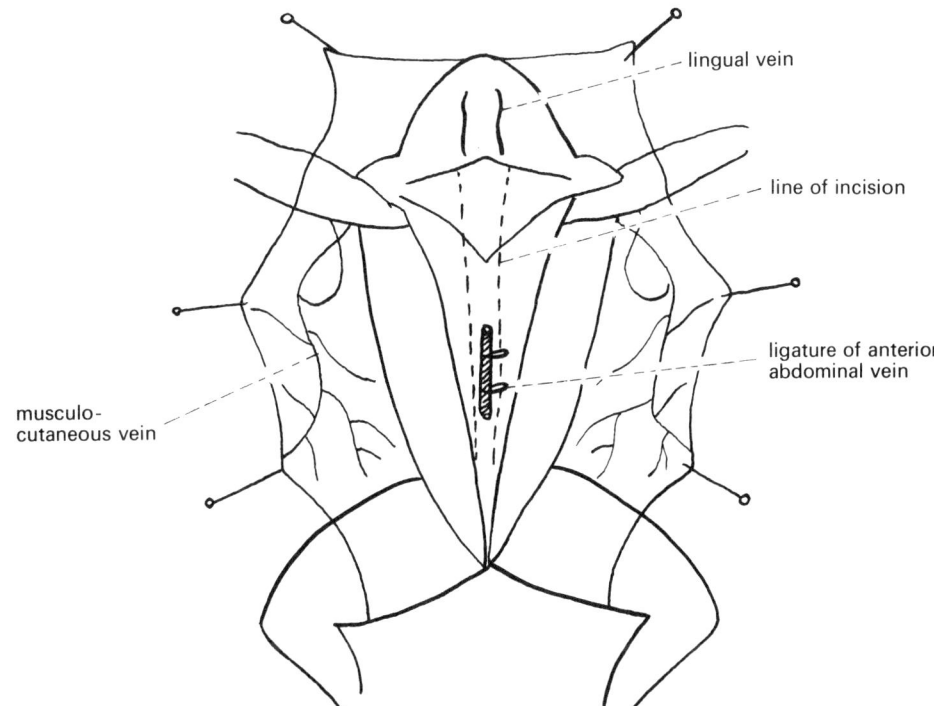

Fig. 152. Initial stage in the dissection of the Frog.

Locate the *anterior abdominal vein* in the mid line of the abdominal muscles. This has to be ligatured in two places. Make two incisions, parallel to, and on either side of, the anterior abdominal vein, as shown in the diagram, and ligature the vein with cotton in the positions shown (Figure 153). Peel back the two flaps of muscle to either side, and remove the piece of the vein between the two ligatures.

To expose the heart, cut through the *pectoral girdle* and the overlying muscle on either side, about $\frac{1}{4}$ in. from the midline. Cut transversely across the *xiphisternum* and lift the resulting loose flap of tissue, and cut away.

exposing the various systems, but you must be sure not to break any nerves and blood vessels.

Clean the muscle from the remaining ends of the anterior abdominal vein.

If the specimen is female, in early spring, the dissection may well be dominated by the large *ovisacs* (which will swell as previously described if wetted). These should be cut open and the eggs removed. In addition, the *fat bodies* may be enlarged. Cut them back to their origin anterior to the kidney.

Lift the ventricle of the *heart* up and forward and pin it through the tip. Display the *lungs* to either side and pin them if necessary. Lift the lobes of the *liver* forward

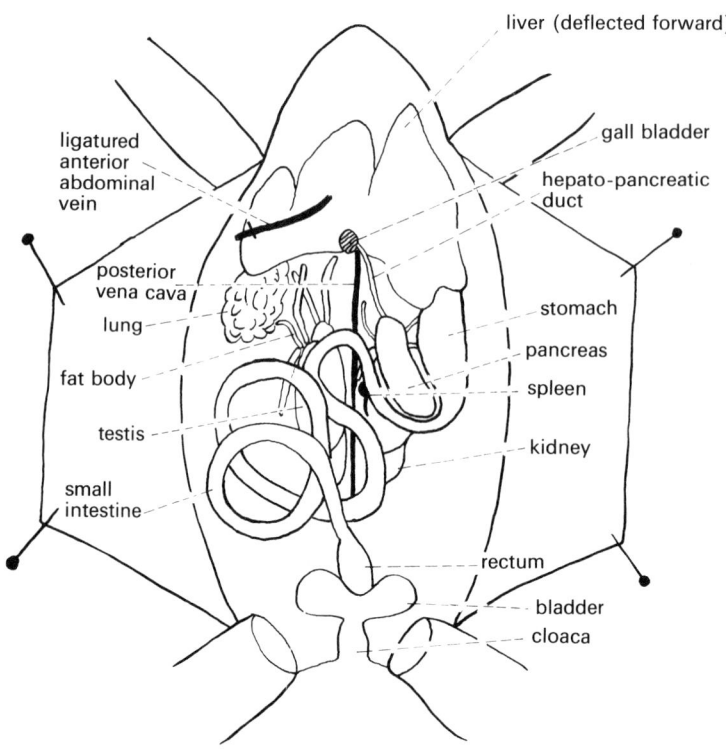

Fig. 153. General ventral dissection of the Frog.

and pin, to display the green *gall bladder* and *hepato-pancreatic duct*. Deflect the stomach to the animal's left and the intestine to the right to display the *kidneys* and *gonads* and the dark red *spleen* near the rectum. Clean the tissues from the bladder to display it.

Draw this dissection.

The venous system (Figure 154)

The veins will be dark red. Remove the pin from the heart, and allow it to fall into its original position. Carefully cut through the *pericardium* and remove it. Pin the heart back as before. Locate the large, suffuse *sinus venosus* with the *anterior vena cavae* arising from it. Working outwards to the right and left, the anterior vena cavae soon branch into the *external jugular veins,* the *innominate veins* and the *subclavian veins*. Trace these to their ultimate origins. Take good care to avoid the arteries and nerves

which lie under and close to these veins.

Move the intestine so that at least one of the kidneys is clearly visible, and locate the *renal portal vein* on its outer side. Trace this until it branches into the *sciatic vein* on the inside of the leg, and the *femoral vein* on the outside. The leg will have to be twisted to display these two. Now trace the posterior stump of the anterior abdominal vein to either side as the *pelvic vein* which joins with the femoral vein.

The next step is to expose the *hepatic portal system*. Trace the anterior stump of the anterior abdominal vein until it joins the hepatic portal vein which soon gives two branches running into the liver. Move the liver as necessary to display this. Posteriorly to the junction of the anterior abdominal vein and the hepatic portal vein, the latter gives off several branches to the various parts of the intestine. Display the intestine to show this.

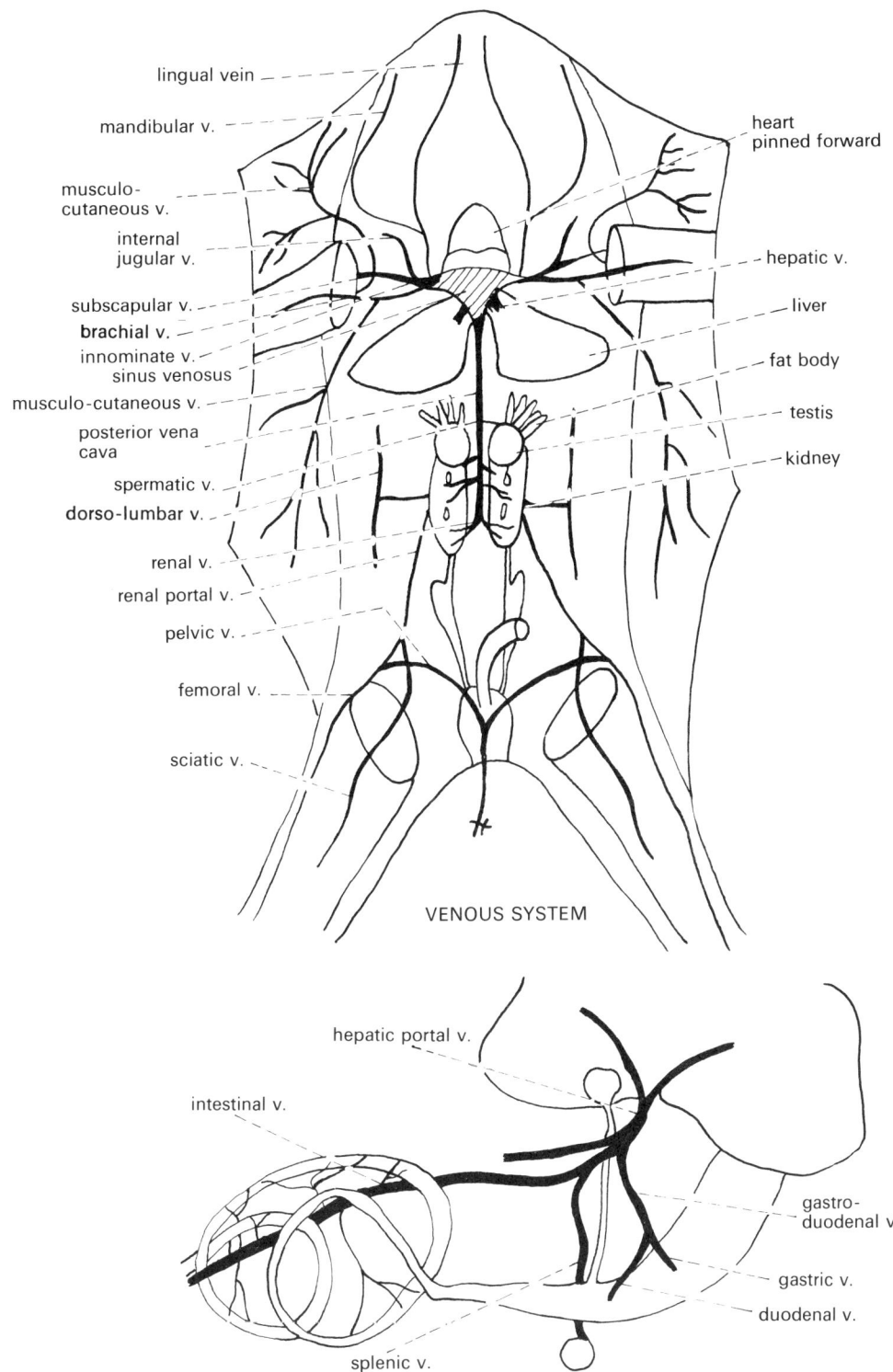

lingual vein

mandibular v.

musculo-cutaneous v.

internal jugular v.

subscapular v.

brachial v.

innominate v.
sinus venosus

musculo-cutaneous v.

posterior vena cava

spermatic v.

dorso-lumbar v.

renal v.

renal portal v.

pelvic v.

femoral v.

sciatic v.

heart pinned forward

hepatic v.

liver

fat body

testis

kidney

VENOUS SYSTEM

hepatic portal v.

intestinal v.

gastro-duodenal v.

gastric v.

duodenal v.

splenic v.

HEPATIC PORTAL SYSTEM

Fig. 154. The veins of the Frog.

Draw your dissection. It may be convenient to do three separate drawings, *viz.* the anterior venous system, the posterior venous system, and the branches of the hepatic portal vein.

The arterial system (Figure 155)

Push a roll of paper through the mouth into the oesophagus. This stretches the blood vessels. Locate the *truncus arteriosus*, and ligature it close to the heart. Place a second ligature on the truncus arteriosus, outside, and as near to the first one as convenient.

These two ligatures are to prevent bleeding from the veins which are now to be removed. Cut the truncus arteriosus between the two ligatures and carefully lift the veins from the underlying arteries. Continue outwards to the fore-limbs, removing the external jugular, innominate, and subclavian veins. This will expose the three visceral arches—the *carotid pulmocutaneous,* and *systemic.* Displace the lungs to display the *pulmonary artery.*

The intestine must now be displaced to the right to display the *dorsal aorta.* Displace

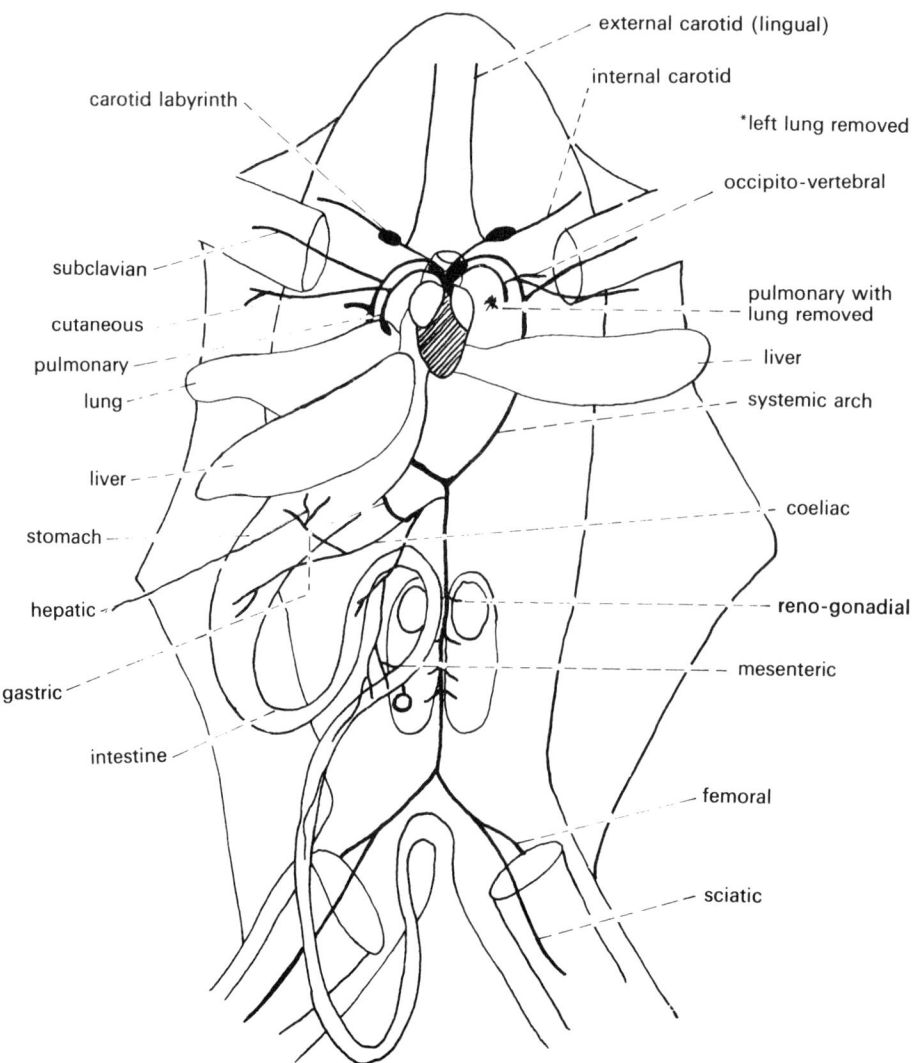

Fig. 155. The arterial system of the Frog.

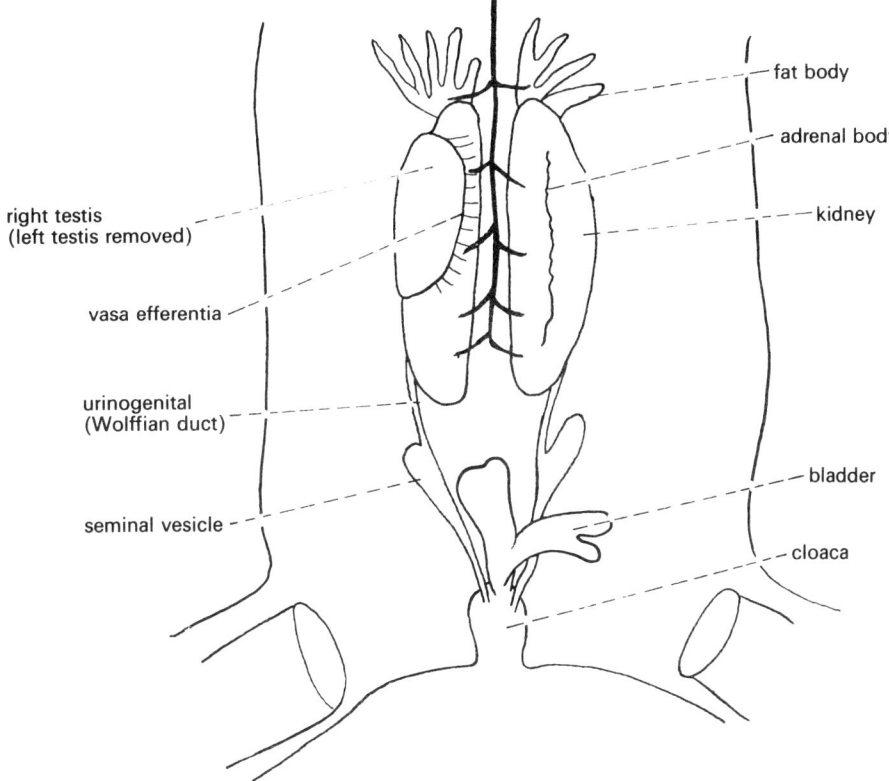

right testis
(left testis removed)

vasa efferentia

urinogenital
(Wolffian duct)

seminal vesicle

fat body

adrenal body

kidney

bladder

cloaca

Fig. 156. Urinogenital system of male Frog.

the left kidney over to the right to display the *renal artery*. Clean away any of the peritoneum left around the heart, and any muscles as necessary to expose the rest of the systemic arch. Draw your dissection. (The *iliac arteries,* and *femoral* and *sciatic arteries* will have to be added to the drawing after the next operation).

Now remove the intestine, by cutting through the rectum and oesphagus, and lifting the whole intestine out from the dissection. Trace the iliac arteries into the limbs, where they branch from the sciatic and femoral arteries. The legs will have to be twisted to display both of them at the same time.

Urinogenital system

With the removal of the intestine, this system is virtually fully exposed. In either sex, open the *cloaca,* and display as shown in

the diagram (Figures 156 and 157). Remove one of the gonads from the kidney to reveal the *adrenal body*.

The Nervous system

Lift the kidneys and gonads free of the body wall, and remove the whole of the urinogenital system. But *do not touch* the dorsal aorta, as the sympathetic nervous system is in intimate contact with it. This will expose the *spinal nerves* as illustrated in the diagram (Figure 158).

The *sympathetic ganglia* lie on either side of the dorsal aorta, between it and the clear white calcareous patches, which are very very obvious in the dissection, lying along the side of the spinal cord.

Notice the IInd and IIIrd spinal nerves fusing to form the *brachial plexus* that runs out to the fore-limb, and the formation of

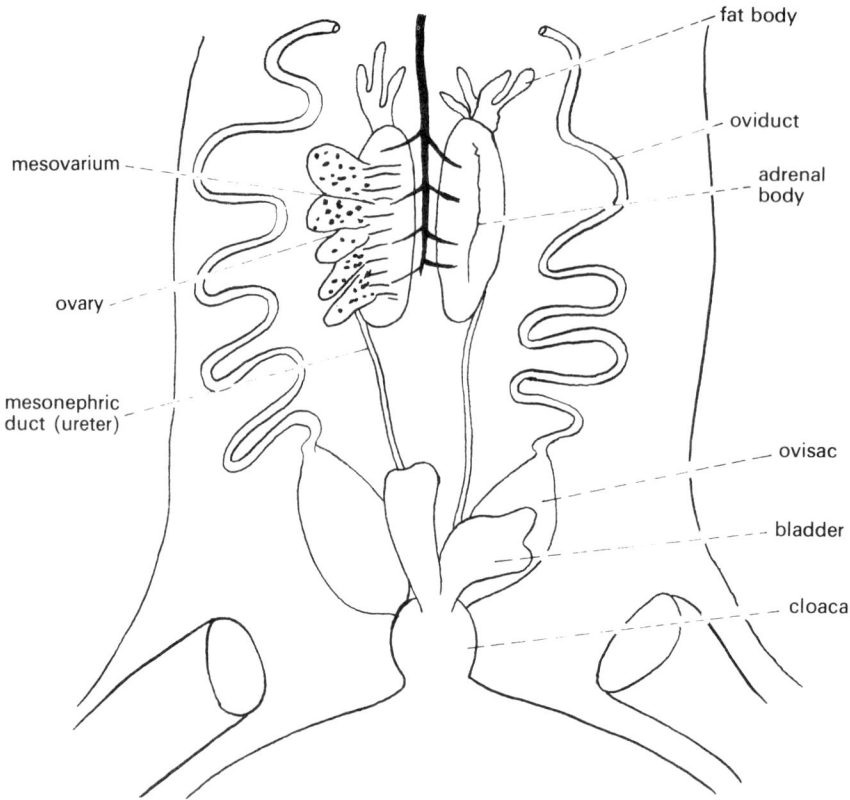

Fig. 157. Urinogenital system of female Frog.

the *sciatic nerve* from the fusion of the VIII and IX. This with the VII and the fine X form the *sciatic plexus*.

The Brain (Figure 159)

Skin the top of the head and back. Remove as much muscle as possible especially at the base of the skull. Refer to your drawing of the dorsal view of the skull, and, in your dissection locate the cartilage dorsal to the exoccipital. Pierce this at the mid point, and split the cranium along the fissure between the two fronto-parietals. Cut laterally at either end of the front-parietals and lift the pieces of bone away. This will expose the top of the brain. Cut away the neural arches from some (or all) of the vertebrae to expose the spinal cord. At this stage it is advisable to soak the dissection in 75% ethanol for 2–3 days to harden the nerve tissue. Cut the

bases of the cranial and spinal nerves, and the brain should be easy to lift, or shake out of the cranium. Avoid damaging the *pituitary gland* on the underside of the brain. It may well need freeing from the skull, before the brain is removed. Draw the brain from the dorsal and ventral aspects.

Use a razor to shave off the dorsal walls of the brain to show the ventricles.
Class: Aves (Birds), e.g. The **Pigeon** or **Dove** *(Columba* spp.)

Examine a complete specimen. Notice the *beak* which is devoid of *teeth* the *nostrils*, the *auditory capsules* behind the eyes, and covered with feathers *(ear coverts)*, the *clawed feet* and legs covered with scales.

The fore-limb, which is of the pterodactyl pattern is modified as a wing. Extend one wing and draw it, comparing the external features with those of the wing skeleton

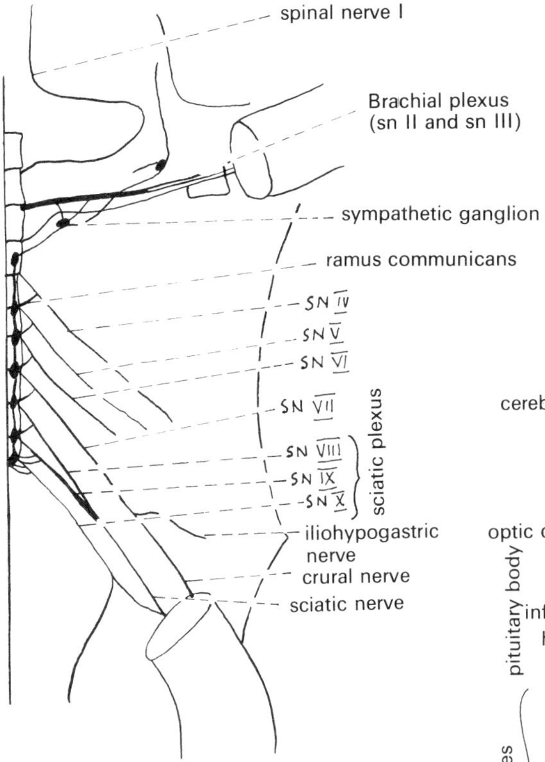

spinal nerve I

Brachial plexus
(sn II and sn III)

sympathetic ganglion

ramus communicans

SN $\overline{IV}$
SN $\overline{V}$
SN $\overline{VI}$

SN $\overline{VII}$

SN $\overline{VIII}$
SN $\overline{IX}$
SN $\overline{X}$

sciatic plexus

iliohypogastric
nerve
crural nerve
sciatic nerve

Fig. 158. Spinal and sympathetic nerves of
the Frog (all other vessels not shown).

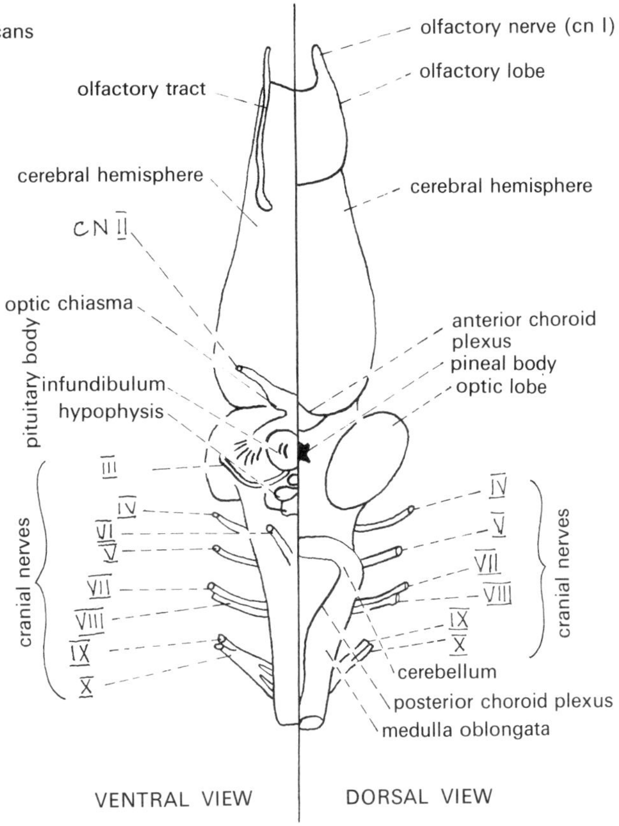

olfactory tract

cerebral hemisphere

CN $\overline{II}$

optic chiasma

pituitary body

infundibulum
hypophysis

cranial nerves

$\overline{III}$

$\overline{VI}$
$\overline{V}$

$\overline{VII}$

$\overline{VIII}$

$\overline{IX}$

$\overline{X}$

olfactory nerve (cn I)

olfactory lobe

cerebral hemisphere

anterior choroid
plexus
pineal body
optic lobe

$\overline{IV}$
$\overline{V}$

$\overline{VII}$
$\overline{VIII}$

$\overline{IX}$
$\overline{X}$

cranial nerves

cerebellum
posterior choroid plexus
medulla oblongata

VENTRAL VIEW   DORSAL VIEW

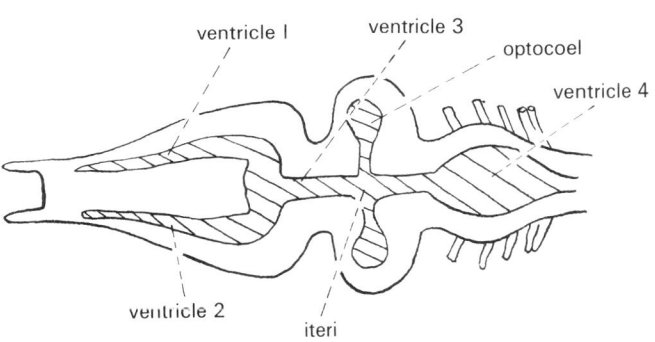

ventricle I   ventricle 3
                              optocoel
                                    ventricle 4

ventricle 2

iteri

Fig. 159. The brain of the Frog. The Ventricles

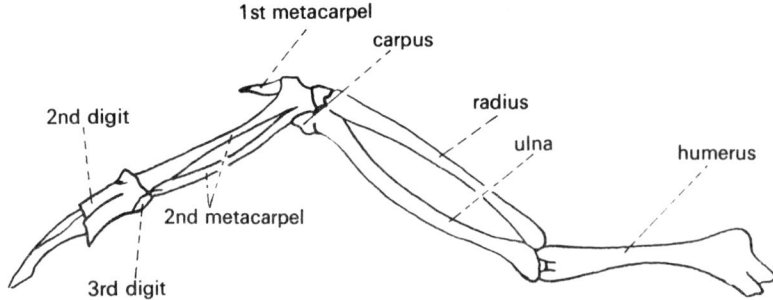

Fig. 160. The skeleton of a bird's wing.

(Figure 160), which should be examined in conjunction with the wing.

Examine the various types of feathers. Remove them from the specimen and draw them (Figure 161). 1. *Quill feathers* on the wing and tail. 2. Contour feathers, include the *covert feathers* with the *down feathers* between them. These cover the whole body, where feathers are found. 3. *Filoplumes* are very small, fine feathers between the contour feathers.

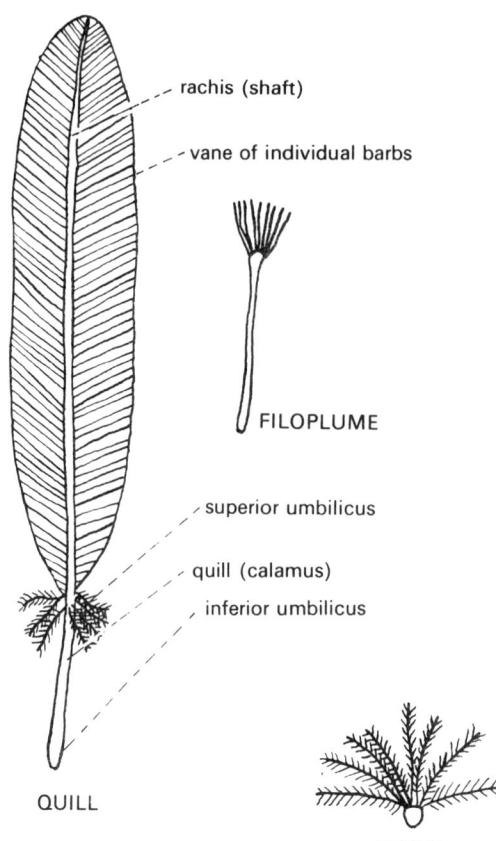

Fig. 161. Types of feathers.

9 The Flowering Plant

MORPHOLOGY AND ANATOMY

In studying the morphology and anatomy of flowering plants, it is essential to be able to recognise the different regions. To do this there are a few simple rules to follow, and these should be learnt. Do not be put off by size, shape or general appearance. Broadly speaking, except for such specialized zones as the hypocotyl, the flowering plant has four features, the *stem, root, leaf* and *flower*. The *dicotyledons* and *monocotyledons* can also be differentiated by their morphology and anatomy. Remember that the following list of differences do not constitute laws; they serve as a guide to recognition, and there may well be exceptions which you might come across.

Differences between monocotyledons and dicotyledons

(a) *Morphology*

Monocotyledons	*Dicotyledons*
1. Seeds have one cotyledon.	1. Seeds have two cotyledons.
2. Leaves usually have parallel veins, with upper and lower surfaces similar, and stomata on both surfaces.	2. Leaves usually have reticulate veins, with the upper and lower surfaces different, and stomata on the lower surface.
3. The flower parts are in multiples of three, without a distinct calyx and corolla.	3. The flower parts are in twos or fives, or multiples, usually with a distinct calyx and corolla

(b) **Anatomy**

Monocotyledons	*Dicotyledons*
1. Bundles closed (no cambium).	1. Bundles open, (cambium between the xylem and phloem, except in the leaves where they are closed).
2. Bundles of the stem scattered.	2. Bundles of the stem in a definite ring.
3. Many protoxylem groups in the root.	3. Few (maximum of five) protoxylem groups in the root.

Differences between the stem, root, and leaf

Morphology

1. The stem bears buds and leaves; there are buds on neither roots nor leaves.
2. Branches of the stem develop from buds in the axils of leaves, i.e. a branch will be above a leaf-scar (Figure 162). The branches of roots develop from an internal meristem, and their arrangement is rarely obvious. Leaves usually have a bud in their axils.

Anatomy

1. Only the root has a piliferous layer.
2. Only the leaf shows a distinct organization of a palisade layer for photosynthesis.
3. The vascular bundles of the stem are scattered, or in a circle around a central large pith. Those of the root are central, with a small pith (if any), the xylem of the dicotyledons usually forming a central star in transverse section.
4. An endodermis is found in the root and many stems.

The typical herbaceous stem

Examine and draw the stem of an annual plant, e.g. groundsel, sunflower, broad bean. Notice that it is divided into *nodes* and *internodes*, the point at which the leaf arises being the node, and the space between two successive leaves being the internode (see Figure 163). Each leaf has a bud in its axil. Find it although it may be very small, and possibly enclosed by the sheathing base of the petiole.

Notice the presence or absence of *stipules*—leaf-like outgrowths at the base of the leaf petiole. Find out the arrangement of the leaves on the stem. They are usually arranged spirally. The arrangement is called the *phyllotaxy,* and is expressed by a fraction,

passed around the stem in moving from one bud to the next in line.

Discover the method of branching of the stem. There are three main types of branching (Figure 164): 1. *Monopodial (racemose)*—the extension growth is by the apical bud,

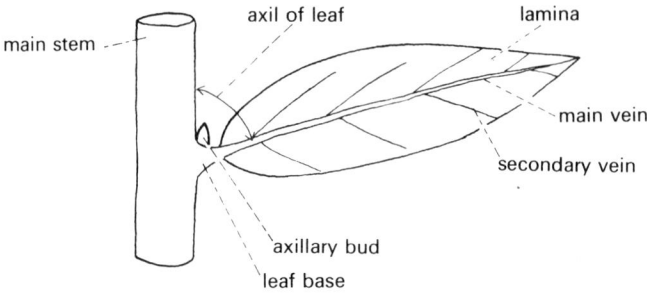

Fig. 162. Diagram of the morphology of a leaf on a stem.

e.g., $\frac{2}{5}$, the numerator being the number of times a point would pass around the stem to pass from one leaf to the next lying in a straight line, and the denominator is the number of leaves passed in this spiral. The phyllotaxy can be observed as follows: Sight along the length of the stem, like sighting a rifle, so that two leaves are in line. Count the leaves between your two 'sights', counting either the first or the last but not both of the 'sights'. This gives you the denominator; the numerator being the number of times you

the lateral buds producing less dominant branches or flower stalks; 2. *Sympodial (cymose)*—the extension growth is by a lateral bud, the apical bud, usually producing a flowering branch. In a specimen, this gives the appearance of a flower (or inflorescence) stalk arising on the opposite side of the stem to the leaf; 3. *Dichasial*—the extension growth is by two lateral buds of equal vigour, the apical bud developing into a flower (or inflorescence) stalk.

The typical dicotyledon stem

Cut transverse sections of the stem and examine them under the microscope. Stain them in aniline sulphate (or aniline hydrochloride), mount them in glycerine and draw. Draw a low power plan of the whole stem, and a sector under high power, to include one bundle and the surrounding tissue.

The diagram (Figure 165), is of a transverse section through the stem of the sunflower (*Helianthus annuus*), which is typical of an unthickened dicotyledonous stem, except that there are *pericyclic fibres* outside the phloem. Depending on the age of your particular piece of stem, *collenchyma* may be

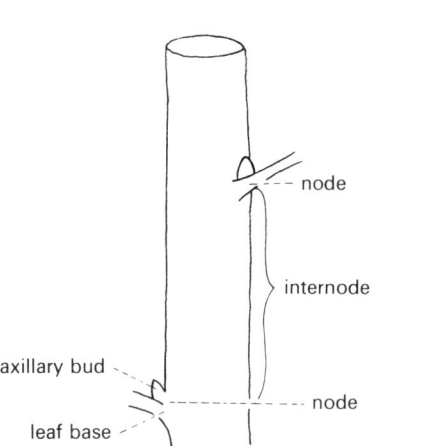

Fig. 163. A figure to show the relationship between node and internode.

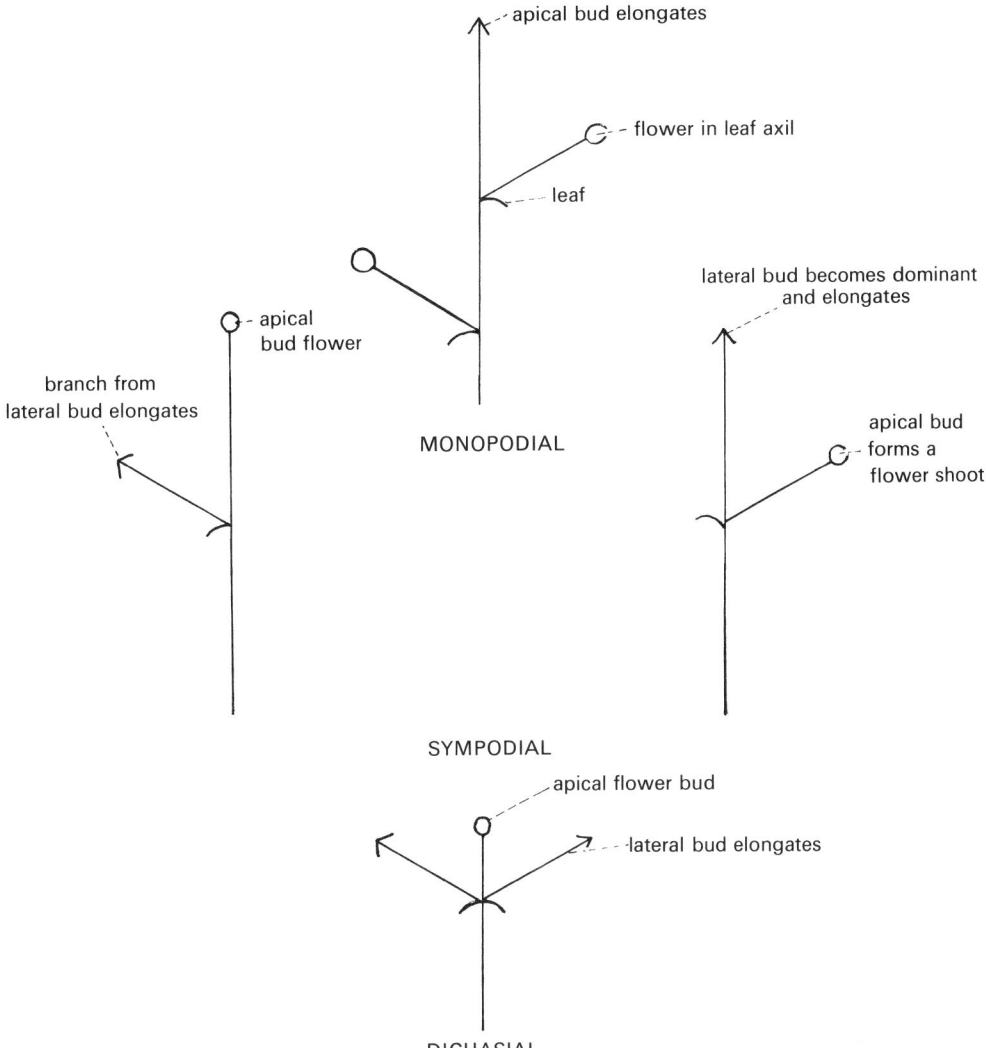

Fig. 164. Types of branching of flowering plants.

well-developed in the outer cortex, and there may be various stages of secondary thickening. The *interfasicular cambium* (that between the primary xylem and phloem) may have begun to divide to produce secondary xylem and phloem, and, at a later stage, the *interfasicular cambium* (that across the primary medullary rays, between the bundles) may have developed.

Examine longitudinal sections of the stem, staining with aniline sulphate (or hydrochloride). Notice particularly the different forms of thickening in the xylem elements (Figure 166)

Cut further transverse sections of your stem, and stain them with iodine solution. Wash them with water and examine them under the microscope, to see the distribution of starch. Draw a plan to show the main tissues and the distribution of the starch-containing areas.

Your stem may well differ from the sunflower considerably in detail, but the basic pattern will be the same.

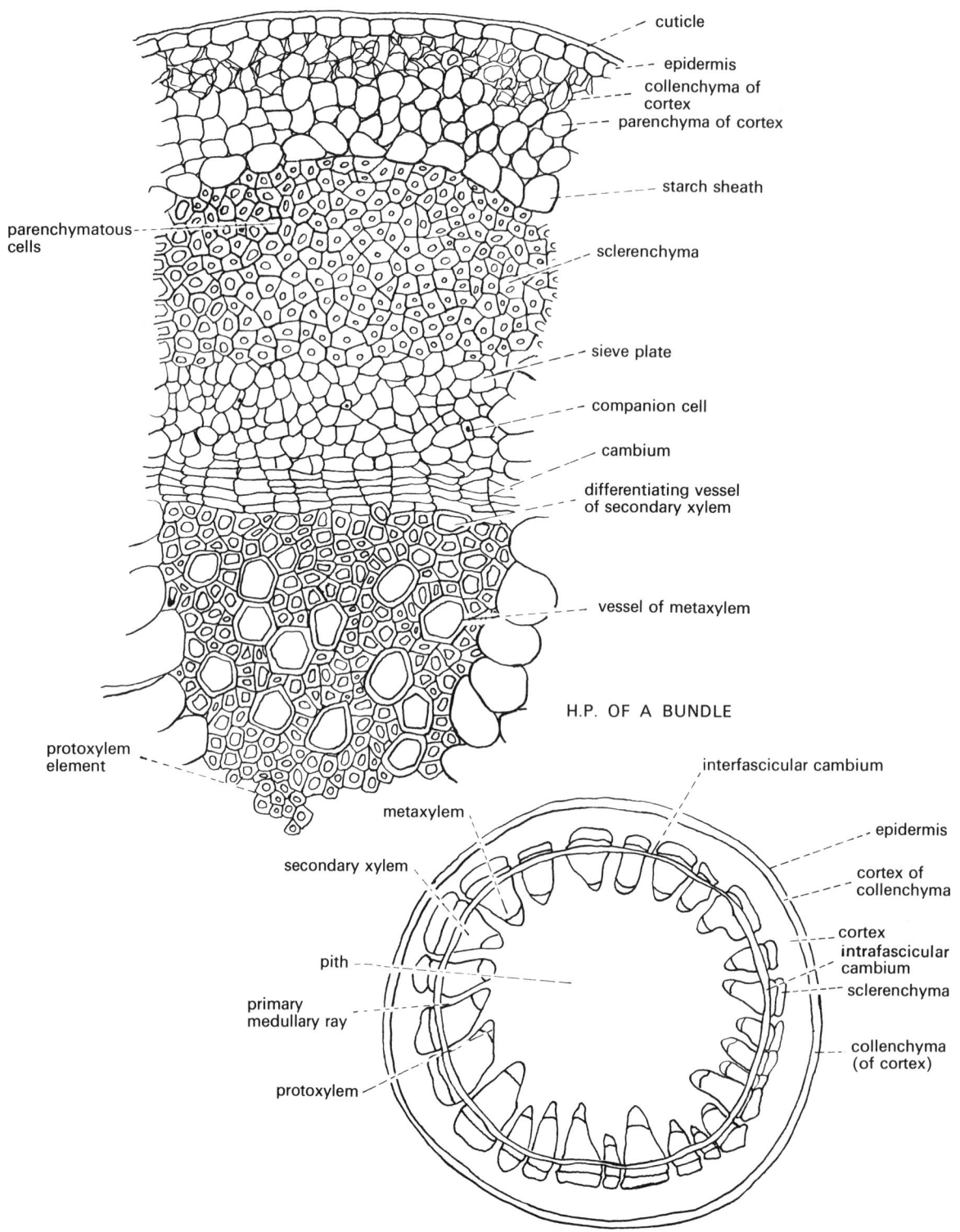

cuticle

epidermis

collenchyma of cortex

parenchyma of cortex

starch sheath

parenchymatous cells

sclerenchyma

sieve plate

companion cell

cambium

differentiating vessel of secondary xylem

vessel of metaxylem

H.P. OF A BUNDLE

protoxylem element

interfascicular cambium

metaxylem

secondary xylem

epidermis

cortex of collenchyma

cortex

intrafascicular cambium

sclerenchyma

collenchyma (of cortex)

pith

primary medullary ray

protoxylem

Fig. 165. Transverse section of a Sunflower stem.

PLAN

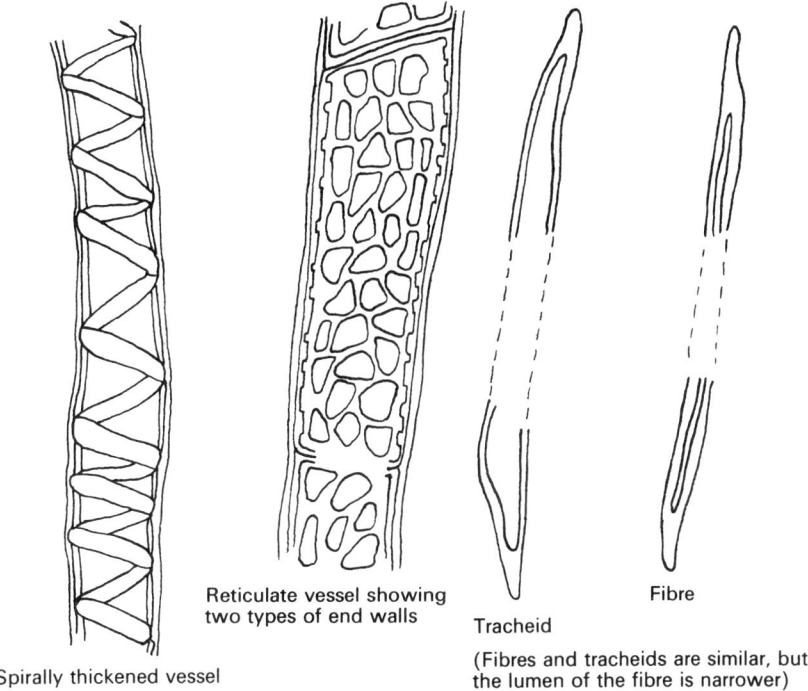

Spirally thickened vessel

Reticulate vessel showing
two types of end walls

Tracheid

Fibre

(Fibres and tracheids are similar, but
the lumen of the fibre is narrower)

Fig. 166. Some of the elements found in the xylem.

The Vegetable Marrow (or other cucurbit) *stem* (see Figures 167 and 168).

This is worth looking at, as it shows the phloem *sieve tubes* and *companion cells* remarkably well. Its other peculiarities include: 1. the second bundle of phloem inside the xylem, 2. the two rings of bundles, 3. the hollow pith (which is not uncommon).

Cut, stain and draw transverse and longitudinal sections as before.

The typical monocotyledon stem

Cut, stain and examine transverse sections of the stem of *Maize (Zea mais)* (Figure 169), or examine prepared slides. Any other convenient monocotyledon may be used, but if you are collecting the material yourself, remember that it is a true stem that you need, and not, for example, a flower stalk.

Notice the scattered bundles, each with relatively few xylem elements, and the large vessels in the metaxylem, with one or two

in the protoxylem; the whole bundle being surrounded by a sheath of fibres.

The woody stem

Examine and draw the winter twigs from about six deciduous trees. Cut, stain (using aniline hydrochloride), examine and draw transverse sections of the stems.

On the specimen notice the *terminal* and *lateral buds*. Work out the phyllotaxy of the latter. Beneath each bud there is a *leaf scar,* and beneath the terminal buds is a group of *ring scars* from the scale leaves of the previous terminal bud. Similar rings of ring scars will be found at intervals down the stem, each having arisen from a terminal bud. The intervals between these groups indicate the growth of successive years. Notice the *lenticels* scattered on the bark.

Use a fairly large bud, e.g. Horse Chestnut, and dissect off the individual bud scales until you find the true foliage leaves. The intermediate forms may have an enlarged base with small lamina apically. This indicates that the bud scales are enlarged petioles, and not the leaf lamina.

Leave twigs from different species in water, and observe the various stages as the buds open.

size of the elements. Distinguish *vessels* and *tracheids* noticing particularly the distribution of the former. There will be isolated cells, or small groups of *xylem parenchyma* and the *primary* and *secondary medullary rays*. These cells will show up quite clearly, as they will not be stained as lignin.

Locate the cambium and notice the undifferentiated cells immediately adjacent

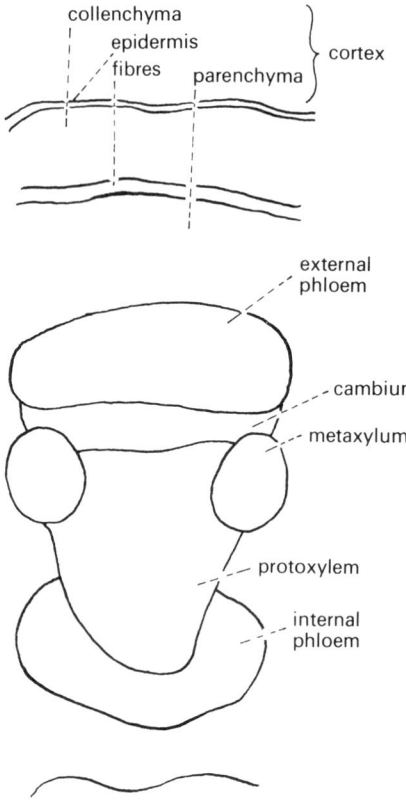

Fig. 167. Plan of a sector through the stem of the Vegetable Marrow.

In examining the transverse sections, remember that you will find a great variety of arrangement, and of structure between individual species, e.g. there may be sclerenchyma in the cortex or phloem, resin canals may be found in different zones, etc. The protophloem will be difficult to see. Look for the *phellogen* (bark cambium), and distinguish the *phelloderm* from the cortex, which itself will be very squashed.

In the xylem notice the *growth rings* are distinguished by the difference in the relative

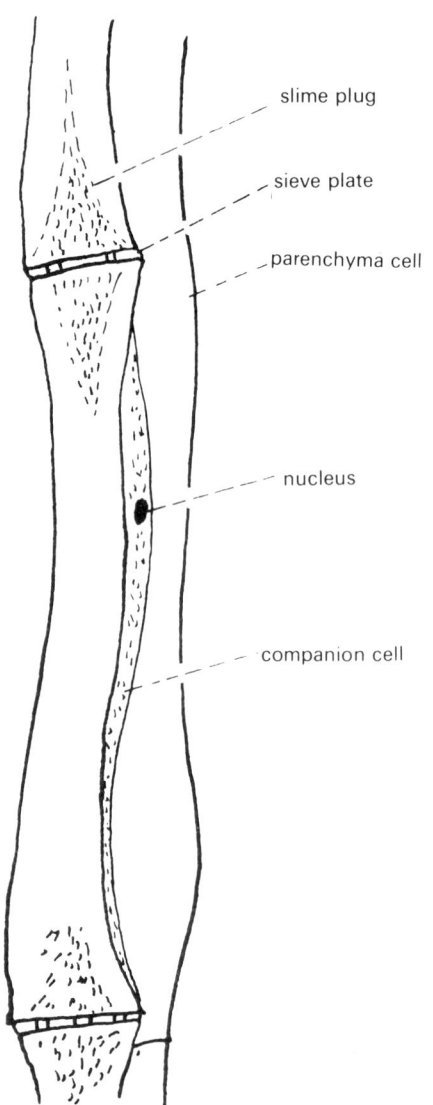

Fig. 168. Longitudinal section through sieve tube of phloem.

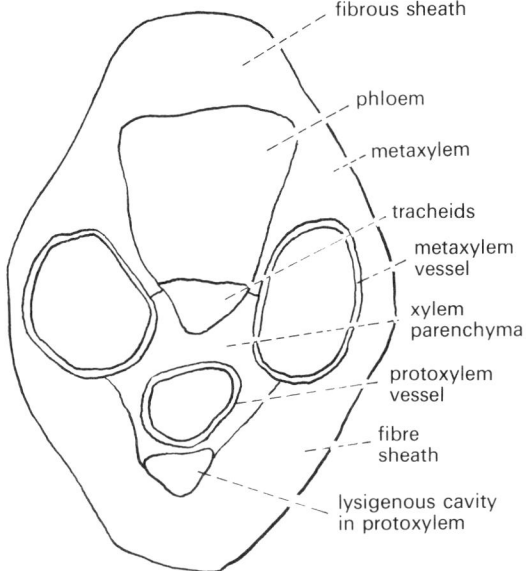

Fig. 169. Plan of a transverse section of a vascular bundle in a Maize stem.

to, and on either side of, the meristematic cells.

The *pith* may be quite small, but find it, and the adjacent *protoxylem* elements. The diagram (Figure 170), shows the basic pattern of the stem, which will always be the same in dicotyledons, and you are not expected to study the various ways in which exceptional monocotyledons are secondarily thickened.

You may have been fortunate and have cut through a *lenticel*. If so, draw it. If not examine a prepared slide of the stem of elder *(Sambucus nigra)*, which shows lenticels very well (Figure 171).

Modifications of the Stem

If the structure is a stem, it will have leaves, scale leaves, or leaf scars, with buds or shoots in their axils. The converse of this statement is true, namely, that if a structure bears leaves

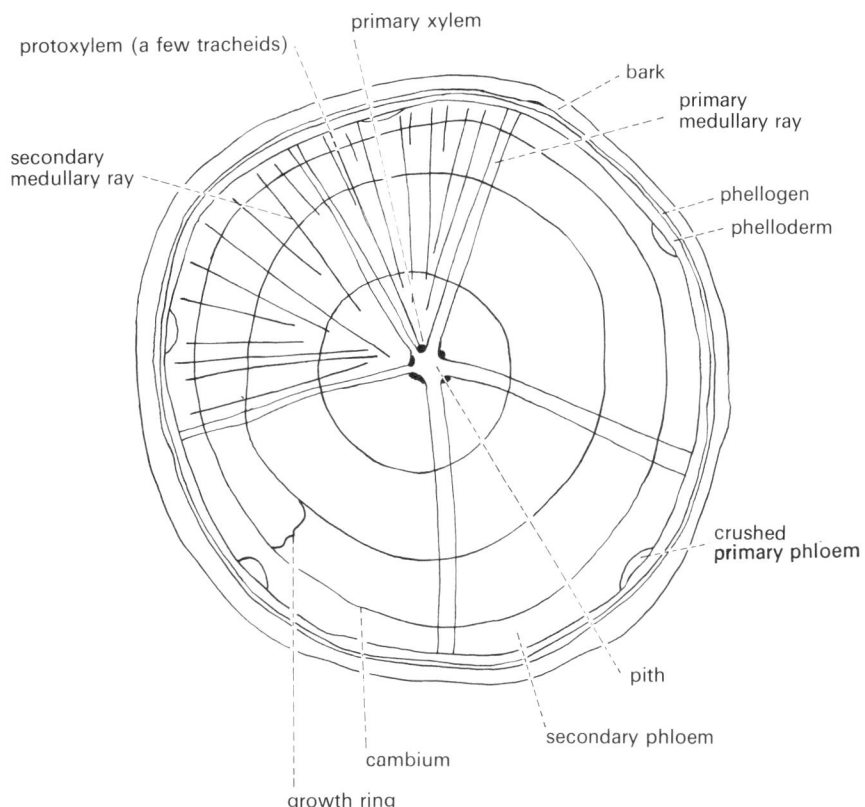

Fig. 170. Transverse section of typical stem with secondary thickening.

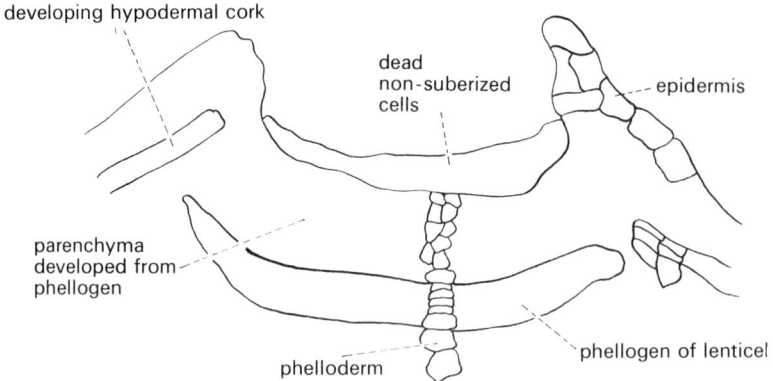

Fig. 171. Plan of a transverse section through a lenticel of Elder.

and buds, it must be a stem, whatever it looks like.

Tuber

Examine a potato *(Solanum tuberosum)* or Jerusalem Artichoke *(Helianthus tuberosus)*. On it you will see a scar of the stem, where it was attached to the parent plant, the 'eyes' which are leaf scars, with buds in their axils. Cut the tuber in half, wash the cut surface and stain it with aniline sulphate. Notice the distribution of the vascular bundles, the xylem of which being stained. Stain the other surface with iodine to determine the nature of the stored food (Figure 172). If you are using a potato scrape a little of the cut surface on to a slide, and examine the starch grains. The reserve material of the Jerusalem artichoke is *inulin* (a polysaccharide) which you can see under the microscope if the material has been preserved in ethanol. Large hemispherical crystals are deposited on the cell walls.

Corm

Examine a corm of Crocus *(Crocus vernus)*,

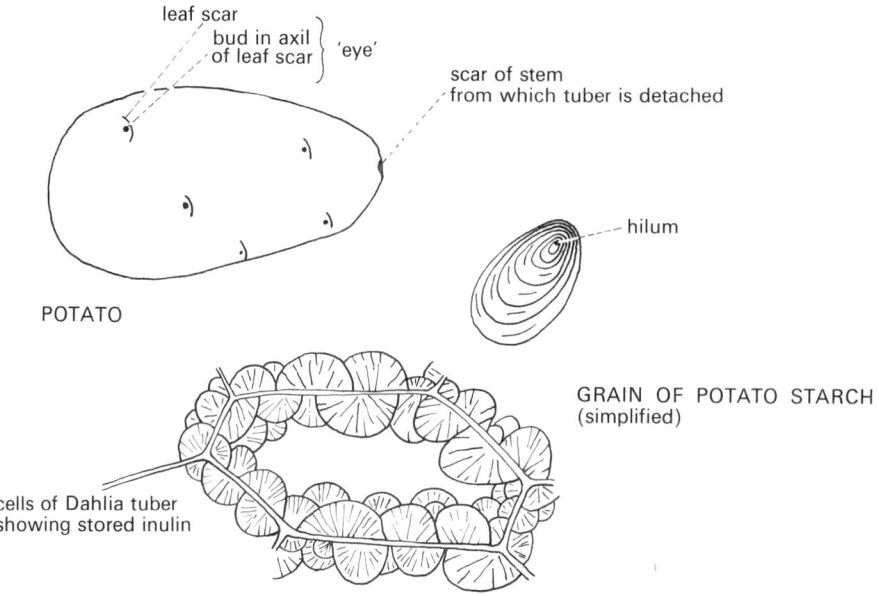

Fig. 172.

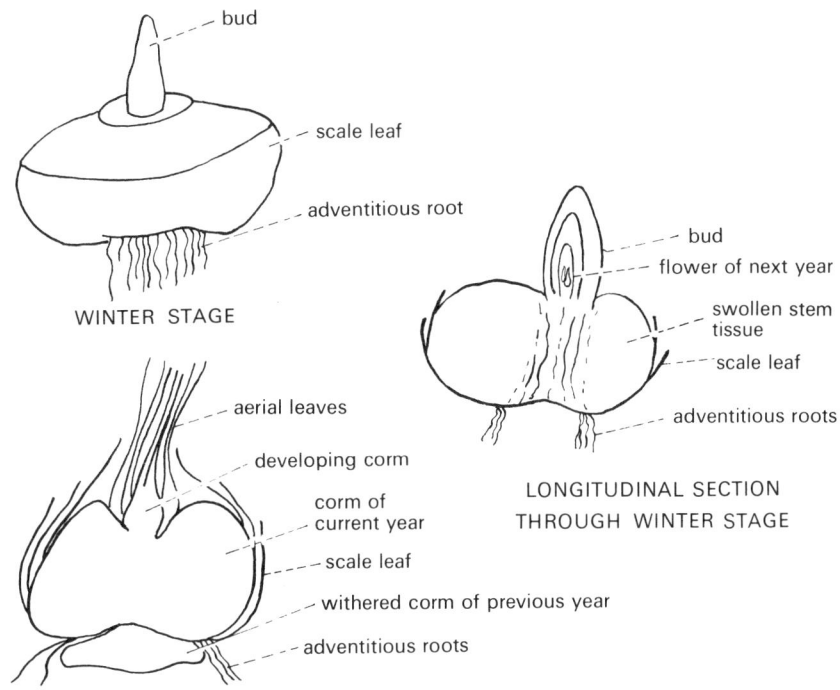

Fig. 173. Crocus corm.

or *Gladiolus (Gladiolus communis)* (see Figure 172). The parts are shown in the diagram. Cut the corm in half longitudinally, and stain one half with iodine to determine the nature of the reserve material. Draw the other cut surface. Cut another corm (or one half of the original one) transversely. Wash and stain with aniline sulphate. Note the distribution of the bundles, recalling that these are monocotyledons.

Rhizome (Figure 174)

Examine the rhizome of Solomon's Seal *(Polygonatum multiflorum)*, Iris *(Iris pseudacorus)* or Lily of the Valley *(Convallaria majalis)* and the thin rhizome of Couch Grass *(Agropyrum repens)*. Notice the scale leaves and the apical bud, or the shoot arising from it, and the *adventitious roots*. The scars of the shoots of the previous years may be present. Is the branching monopodial or sympodial?

Bulb (Figure 175)

Examine the bulbs of Tulip *(Tulipa* spp.) or Hyacinth *(Hyacinthus* spp.), and Daffodil *(Narcissus pseudonarcissus)*. There are adventitious roots, and papery scale leaves.

Cut a median longitudinal section through each of the bulbs. Both will show the flattened condensed stem, from which the adventitious roots arise in a circle. The fleshy scales of the bulb are morphologically different in the tulip (and hyacinth) than in the daffodil. In the former they are really swollen bud scales, and in the latter, the swollen leaf bases of the previous year's foliage leaves. In your longitudinal section, you may see the developing flowering axis. Having discovered this, you can decide the method of branching of the bulb.

Test the bulb with iodine solution and Fehling's solution. What are the storage substances?

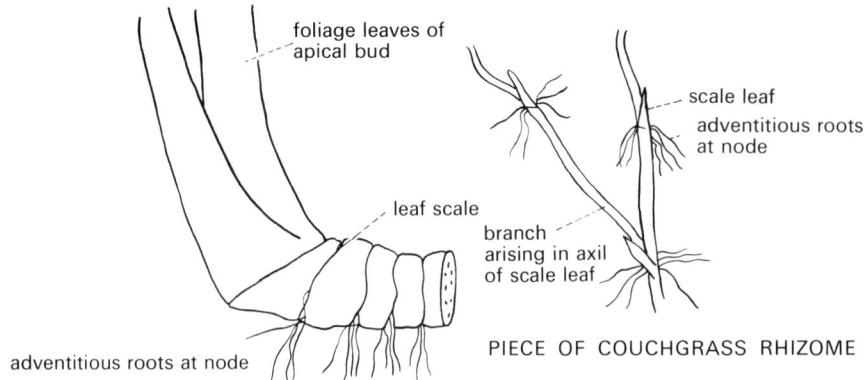

Fig. 174. Rhizomes.

Runners (Figure 176)

Examine runners of any of Creeping Buttercup *(Ranunculus repens)*, Yellow Archangel (Yellow Deadnettle) *(Galeobdolon luteum)*, Strawberry *(Fragaria vesca)*, or Cinquefoil *(Potentilla reptans)*.

The adventitious roots arise from the nodes only. Notice any branching and where the branches arise. Do not confuse the stipules of the strawberry with leaves.

Twining stems

Examine specimens of the Runner Bean *(Phaseolus multiflorus)* and Bindweed *(Convolvulus* spp.).

Stolons

Examine the stolons developed by the Blackberry *(Rubus* spp.). Here the tips of the stem develop the adventitious roots, and the new shoots develop from lateral buds in the axils of leaves near the shoot tip.

Stem tendrils (Figure 177)

Examine the tendrils developed on the Passion Flower (*Passiflora* spp.), Vine *(Vitis vinifera)*, and Virginia Creeper *(Ampelopsis vetchii)*.

The tendrils of the Passion flower clearly arise from the axil of a leaf and are hence obviously branches, but those of the Vine and

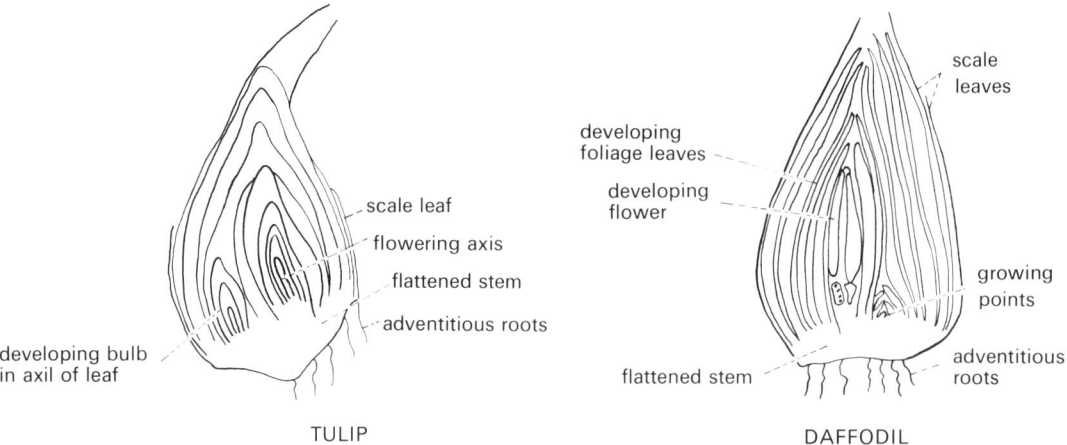

Fig. 175. Bulbs of Longitudinal section.

Virginia Creeper appear to arise opposite the leaf. This is a good example of sympodial branching, where the main branch has

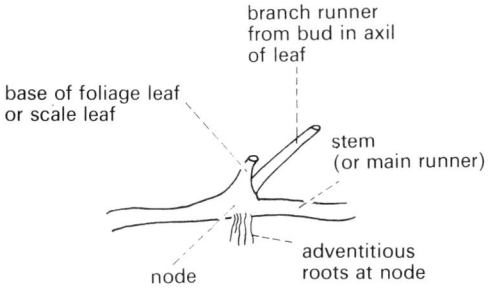

Fig. 176. Origin of branch runner.

developed into a tendril, and the axillary bud has grown to form the elongating branch.

Cladode (Figure 178)

Examine the cladode of Butcher's Broom *(Ruscus aculeatus)*. The flattened branch arises in the axil of a scale leaf, and must therefore be a stem. It also bears a scale leaf with a bud in its axil.

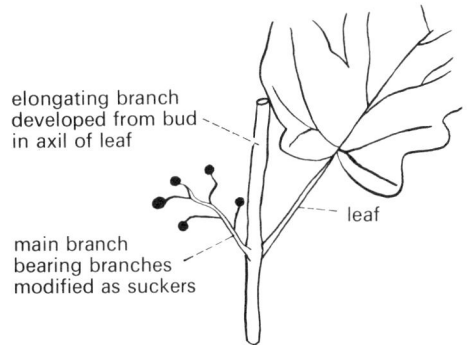

Fig. 177. Virginia creeper.

Stem spines (Thorns)

Examine the stem-spines of the Sloe (Blackthorn) *(Prunus spinosa)* or Hawthorn *(Crataegus* spp.). The spines are obviously stems, as they arise in the axils of leaves, and themselves bear buds and leaves. The apical bud has become replaced by the thorn point.

Prickles

Examine the stem of *Rosa* spp. or Blackberry (*Rubus* spp.). Prickles are not modified stems, as only the outer tissues are involved in their formation. This is shown by their random distribution on the stem, bearing no definite

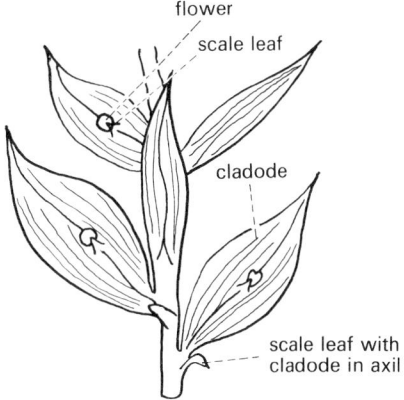

Fig. 178. Cladodes of Butcher's Broom.

relation to the leaves, and the absence of vascular tissue. This latter point is illustrated by the ease with which the prickle breaks from the stem, and their not ripping the 'bark' when they are removed. If you try to break a stem spine from the main branch, the bark invariably rips.

Aquatic stems

Examine transverse sections of the stem of Mare's Tail (Figure 179) *(Hippuris vulgaris)* or Pondweed (*Potamogeton* spp.). Both these genera show the characteristics typical of aquatic stems, namely (1) thin-walled epidermis, (2) the extremely large cortex with large air spaces, (3) the small stele with the vascular elements reduced. The xylem of *Hippuris* is represented only at the internodes by a few isolated tracheids, so you may not see evidence of xylem in your section.

The Root

The Morphology of the Primary Root

Germinate some Cress (or other suitable) seed on some moist filter paper, and when

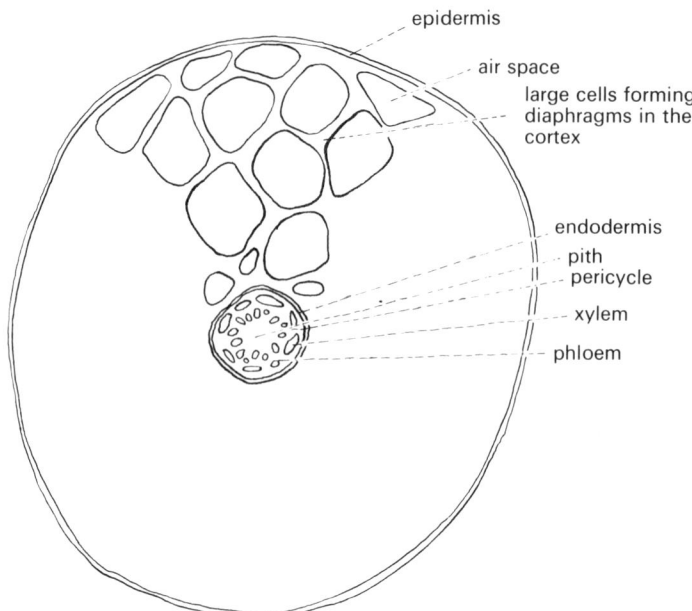

epidermis

air space

large cells forming
diaphragms in the
cortex

endodermis
pith
pericycle
xylem
phloem

Fig. 179. Plan of a transverse section of Mare's Tail stem.

the young root is suitably grown, examine
the external features. You will see the *root
hairs;* notice that they are formed only in a
limited zone, at some distance from the tip
of the root. Examine the root microscopically.
Have the root hairs any crosswalls?

Examine a prepared slide of a longitudinal
section through an unthickened root of Broad
Bean *(Vicia faba)* or Barley (*Hordeum* spp.).
This will show the *root cap* and the
meristematic zones which will be examined
in more detail later.

Anatomy of the dicotyledonous root

Cut, stain and examine transverse sections
of the root of Buttercup (*Ranunculus* spp.),
Broad Bean *(Vicia faba)*, or Sunflower
(Helianthus annus), cutting in the region of
the root hairs. Draw a low-powered plan and
a detailed high-power drawing of the bundle.
Notice the relative diameter of the stele to
the rest of the root, and the few groups of
protoxylem (maximum of five). Look carefully
at the *endodermis* and locate the *Casparian
strip.*

Cut further sections and stain them with
iodine solution. Notice the distribution of
the starch grains.

Anatomy of the monocotyledonous root

Cut and examine roots of Barley (*Hordeum*
spp.), Iris (*Iris* spp.), Onion (*Allium* spp.)
or Wheat *(Triticum* spp.) as above. Notice
that the basic plan is similar to that of your
dicotyledon root (Figure 180), but that there
are many more groups of protoxylem. Such
a vascular bundle is said to be *polyarch.*

Secondary thickening of the root (Figure 181)

Cut similar sections of the root of your
dicotyledon.

You may see any stage in the secondary
thickening, but the following points will help
you to recognise the thickening process:
(i) In the unthickened root the xylem and
phloem alternate around the stele, (ii) The
cambium lies internal to the groups of
phloem, and when it becomes active, it cuts
off xylem elements on the inside and phloem
elements on the outside, (iii) This means that

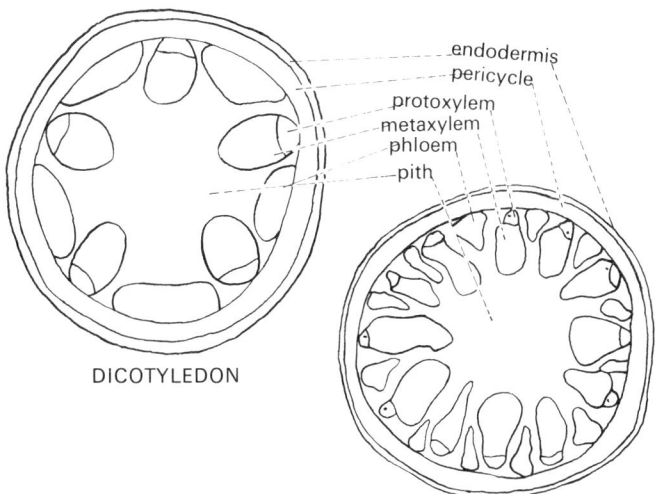

Fig. 180. Diagrams of transverse sections through root steles.

the primary phloem becomes pushed to the outside of the root, with the secondary phloem immediately adjacent to it. The cambium cuts off secondary xylem internally. The primary xylem now appears as a star in the centre of the stele, with the protoxylem at its points, which point into the primary medullary rays.

At this stage a *phellogen*, with bark and phelloderm will have developed.

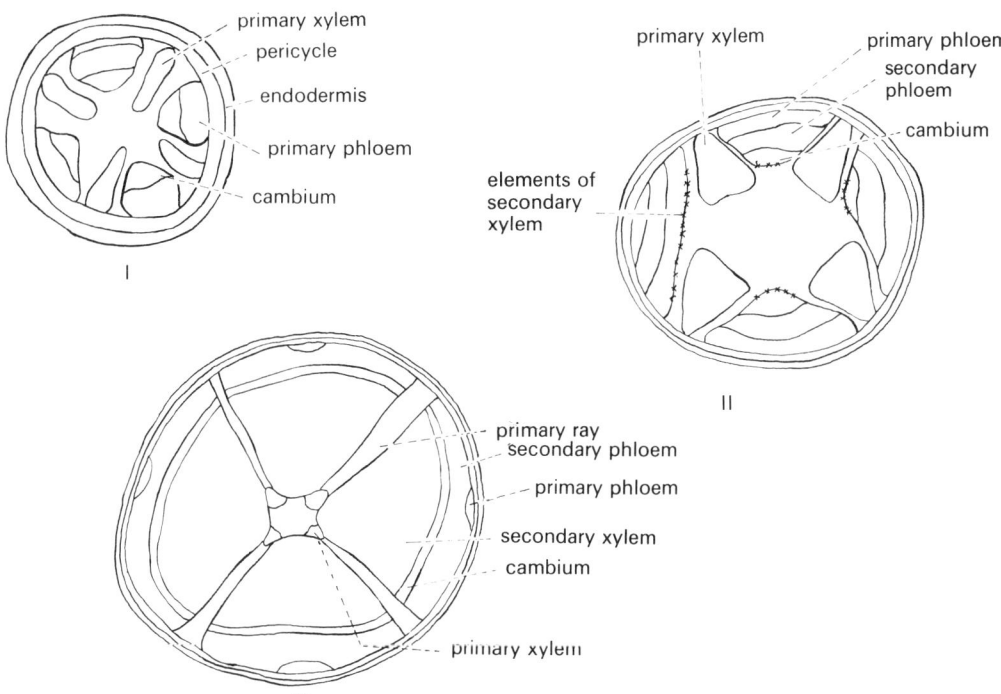

Fig. 181. Stages in the secondary thickening of a tetrarch root (not to same scale).

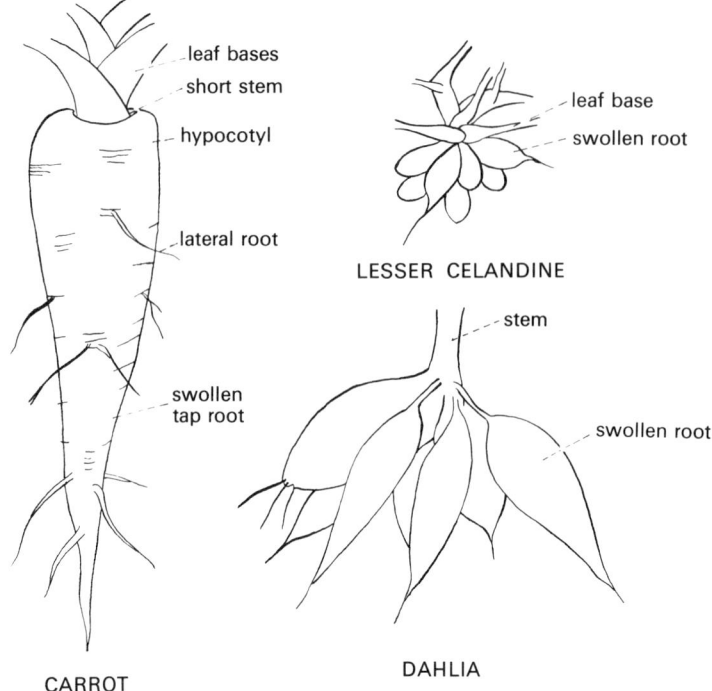

Fig. 182. Various storage roots.

Origin of lateral roots

Examine prepared slides of the development of lateral roots. Draw a plan of the slide to show the internal *(endogenous)* origin of the branch root.

Modification of the root

Tap roots

This type of root system is typical of the dicotyledons. Look at the root system of any typical dicotyledon, e.g. Pea, Bean. It consists of a main root or *primary root* bearing a series of branch roots or *lateral roots,* the former having been derived from the radical of the seedling.

Fibrous roots

These are typical of monocotyledons.

It is difficult to generalize about this type of root system, but it is typified by there not being one root larger than, or dominant to, the others. Usually, the main root (i.e.

that developed from the radicle) has stopped growing, and the fibrous roots are *adventitious*, having developed secondarily from the base of the stem.

Storage roots (Figure 182)

Examine and draw a Carrot *(Daucus carota),* Radish *(Raphanus sativus),* or Parsnip *(Pastinaca sativa).*

It is only the basal part that is root. The upper part bears leaves, so that it must be a stem, while the region immediately below it is the hypocotyl.

Examine and draw the tubers of *Dahlia* or Lesser Celandine *(Ranunculus ficaria).* These are swollen lateral roots.
Test the roots for the presence of starch, inulin and sugars. Tabulate your results.

Prop roots (Figure 183)

Examine a prepared slide of the prop root of Maize *(Zea mais).* Notice that the thickened vascular tissue is distributed nearer to the periphery of the root than is normal. Can you account for this?

Adventitious roots

These are defined as roots that are formed from some organ, other than the radicle.

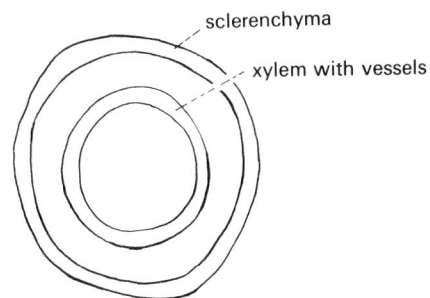

Fig. 183. Plan of a transverse section of a prop root of Maize.

You will have already seen them in studying the modifications of the stem.

Examine a piece of the climbing stem of Ivy *(Hedera helix)*.

Root nodules

The formation of root nodules, which contain nitrogen-fixing bacteria is common among the Leguminosae. Examine the root system of Pea, Bean (any) or Vetch and see the small nodules. Notice that they do not occur at regular intervals.

Mycorrhiza

The fungus-spermatophyte association is found in a large number of plants of a wide

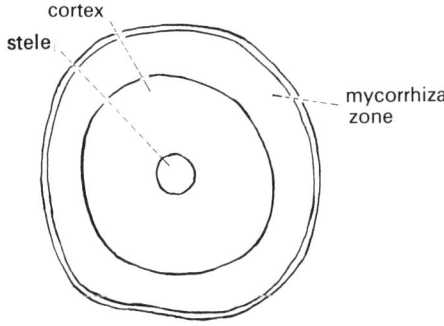

Fig. 184. Plan to show the endotrophic mycorrhiza of Bird's Nest Orchid.

range of families. There are two main types of mycorrhiza: (1) *Ectotrophic*—where the fungus forms a sheath of mycelium on the surface of the root, (2) *Endotrophic*—where the fungus is found internally, and is not visible on the root surface.

Examine prepared slides of transverse sections through the roots of Heather *(Erica* spp.), or Bird's Nest Orchid *(Neottia nidus—avis)* to see the fungus in the cortex. This is an endotrophic mycorrhiza (Figure 184).

Examine similar slides of Beech *(Fagus sylvatica)* to see the ectotrophic mycorrhiza. If you cut your own sections, stain them with lactophenol blue (cotton blue), when the fungus will take up the stain.

The leaf

The Typical leaf

Examine a Privet *(Ligustrum vulgare)* leaf, and draw it. This is a *simple leaf* and typical of the dicotyledons (Figure 185). The *lamina* is not divided into separate lobes, there are no stipules and the veins are formed like a net *(reticulate venation)*.

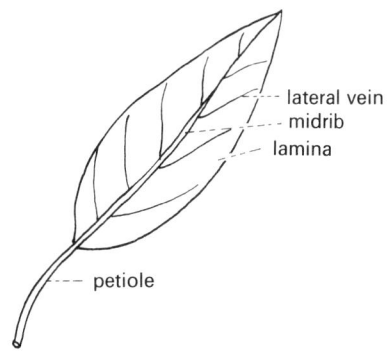

Fig. 185. The parts of a typical dicotyledonous leaf.

Cut and draw a section across the midrib of a privet leaf (Figure 186), and examine the structure. This is typical of the dicotyledons, except that in privet, the palisade layer is multilayered. A leaf like this has a distinct upper and lower surface, both morphologically and anatomically. It is therefore a *dorsiventral leaf.*

As your section is taken across the midrib,

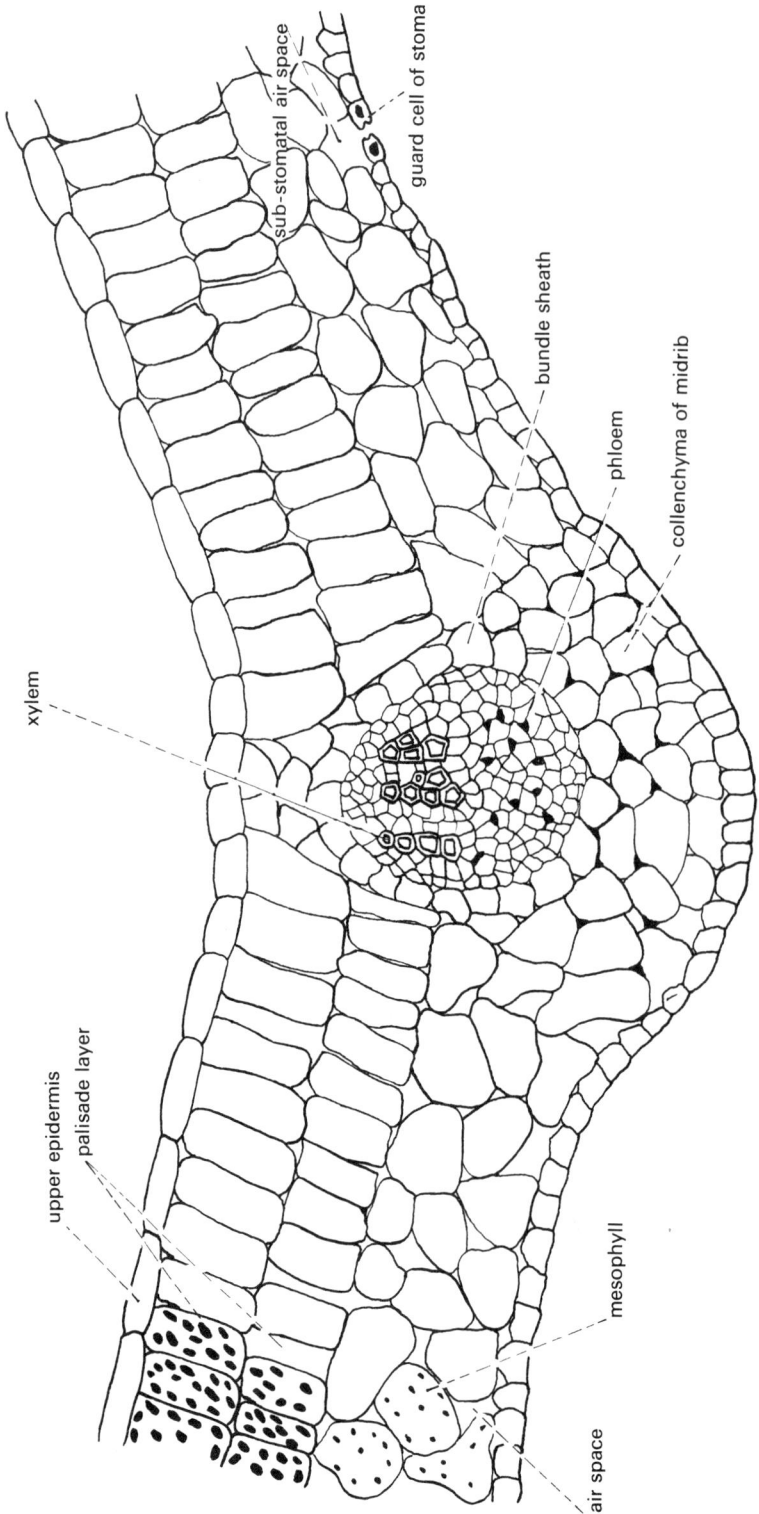

Fig. 186. Transverse section through a Privet leaf, passing through the midrib.

only that will appear in true transverse section; all other veins will be cut through obliquely.

Similarly examine the leaf of a grass or Daffodil *(Narcissus pseudonarcissus)*, or other convenient monocotyledon. Notice the *parallel venation,* and the absence of a distinct petiole, which are features of the monocotyledonous leaf.

Cut and draw a transverse section through the leaf. Now several or all the bundles will appear in true transverse section, and the upper and lower leaf-surfaces are not clearly distinguishable.

The xylem is on the upper side of the bundles in leaves. This is an easy way of orientating your section, when the leaf is not organized dorsiventrally.

The leaves so far examined are simple, i.e. the lamina is not divided into separate lobes. If the lamina is so divided, the leaf is *compound.* Compound leaves fall into two broad catagories (1) *Pinnate*—where the leaflets arise in pairs on opposite sides of the petiole, and (2) *Palmate*—where the leaflets appear to arise from the same point at the apex of the petiole. (Anatomically

they do not in fact arise at the same point, but very close together.)

Examine and draw the leaves of Ash *(Fraxinus excelsior),* with pinnate arrangement, and Horse Chestnut *(Aesculus hippocastanum)* with a palmate arrangement.

At the base of the petiole there are often accessory structures called *stipules.* These may be leaf-like or scales, and may be present throughout the life of the leaf *(persistent),* or fall soon after the leaves have opened *(caducous).*

Examine the leaf of a Pea *(Pisum sativum)* and draw it (Figure 187). The stipules are large and have largely taken over the photosynthetic function of the leaflets, which are modified as tendrils.

Look at the deeply dissected stipules of the Pansy *(Viola* spp.). Look at the stipules of the Rose *(Rosa* spp.). They are fused with the petiole of a compound pinnate leaf.

Leaf arrangement

If only one leaf arises at each node, the leaves are *alternate,* giving a *spiral* phyllotaxy.

If two leaves arise opposite each other at one node, they are said to be *opposite,* and

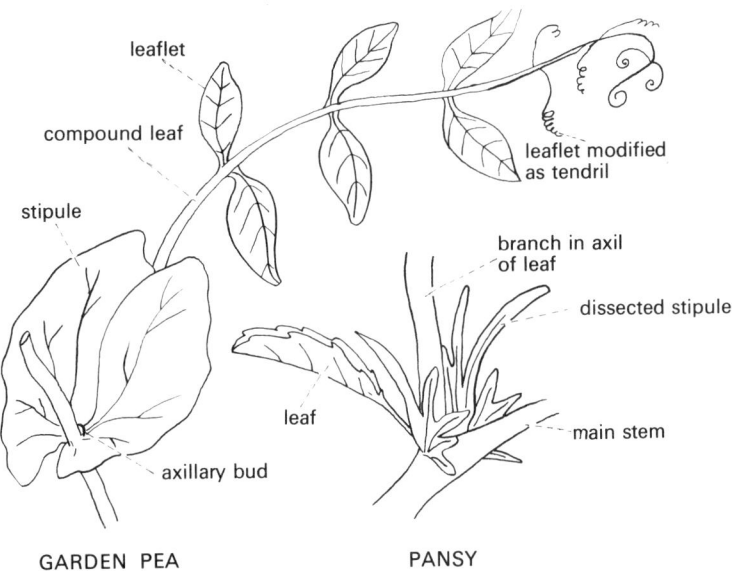

GARDEN PEA   PANSY

Fig. 187. Forms of stipules.

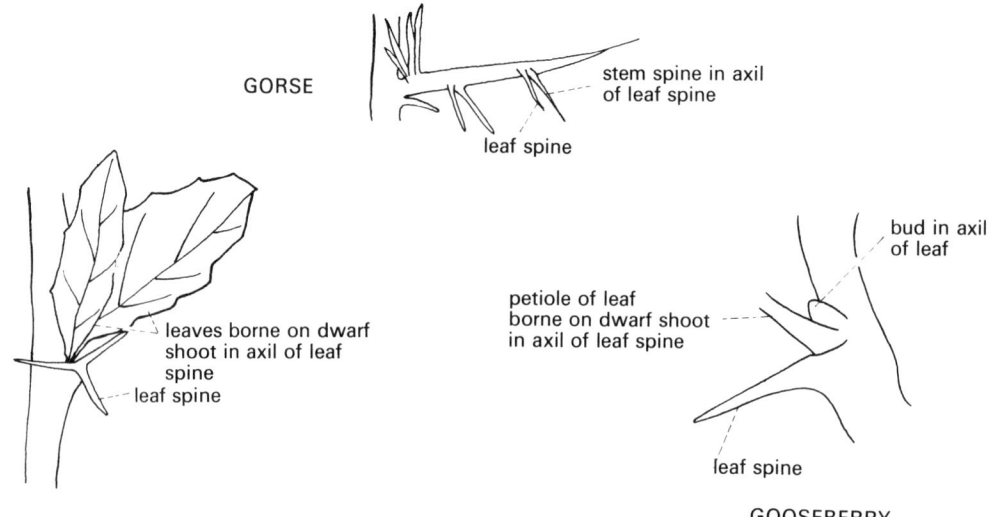

Fig. 188. Leaf spines.

if the leaves at successive nodes are at right angles to those above and below them, they are *decussate* e.g. Horse Chestnut *(Aesculus hippocastanum)* or Dead Nettle *(Lamium album)*. The phyllotaxy is then *cyclic*.

Frequently the leaves arise very close together on a short stem, e.g. Dandelion *(Taraxacum officinale)* when they are in a *rosette*.

Examine as many species as convenient, and work out the leaf arrangement.

Modifications of the leaf

Leaf spines (Figure 188)

The morphology of the structure is apparent, as a bud or branch arises in the axil. The spine is therefore a modified leaf. Examine a shoot of Barberry *(Berberis* spp.). The leaf is modified as a three-pointed spine with the foliage leaves borne on a short shoot in its axil.

Similarly look at the Gooseberry *(Ribes grossularia)*, where the spine is an outgrowth of the leaf base; and Gorse *(Ulex* spp.) where both the leaves and stipules are modified as spines.

Leaf tendrils

This modification is well-illustrated by the

Sweet Pea *(Lathyrus oderata)* where the leaflets are modified as tendrils. The stipules are enlarged, and the stem winged, presumably to counteract the loss of photosynthetic surface.

Phyllodes

These are leaf petioles which are flattened, taking over the photosynthetic function of the leaf blade (see Figure 18a). Look at specimens of *Acacia*, which show the development of the phyllode at successive stages. (This is a tropical plant, so your material will probably be preserved, but it is sometimes to be seen in glasshouses in parks.)

Hydrophytes (Figure 190)

The dimorphism of leaves in aquatic habitats is shown very well by the Water Crow-foot *(Ranunculus aquatilis)*, and the Arrowhead *(Sagittaria* spp.). In the former the submerged leaves are finely divided, while in the latter they are simpler in form than the aerial ones. Can you account for this?

Xerophytes

These plants live in conditions of drought, caused by the absence of water, or the water being unavailable (physiological drought).

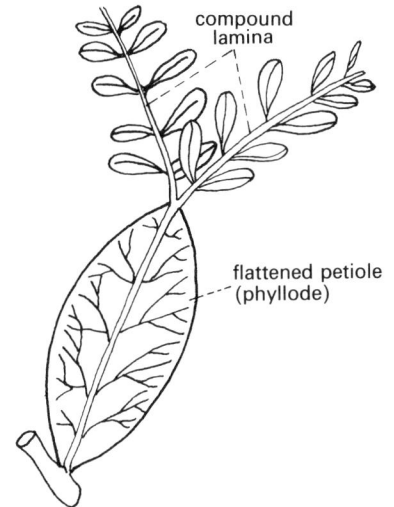

Fig. 189. Phyllode of Acacia.

Examine transverse sections of the leaves of Marram Grass *(Ammophila (Psamma) arenaria)*. Cross-leaved Heather *(Erica tetralix)*, *Hakea pectinata*, and Scot's Pine *(Pinus sylvestris)*. What features do they have in common, and how will they reduce loss of water by transpiration?

The structure of buds and meristems (Figure 191)

A Brussel's Sprout *(Brassica oleracea* var.

gemmifera) is an enlarged bud, which shows the typical structure very well. Cut one longitudinally to see the arrangement of the leaves, and the buds developing in their axils. A bud is a compressed shoot, and will have all the structures you would normally find on a shoot.

From a prepared slide, draw plans of longitudinal sections through the growing apex of a stem and a root.

The flower

The anatomy of the flower

Draw prepared slides of transverse sections of *Lilium* flowers (Figures 192 and 193), to show the different stages in the development of anthers and ovules.

Germination of pollen grains

Make a 5% solution of sucrose. Place a drop of the solution on a microscope slide, and on to it sprinkle some ripe pollen. Cover the slide suitably to prevent evaporation, and leave it for a few hours. After this time some of the pollen will have germinated, and the pollen tubes can be seen under the microscope. Stain the slide with a drop of Neutral Red to see the three nuclei.

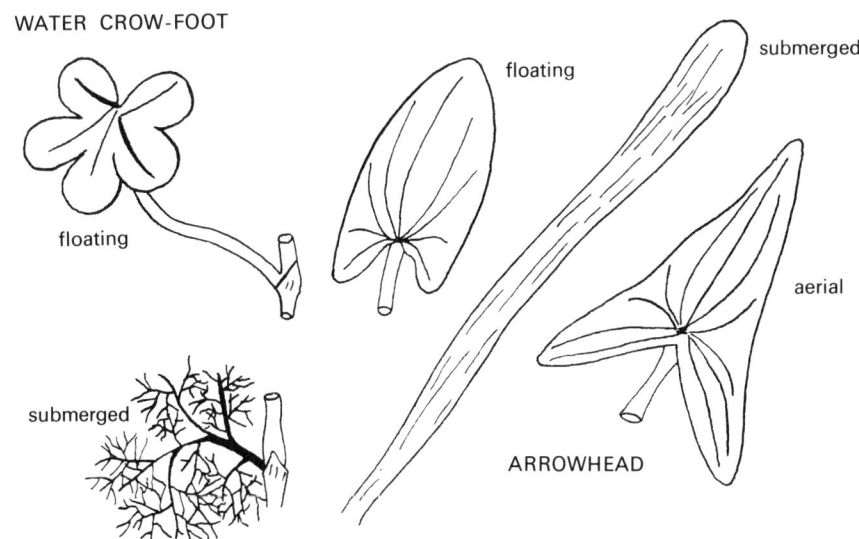

Fig. 190. Leaves of hydrophytes.

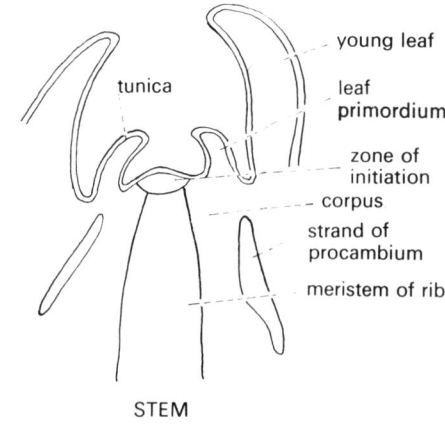

STEM

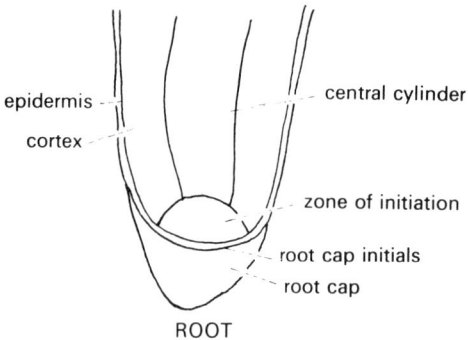

ROOT

Fig. 191. Apical meristems.

The Inflorescence (Figure 194)

There are two basic forms of inflorescence:
(1) the *racemose type* where the apical bud
remains active, and the lateral buds are
flower buds. Hence branching is monopodial.

(2) the *cyme* where the apical bud is a flower
bud, and the extension growth is brought
about by the lateral buds. Hence the
branching is *sympodial*.

These two basic forms are further sub-
divided depending on the length of the
flower-stalk *(peduncle)*, and the final relative
position of the flowers. The diagrams in
Figure 194 will make this clear. The examples
shown in brackets should be examined, but
be on the look-out for further ones.

A *cymose umbel*, e.g. Onion *(Allium)*,
appears like a racemose umbel, but the
oldest flowers are at the centre, similarly with
a *cymose corymb (Viburnum)*, and a *cymose
capitulum (Scabiosa)*.

The use of the flora

Various floras are recommended by
Examining Bodies, but the most popular
these days is probably the 'Excursion Flora
of the British Isles' (Clapham, Tutin, and
Warburg).

A key is used to identify the plant, by
using one of two or three alternative
characteristics. This then leads to the next
step—the number on the right hand side of
the page. The first key identifies the family
to which the plant belongs. You repeat the
procedure with the key of the family, which
leads to the genus, which again has its own
key giving the species. There are three points
to remember when using the flora:
(1) You will need the whole plant to

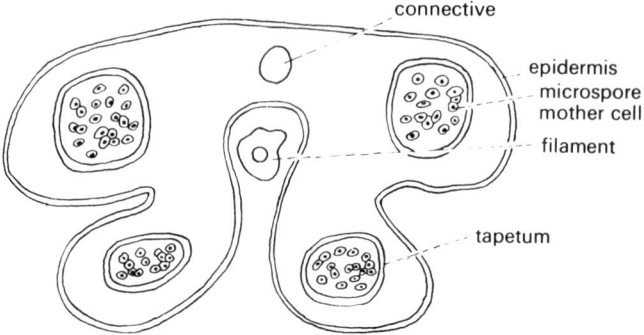

Fig. 192. Transverse section through a developing Lily anther (varying stages in the development of
the microspores and pollen grains are in different sections).

identify it properly. The specimen should have flowers, leaves, roots, and fruits. It is often difficult, if not impossible to find a plant with both flowers and fruits, but the ovary is an immature fruit, and close examination of this, possibly by examination of sections microscopically, will show you the structure of the mature fruit.

(2) There are a number of technical terms used in a flora, most of which you will find defined in the glossary. Make sure that you are familiar with the precise meanings of these terms. This does not mean that one has to learn the glossary. You will learn most of the terms by frequent use of the flora.

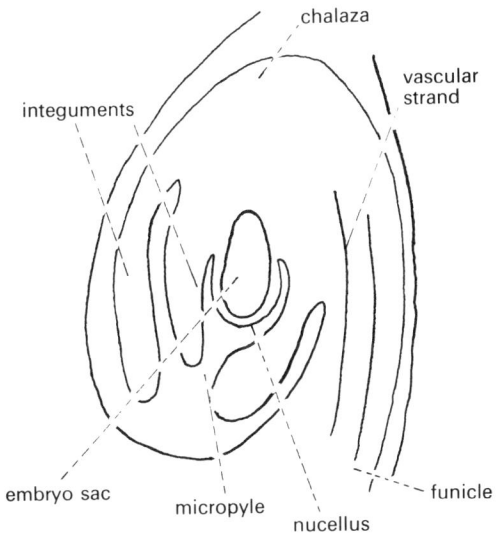

Fig. 193. Longitudinal section of an ovule of Lilium (various stages in the development of embryo sac will be seen).

(3) Use your common sense. If you pick a flower in the hedgerow and identify it as one which is found only locally in the Scilly Isles, you have made a mistake. Go back and start again. You are not likely to discover rare species, and less likely to discover new ones!

Describing the flower

There are four ways of describing and illustrating a flower. These are: (1) the floral

formula; (2) the floral diagram; (3) the half flower; (4) the longitudinal section. Do not necessarily confine yourself to these four ways. If there are any other points of interest, draw them separately. Always use a hand-lens in examining flowers. If you do not, you will miss a lot of detail.

(1) *The floral formula*

This is an abbreviated way of writing the structure of the flower, using the following symbols:

$\dagger$ = zygomorphic
K = calyx
$\oplus$ = actinomorphic (regular)
C = corolla
P = perianth (i.e. in plants where the calyx and corolla whorls cannot be morphologically distinguished)
A = androecium (stamens)
G = gynaecium with perigynous floral leaves
$\underline{G}$ = superior ovary (hypogynous flower)
$\overline{G}$ = inferior ovary (epigynous flower)
( ) around number indicates that the parts are fused, e.g. C(5) means five fused petals.
$\frown$ between letters means that the whorls are fused to each other.
The absence of brackets means that the parts are free (not fused).
∞ means a large (usually more than about 12) and indefinite number of parts.

A few examples will make this clear:

(a) $\oplus$ K5 C5 A∞ $\underline{G}\infty$ *(Ranunculus)*. The $\oplus$ shows that the flower is actinomorphic; K5—five free sepals; C5—five free petals; A∞ many free stamens; $\underline{G}\infty$ many carpels, not united, and superior.

(b) $\oplus$ K2+2 C4 A2+4 $\underline{G}$(2) (Wallflower *(Cheiranthus)*). +—regular flower; K2+2—the sepals are free but in two whorls of two each; C4—the four petals are free; A2+4—the stamens are free, but in two whorls, the outer one of two, and the inner one of four; $\underline{G}$(2)—the gynecium is superior, consisting of two fused carpels.

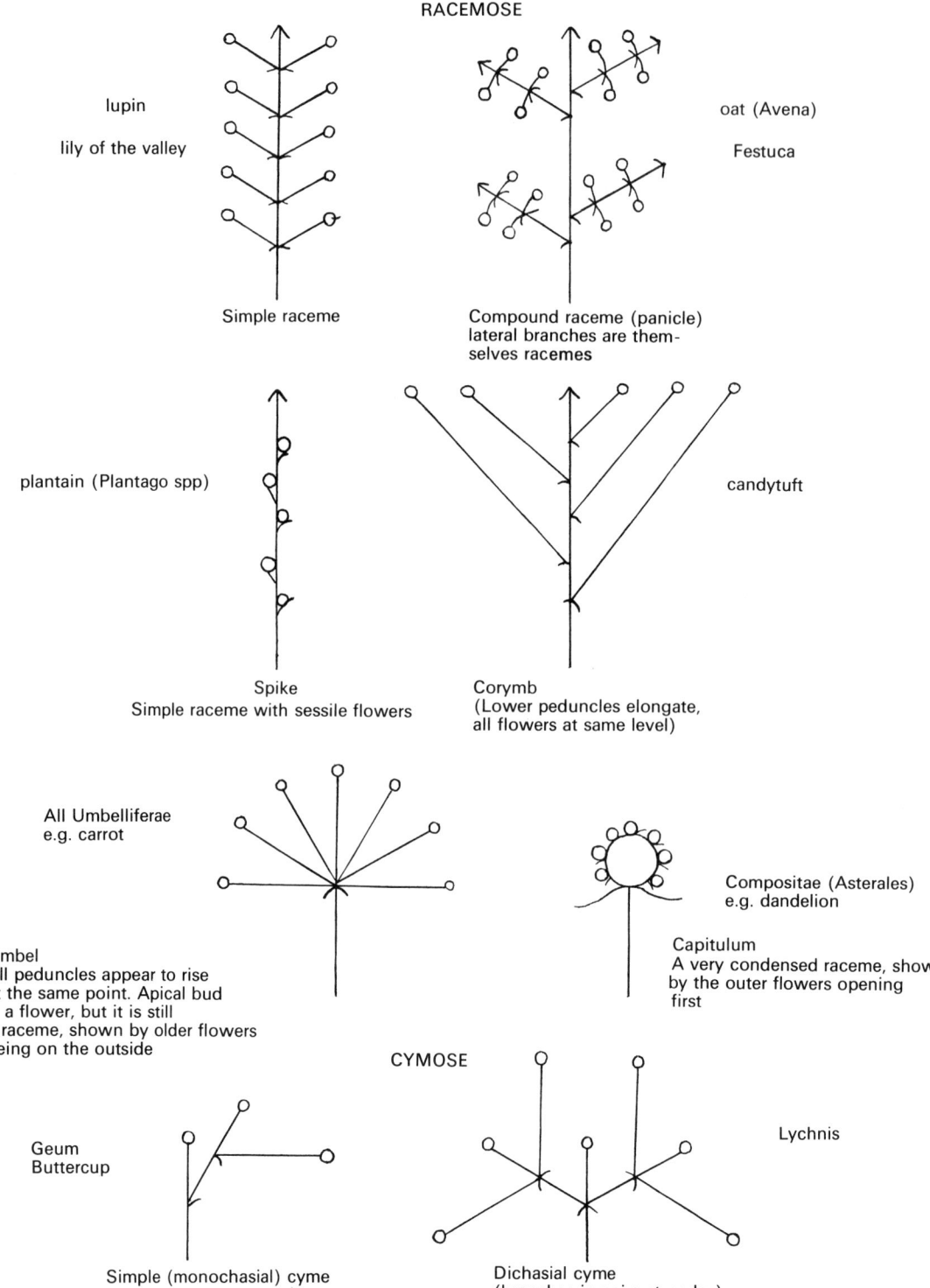

Fig. 194. Types of Inflorescence. O = Flower, ↑ = Growing apex.

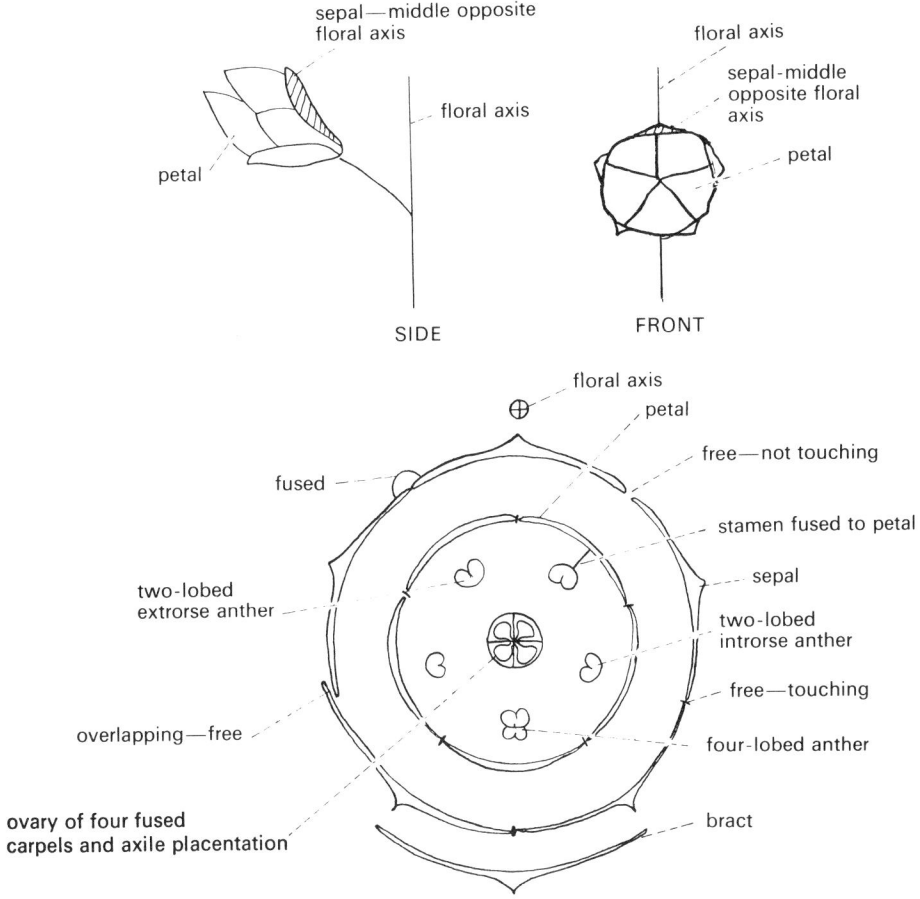

Fig. 195. Construction of, and terms used in, floral diagrams.

(c) †K(5) C(5) A4 G(2) (Dead Nettle (*Lamium*))

†—zygomorphic flower; K(5)—there are five sepals fused to each other; C(5)—the five petals are fused to each other. A4—the four stamens are free from each other, but ⌢ means that they are fused to the petals; G(2)—the two carpels are fused to form a superior ovary.

Floral diagram

This is a diagrammatic plan of the flower, using various conventions. These are shown in the diagram (Figure 195). To get the correct orientation of the diagram it is essential to place the floral axis correctly. The diagram will make this clear.

The half flower and longitudinal section

The flower is prepared by cutting it in half longitudinally. The cut should go through the mid-line indicated by the floral axis. The best way to do this is with a *sharp* razor blade. Remove the flower from the stalk and hold it by the petals. Cut through the middle of the gynaecium, leading into the petals. This gives a much cleaner cut than that in trying to cut through the petals first.

The difference between the half flower and the longitudinal section is in the method of drawing. For the former, you draw the whole half flower looking at the cut surface, and for the latter, you draw only the cut surfaces. Whichever you choose to do is a matter for personal preference, and, more important, which shows the essential detail of the flower. The longitudinal section usually gives sufficient detail, and is clearer than the half flower, but cannot show differences between structures in different planes.

The only way to learn the characteristics of the different families is by constant practice. The differences between the important families will be found in your text-book or flora. Pay particular attention to the following: *Ranunculaceae, Rosaceae, Papilionaceae, Compositae, Scrophulariaceae, Labiatae, Liliaceae,* and *Graminae.*

Fruits and Seeds

Fruits are derived from the ovary by its enlargement. They therefore have the following characteristics: (1) they contain seed(s), (2) there are two external scars, or similar structures, one from the flower stalk, and one at the opposite end from the style, (3) The wall of the ovary has three layers, although these are not frequently clearly visible. (See Figure 196).

The seed has (1) the single embryo and the food reserve inside it; (2) One scar derived from the attachment with the inside of the ovary wall; (3) A single wall—the *testa.*

These are the features to look for in a specimen. Do not be put off by appearances—several so-called seeds are in fact fruits.

Types of fruits

There are a wide variety of fruit types as is shown in the following classification. Draw as many of them as time permits, noting especially, any particular methods of dispersal.

Simple fruits

These are formed from a single gynaecium, which may contain one or more carpels. These latter may be shown as partitions in the fruit, or may have been broken down before maturing.

Dry indehiscent simple fruits

Achene e.g. Buttercup (*Ranunculus* spp.). The achenes are usually found on a common receptacle in groups, forming an aggregate fruit (see later). The wall of the individual achene is hard and leathery, having developed from a superior ovary.

Cypsela e.g. Sunflower *(Helianthus annuus),* Dandelion *(Taraxiacum officinale).* This type of fruit is typical of the Compositae, and is similar to the achene, but is derived from an inferior ovary. Frequently the cypsela has attachments associated with wind-dispersal, e.g. the 'parachutes' of the dandelion which are derived from the calyx, and that of *Clematis* derived from the style.

Nut e.g. Hazel (*Corylus* spp.), Beech (*Betula* spp.), Acorn (*Quercus* spp.). The pericarp is hard or leathery. The fruit is similar to, but larger than, an achene.

Caryopsis e.g. any grass. This type of fruit is typical of the Graminae. Examine Wheat (*Triticum* spp.), Oat *(Avena),* or Barley *(Hordeum).* This is similar to an achene, but the pericarp is thinner, and fused with

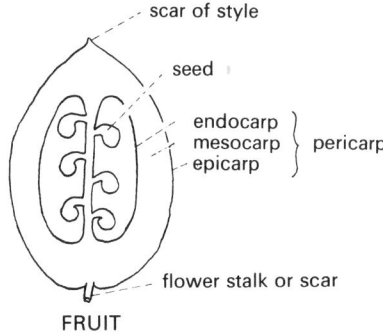

FRUIT

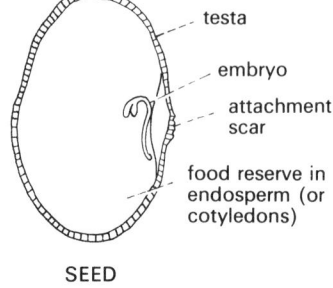

SEED

Fig. 196.

the testa, so that if the fruit is peeled, both are removed to show the endosperm.

Samara e.g. Ash *(Fraxinus excelsior)*, Elm *(Ulmus* spp.). These are essentially achenes, in which the pericarp has become flattened and enlarged to aid in dispersal by wind.

Dry dehiscent simple fruits

Follicle e.g. Larkspur *(Delphinium)*, Columbine *(Aquilegia)*, Monk's hood *(Aconitum)*. These examples have groups of follicles, forming aggregate fruits. Single follicles are rare. Each follicle has developed from a single carpel, and dehisces by a longitudinal split down *one* suture viz. the ventral one.

Legume e.g. Pea *(Pisum sativum)*, Broad Bean *(Vicia faba)*. This is typical of the Leguminosae. Each legume develops from an ovary of one carpel, and it splits longitudinally down *both* sutures. Some legumes e.g. Gorse *(Ulex* spp.) split violently and disperse the seeds by this explosive mechanism.

Capsule e.g. Poppy *(Papaver* spp.). This is made up of several carpels and contains several light seeds. On ripening, the seeds dry, and pores develop at the top of the capsule. The seeds are shaken out of the fruit as it is swayed in the wind—a censer mechanism.

Siliqua e.g. Wallflower *(Cheiranthus)*. This an elongated ovary of two carpels, separated by a *false septum*. On ripening, the two walls of the ovary (valves) separate from the base upwards to expose the seeds attached to the replum, and lying on the false septum.

Silicule e.g. Honesty *(Lunaria)* and Shepherd's Purse *(Capsella)*. This is similar to the siliqua, but shorter and broader.

Both siliqua and silicule are typical of the Cruciferae. Both bear the remains of the stigma at the apex.

Schizocarpic fruits

These are fruits formed from a single gynaecium, but which split into distinct segments during ripening.

Cremocarp e.g. Carrot *(Daucus carota)* or several of the Umbelliferae. The fruit splits longitudinally into two halves *(mericarps)*, which remain attached at the apex.

Double samara e.g. Sycamore *(Acer* spp.). These are two fused samaras which split at dehiscence.

Lomentum e.g. Radish *(Raphanus sativus)*. This is a siliqua which has become constricted into one-seeded portions which separate at dehiscence. A lomentum is sometimes formed as a modified legume.

Regma e.g. Geranium *(Pelagonium)*. The fruit splits into one-seeded units called *cocci*, remaining attached to the style at the apex. The violence of the splitting aids in the dispersal of the seeds.

Carcerulus e.g. White Deadnettle *(Lamium album)*. There are two carpels, each of which is divided by a false septum to form four mericarps.

Simple succulent fruits

The fleshy part of the fruit is the mesocarp, while in the drupe, the endocarp is thickened to form a wall around the seed (the 'stone').

Although there are exceptions, a *berry* has many seeds, e.g. Tomato *(Lycopersicum esculentum)*, Gooseberry *(Ribes uva-crispa)*, and a *drupe* has a single seed, e.g. Plum, Cherry *(Prunus* spp.).

Aggregate fruits

Collection (etaerio) of drupes e.g. Raspberry, Blackberry *(Rubus* spp.).

Collection of achenes e.g. Buttercup *(Ranunculus* spp.), Strawberry *(Fragaria vesca)*. (See *Strawberry* as a false fruit.)

Collection of follicles e.g. Larkspur *(Delphinium* spp.), Monkshood *(Aconitum* spp.), Colombine *(Aqualegia* spp.).

False fruits (Pseudocarps)

These are structures, commonly called fruits, but which consist of parts other than a swollen ovary.

Strawberry *(Fragaria vesca)*—the fleshy part is a swollen receptacle, and therefore not a part of the true fruit. It does not have the two scars, one at each end.

Pome e.g. Apple, Pear *(Pyrus* spp.)—the pericarp is the core, and the fleshy part the swollen receptacle. The skin is epidermis. There are two scars, but one is the flower stalk, and the other a complex scar of the remains of the stamens, petals and sepals. Cut a longitudinal section and draw it.

Structure of seeds and seedlings

Always soak seeds in water for 24 hours before examination.

Dicotyledonous seeds

Non-endospermous seeds

Broad Bean (Vicia faba)

Examine the seed and note the points shown in Figure 197. Remove the testa and draw to show the two large cotyledons which contain the reserve food materials.

Cut thin slices of the cotyledon, and place them on separate slides. Stain them with iodine solution, Fehling's solution, and Millon's reagent separately. Deduce the distribution of starch, reducing sugars and protein in the soaked seed.

Separate the cotyledons and note the points shown in the diagram.

Place some soaked seeds on damp cotton-wool, or in a beaker between damp filter-paper and the side, and allow them to germinate. Draw a few of the stages. Is the germination epigeal or hypogeal?

Sunflower (Helianthus annus)

This is a cypsela, not a true seed.

Examine it as you did the broad bean seed, but note the two scars on the pericarp, and

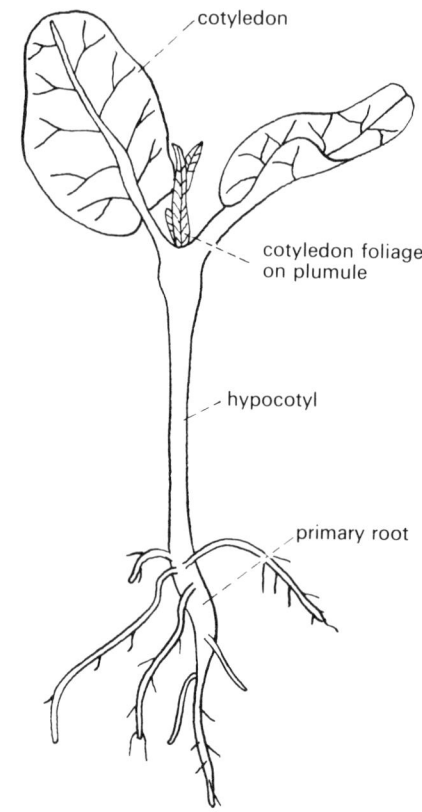

Fig. 198. Castor-oil seedling.

the thin testa visible when the pericarp is removed.

Germinate some of these 'seeds' as described above.

Endospermous seeds

Castor Oil (Ricinus communis)

Examine the external features, noting the *aril* at one end. If this is removed, the hilum and micropyle is revealed. Remove the testa, and this will expose the papery inner layer—the

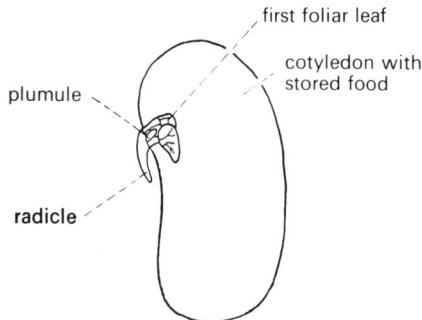

Fig. 197. A single cotyledon of germinated Broad Bean seed, with the embryo attached.

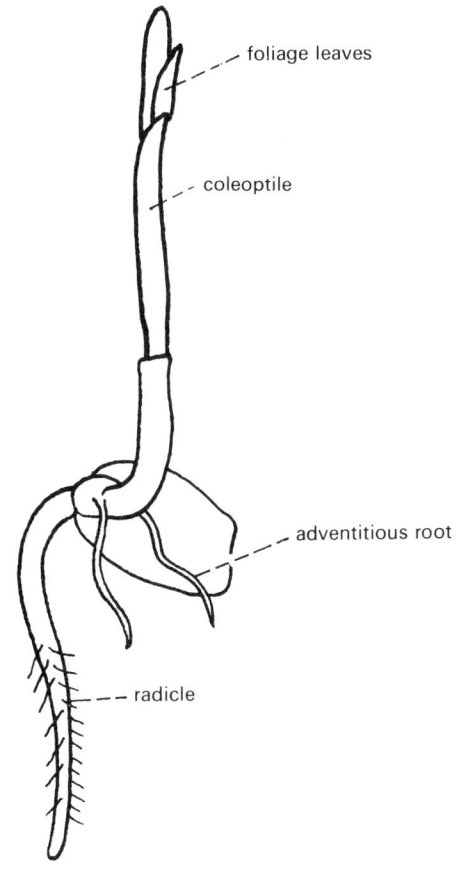

Fig. 199. Maize seedling.

tegmen. Test thin slices of the endosperm for food reserves as for the broad bean, but also include test for fats and oils (Sudan III).

Cut longitudinal sections of another seed to show the embryo. From yet another seed try and dissect out the embryo, to show the fine papery cotyledons, lacking stored food. This can be done by removing the tegmen, and gently squeezing the endosperm to reveal a slit, which will be the space between the cotyledons.

Germinate some castor-oil seeds and draw the various stages (Figure 198). Is this method of germination epigeal or hypogeal?

Monocotyledonous seeds

Maize (Zea mais)

This is a caryopsis.
Draw the external features and a longitudinal section as shown in Figure 199. Test thin pieces of the 'seed' as before for starch, reducing sugars and proteins.

Germinate some maize 'seed' and decide the method of germination. Draw the various stages.

Study the germination of Wheat *(Triticum)*, Onion *(Allium)*, and Cress *(Lepidium)*.

10 Plant Growth

Conditions necessary for germination

General conditions

Experiment: To find the water content of dry seeds, and the amount that they absorb before germination.
Apparatus: Balance, steam (or other suitable) oven, beakers, about 100 pea, or other suitable seeds.
Method: Divide your sample of peas into two approximately equal amounts and weigh them separately. The two weighings should be approximately the same.

Dry one sample in the steam oven, weighing repeatedly until the weight is constant. Soak the other sample and weigh them at 12 hourly intervals for three days. Before weighing, dry the peas carefully on filter-paper.

Calculate your results as precentages of the original weights. Graph your results against the time of soaking.

What conclusions can be drawn about the water content of 'dry' seeds, and the absorption of water before germination?

Experiment: To calculate the percentage germination of seeds.
Apparatus: Petri dishes (or other covered dishes), filter-paper, any small seeds, e.g. Cress.
Method: Place a piece of moist filter-paper on the bottom of the Petri dish, and sprinkle about 100 of each type of seed in each dish. Count the number of seeds germinated each day. From these figures you can calculate:
(a) the percentage germination for each species; (b) the day on which most seeds germinated; (c) the mean time taken for germination.
N.B. A seed has germinated as soon as the radicle has appeared through the testa.

Need for water

Experiment: To show that water is necessary for germination.
Apparatus: Two Petri dishes, filter-paper, any small seeds e.g. Cress, Barley.
Method: Place a soaked filter paper in one dish, and a dry one in another. Place approximately the same number of seeds in each dish, and leave them for a few days.

(The approximate length of time of the experiment can be deduced from the previous experiment.) Calculate the percentage germination in each dish.

Experiment: To show the effect of extremes of temperature on the germination of soaked and unsoaked seeds.
Apparatus: Pea seeds, boxes or six large beakers, filled with damp sawdust, muslin, boiling water, and freezing mixture (or the cold compartment of a refrigerator).
Method: Divide the peas into two equal groups, and soak one half for 24 hours, leaving the other half dry. Divide each of the two groups of seeds into three further groups. Plant one set of dry, and one set of soaked peas in the sawdust, and then place the other four sets in muslin bags. Place one dry set, and one soaked set in the freezing mixture for about five minutes, and the other two sets into the boiling water for half a minute. Plant each set separately and record the percentage germination.

Temperature

Experiment: To show the effect of extremes of temperature on germination.
Apparatus: Petri dishes (or other small dishes), filter paper, thermometer, refrigerator, seeds, e.g. wheat, cress.
Method: In this experiment it is important that there is a proper control. As one group of seeds will be inside the refrigerator, i.e.

in the dark, the other treatment must also be covered with a dark box.

Sow approximately the same number of seeds in each of two Petri dishes which are lined with damp filter-paper. Place one in the refrigerator, but not in the freezer, and the other near a radiator, having previously placed a thermometer near each dish. Count the number of seeds that have germinated, and calculate the percentage germination.

Oxygen

Experiment: To show that oxygen is necessary for germination.
Apparatus: Two covered gas-jars, cotton wool, wire, alkaline pyrogallol, and seeds.
Method: About quarter fill one of the gas-jars with alkaline pyrogallol. Thread some moist cotton wool on the wire and sow it with seeds. Suspend this above the alkaline pyrogallol and cover the gas-jar. Set up a similar gas-jar as a control, replacing the alkaline pyrogallol with water. Leave both jars in the warm, and calculate the percentage germination.

Experiment: To show that water and oxygen are necessary for germination.
Apparatus: Gas-jar, pea seeds, piece of stick.
Method: Three-quarters fill the gas-jar with boiled water (i.e. contains little dissolved oxygen). Pin the peas at intervals along the strip of wood and place it vertically in the cylinder of water. Thus some of the peas will be in water with little oxygen, some will be in air with no water, and the others, at the surface of the water will have plenty of water and dissolved oxygen. Observe what happens.

Light

Experiment: To show the effect of light on germination.
Apparatus: Petri dishes, filter-paper, seeds of onion (*Allium* spp.), Willowherb (*Epilobium hirsutum*), buttercup (*Ranunculus sceleratus*), tobacco (*Nicotiana tabacum*), and any other suitable seeds.

Method: Divide each group of seeds into two approximately equal portions, and place them on damp filter-paper in the Petri dishes. Place one dish of each pair in the dark, and the other in the light. Observe the number of seeds germinated in each dish after a few days and notice the condition of the young seedlings.

This is a very simple experiment, but you may like to devise your own examining the effect on germination of length of exposure to light, and the wave length of the light (using coloured filters).

Vegetative reproduction

Vegetative reproduction takes place widely in flowering plants, and usually depends on the development of adventitious roots from a node, the bud of which grows into a new shoot, thus producing a new individual which becomes separated from the parent plant.

Using horticultural methods, one depends on the adventitious roots developing from the meristematic callus which forms over a wounded surface.

Cut a potato so that each piece contains an 'eye' (bud), and plant them separately in a box of soil. Keep the box in a frost-free place and observe the number of plants which develop. Similarly cut up and plant the roots of dock (*Rumex* spp.), and dandelion (*Taraxacum officinale*), (which, being roots, will not have nodes), and rhizomes of couch grass (*Agropyrum repens*).

Try rooting cuttings of as many perennials as you can.

Take the cutting by cutting a young branch across, just below a node. Remove all but the top pair of mature leaves, and place the cut end in damp sand in a flower pot or box. (Instead of sand you can use 'Vermiculite'.) Make sure that the sand is packed firmly around the cutting, and that there is no air-space between the bottom of the cutting and the base of the hole that you made in the sand. Water well, and cover the pot and cuttings with a sheet of polythene to reduce water loss.

If you have time you can elaborate your experiments with cuttings by trying the following treatments: (a) the effect of rooting 'hormones'; (b) the effect of light; (c) the effect of temperature.

To keep the plants after they have rooted, you will have to pot them in good garden soil or John Innes Compost No. 2. There are no nutrients in sand or 'Vermiculite'.

Some plants will reproduce vegetatively from leaves. Try the following method with any leaves you have available, especially *Begonia* and African violet *(Saintpaulia)*. Both these are fairly popular house plants, and you should have no difficulty in obtaining material. Snick through the leaf veins on the lower surface. Do not cut right through them. Pin the whole leaf down on damp sand in a flower pot, and cover it with polythene or a piece of glass. Make sure that the leaf is in continuous contact with the sand.

A great deal of patience is required when using these methods of propagation. Some of these cuttings take several weeks, maybe months to root.

Remember that the asexual spores produced by other groups of plants are asexual methods of reproduction.

Photosynthesis

A great many of the modern experiments on photosynthesis are of great importance, but unfortunately they cannot be repeated in a school laboratory. Try to interpret your results in the light of modern findings, and make full use of electron microscope photographs e.g., of chloroplasts.

Evidence that photosynthesis has taken place in leaves is usually shown by testing the leaf for the presence of starch. This should be carried out in the following way:
Remove the leaf from the plant and kill it by dipping in boiling water. Do not leave the leaf in the boiling water any longer than it takes to make it limp. Transfer it to boiling ethanol (ethyl alcohol) or methylated spirit to remove the chlorophyll. Boil the alcohol in a water-bath and remember that it gives

off a heavy inflammable vapour. Remove the leaf from the alcohol and wash it in water to soften the tissue and remove the alcohol. Stain the starch by dipping the leaf in iodine solution. The presence of starch is shown by a blue-black coloration. Iodine will stain the leaf brown whether starch is present or not, as it reacts with cellulose.

Light

Experiment: To show that light is necessary for photosynthesis.
Apparatus: Potted plant (e.g. nasturtium *(Tropaeolum)*, runner bean *(Phaseolus)*), iodine solution, apparatus for removing chlorophyll, tin foil, paper clips.
Method: Place the plant in the dark, e.g. in a cupboard, for 24 hours. Remove it, and immediately test one of the leaves for starch. Cover one of the other leaves with a stencil made of the tin foil (see Figure 200). Leave the plant in the light during the day, removing the partially covered leaf during the afternoon. Test it for starch.

From these tests draw your own conclusions about the necessity for light in photosynthesis.

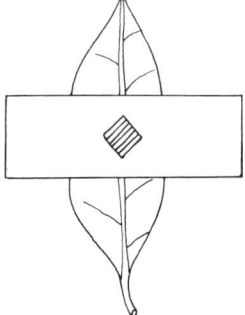

Fig. 200.

Apparatus: *Spirogyra,* microscope, Petri dishes, apparatus for testing for starch, slides.
Method: Place some filaments of *Spirogyra* in two Petri dishes of water. Leave one dish in the dark and the other in the light for 24 hours. Remove the chlorophyll and test each for starch. Mount the specimens and

examine them under the microscope. Which specimen contains starch, and where is it located?

Experiment: To show the wavelength of the light necessary for photosynthesis.
Apparatus: Two or more large plant pots, potted plants, such that the pot and plant will fit inside the large pot. Various light filters.
Method: Set up the apparatus as shown in Figure 201, having de-starched the plants in the dark for 24 hours. Leave the apparatus

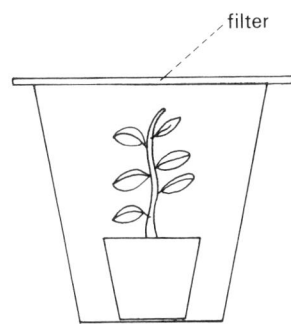

Fig. 201.

in the light for 2–3 days. Remove leaves from each of the plants and test them for starch. What do you conclude about the wavelength of the light necessary for photosynthesis?

More complicated light-tight boxes can be used, but this simple apparatus works effectively. If you have transparent dishes large enough to cover the tops of the large flower pots, these filled with concentrated solutions of potassium dichronate or ammoniacal copper sulphate make, respectively, effective yellow and blue filters.

Experiment: To estimate the effect of light intensity on the rate of photosynthesis.
Apparatus: Light source, four test-tubes, clamps (or other means of holding the test-tubes), pondweed *(Elodea)*, sodium bicarbonate, watch.
Method: This method depends on the evolution of oxygen from the water plant during photosynthesis. You have not, as yet, proved that oxygen is evolved during photosynthesis, so you may like to leave this until you have carried out those experiments which prove the evolution of oxygen.

Prepare a 0.25% solution of sodium bicarbonate. (This is used to ensure that there is plenty of carbon dioxide available.) Partly fill the test-tubes with the solution, and arrange them so that they are 3 in., 6 in., 9 in., and 1 ft. from the light source, and are not in each others shadow.

Cut off lengths of *Elodea* (3–4 in.) under water, and keep them moist. Place a piece, stalk upwards, in each test-tube. Allow the plants to acclimatize themselves. This can be judged to have happened when bubbles of oxygen are given off at an even rate.

Compare the rate at which the bubbles of oxygen are evolved. Draw your conclusions regarding the light intensity and the rate of photosynthesis; remembering that the intensity of the light on a surface varies inversely as the *square* of the distance of the surface from the source.

Experiment: To show that light is not necessary for the production of starch, if the leaf is provided with glucose.
Apparatus: Beakers, potted plant, glucose, materials to test for starch.
Method: De-starch the plant by leaving it in the dark for 24 hours. Make a 3% glucose solution in one of the beakers, and use water in the other as a control. Remove two de-starched leaves and float them on the liquid in the beakers. Place both the beakers in the dark for a week, after which time test the leaves for starch.

To reduce the risk of bacterial contamination, use boiled water, and sterilise the beakers with formalin.

Carbon dioxide

Experiment: To show that carbon dioxide is necessary for photosynthesis.
Apparatus: Conical flask, suitable clamp,

sodium hydroxide solution, cork, 'Vaseline', cork-borer.

Method: Select a leaf on the plant, and bore a hole in the cork which will fit around the petiole. Split the cork longitudinally so that it can be put around the petiole, and so that the leaf will fit into the flask as shown in Figure 202.

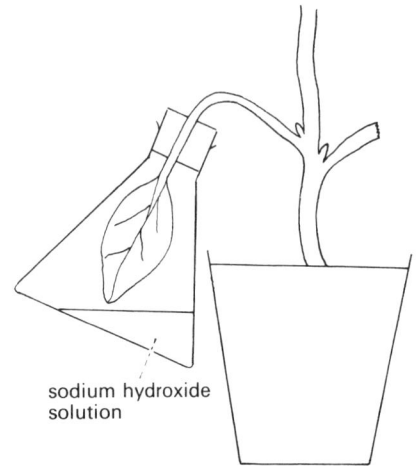

sodium hydroxide solution

Fig. 202.

Place the plant in the dark to de-starch it, then set up the apparatus as shown in the diagram. Make sure that the leaf does not come in contact with the caustic soda solution, and that all the joints are air-tight. Leave the apparatus in the light for a day, and test the experimental leaf and a comparable leaf of the same plant for starch during the early afternoon.

Experiment: To show the relative efficiency of photosynthesis with differing concentrations of carbon dioxide.
Apparatus: Test-tubes, *Elodea*, sodium bicarbonate, pond water, freshly boiled water, watch.
Method: Make sodium bicarbonate solutions of the following concentrations in tap water: 1%, 0.5%, 0.25%, 0.125%. (These may be made up as a dilution series). Partly fill the seven test-tubes with one of these solutions,

the pond water, tap water, and boiled water. Cut pieces of *Elodea* of equal length and place a piece into each tube, remembering to cut the *Elodea* under water. Place the tubes in identical conditions, and count the number of oxygen bubbles evolved from each in a given time. Before beginning to count the bubbles wait until they are being evolved at a steady rate.

Chlorophyll

Experiment: To show that chlorophyll is necessary for photosynthesis.
Apparatus: Variegated leaf, e.g. privet *(Ligustrum)*, or any other soft variegated leaf, apparatus for testing for starch.
Method: Pick the leaf during the afternoon if possible. Draw a plan of the leaf to show the green and yellow areas. Test the leaf for starch, and compare the areas which contain starch with those on the original plan. Is there any relation between the starch-containing areas and the green areas?

Experiment: To investigate the relation between light and the production of chlorophyll.
Apparatus: Cress seed, pea and pine *(Pinus)* seeds, seed trays, sawdust or 'Vermiculite', razor, slides and microscope.
Method: Sow the seeds on damp sawdust in the seed trays, and place them in the dark. When the seeds have germinated, observe the colour. Is light necessary for the development of chlorophyll? If no chlorophyll has developed, cut transverse sections of the leaf, and compare them with the microscopic structure of a normal leaf. Draw the two slides, paying especial attention to the palisade layer.

If no chlorophyll has developed, place the plants in the light and observe how long it takes for the pigment to appear.

Experiment: To investigate the stage in carbon assimilation which requires chlorophyll.
Apparatus: Leaf from a bean or pea plant which has been grown in the dark so that it

has no chlorophyll (a leaf from the previous experiment will do), 3% glucose solution, beaker, apparatus for testing for starch.
Method: Remove one leaf from the plant and float it on the glucose solution. Place the beaker with the floating leaf in a dark cupboard and leave it for a week, after which time test for starch.

Leave the plant, from which the leaves were removed, in the light for 24 hours, remove a leaf and test it for starch. What conclusions can you draw regarding the necessity for chlorophyll for the production of starch? This is far from being a perfect experiment. What criticisms of it can you make?

Experiment: To separate and examine the pigments in leaves.

The separation described below depends on a technique called paper chromatography. In essence this depends on the fact that if you have a mixture of substances in solution, they will diffuse through a piece of filter-paper at different rates, and so become separated into different bands. These bands can then be identified by different techniques. In the case of the chlorophyll pigments they can be seen as separate bands without further treatment.
Apparatus: Soft green leaves, ethanol, 6 in. by 1 in. test-tube with a cork to fit, fine wire hook made from a pin, strip of filter-paper to fit inside the tube without touching the sides, fine pipette, solvent solution consisting of 8% acetone and 92% petroleum ether, (both of these are highly inflammable; do not make up more than your minimum requirement).
Method: Crush the leaves slightly, and boil them in ethanol in a water-bath, to extract the chlorophyll. Continue the extraction, or add more leaves until the solution is a very dark green.

About 1 in. from one end of the filter-paper draw a light line in pencil (*not* ink). Suck some of the chlorophyll extract into the fine pipette, and with it draw a line on the pencil line. Allow it to dry, and repeat until the paper is clearly marked. Pour some of the

solvent into the test-tube, and set up the apparatus as shown in Figure 203. Take care that the level of the solvent in the tube is below the line drawn on the paper. Leave the experiment until the solvent reaches the top of the paper. The pigments will separate out in distinct bands across the paper. How many different pigments are there in the chlorophyll?

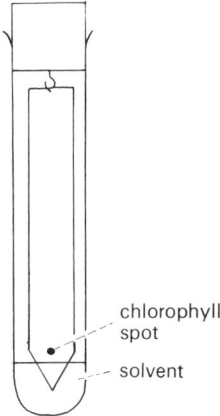

chlorophyll spot

solvent

Fig. 203.

Repeat the experiment using different types of plant, e.g. green algae, seaweeds, mosses, ferns, etc.

If there is a spectrometer available, examine the absorption spectrum of a chlorophyll extract.

Oxygen

Experiment: To show that oxygen is evolved during photosynthesis.
Apparatus: Piece of pond weed *(Elodea)*, large beaker, filter funnel, test-tube, sodium bicarbonate.
Method: Dissolve 2–3 g. of sodium bicarbonate in about 500 ml. of water in the large beaker, filling the beaker so that the stem of the filter funnel will be covered when it is inverted in the beaker. This gives a high concentration of carbon dioxide in the water. Place the pond weed in the beaker and cover it with the inverted funnel. Fill the test-tube with water, and invert it over the stem of the funnel as shown in the diagram (Figure 204).

Stand the apparatus in a sunny place until a large bubble of gas has accumulated in the test-tube.

Remove the test-tube and insert a glowing split into it. What does this test indicate?

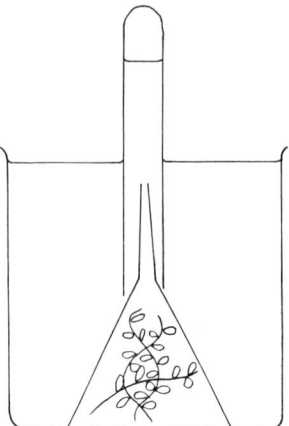

Fig. 204.

Rate of photosynthesis

Experiment: To compare the rate of photosynthesis under different conditions.
Apparatus: Two 6 in by 1 in. test-tubes, 2 two-holed rubber bungs to fit the test-tubes, glass tubing, rubber tubing, 2 clips, sodium bicarbonate, pond weed *(Elodea)* or fresh-water green algae, stop watch.
Method: Construct a manometer as shown in Figure 205. Fill each tube at least three-quarters full of 1% sodium bicarbonate solution. Place the plant material in one tube. Open the two clips to zero the water in the U-tube. Place the tubes together in a warm place, or in ice, or in varying light intensities, and measure the rate of change in the difference of levels in the arms of the manometer. Graph your results. Before each test zero the apparatus by opening the two clips.

If you are using a unicellular green algae for this experiment, the cells will have to be centrifuged out of the culture solution, or allowed to settle overnight. Make up a suspension of the cells in the sodium bicarbonate solution, and proceed as above.

The apparatus measures the difference in pressure between the two tubes caused by the release of oxygen during photosynthesis.

Special methods of nutrition

Nitrogen fixation

Dig up and draw the roots of a legume, e.g. Pea *(Pisum)*, Lupin *(Lupinus)*, Clover *(Trofolium)*, to show the root nodules.

Experiment: To show the effect of the root nodules on the growth of a legume, and on plants grown in association with it.
Apparatus: Four plant pots, sand, soil sterilizer, legume plants, legume seeds, wheat *(Tritcum)* seed, nutrient solution lacking nitrogen, as described in a later experiment.

Method: Steam sterilize the sand and the flower pots to ensure that there are no nitrogen-fixing bacteria present. Fill the flower pots with the sand. Wash the roots of the legume plants in sterile water, and remove the root nodules with a sterilized scalpel. Grind the nodules in sterile water to form a milky suspension. Pour this on two of the flower pots. In one of these pots sow legume seed alone, and in the other legume and wheat seed (so that there are about 10 seeds in an

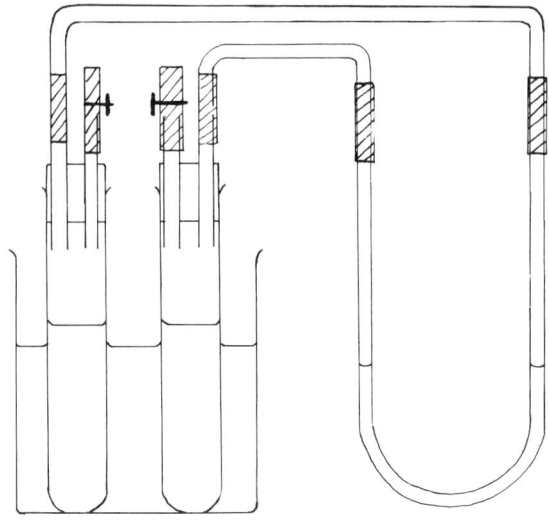

Fig. 205.

8-in. pot). Similarly sow the non-inoculated pots. Water all the pots with the nutrient solution, and moisten them with the same solution if they dry out. Allow the plants to germinate and grow, observe any differences between the treated and non-treated plants.

Examine and draw prepared slides of root nodules.

Parasitic plants

Dodder *(Cuscuta)* is found on gorse *(Ulex)*, nettle *(Urtica)*, and clover *(Trifolium)*. Draw a specimen. What characteristics make it evident that it is a parasite?

Draw a prepared slide of a haustorium of Dodder.

Broomrape *(Orobanche)* is found parasitizing the roots of gorse *(Ulex)*. Dig up a specimen and draw it, noting any parasitic characteristics. Similarly examine Toothwort *(Lathraea)* which is a root-parasite of Hazel *(Corylus)*.

Mistletoe *(Viscum album)* is a partial parasite. What evidence is there to support this statement? Examine a piece of apple wood with mistletoe growing on it. Cut through the branch to see the haustorium of the mistletoe. Draw the cut surface (see Figure 206).

Other partial parasites are Lousewort *(Pedicularis sylvatica)* found in marshy situations, and Yellow Rattle *(Rhinanthus* spp.) in drier situations. Both are root parasites.

Insectivorous plants

Examine whole plants of Sundew (*Drosera* spp.) and Butterwort (*Pinguicula* spp.) noting the glandular hairs on the leaves. Examine

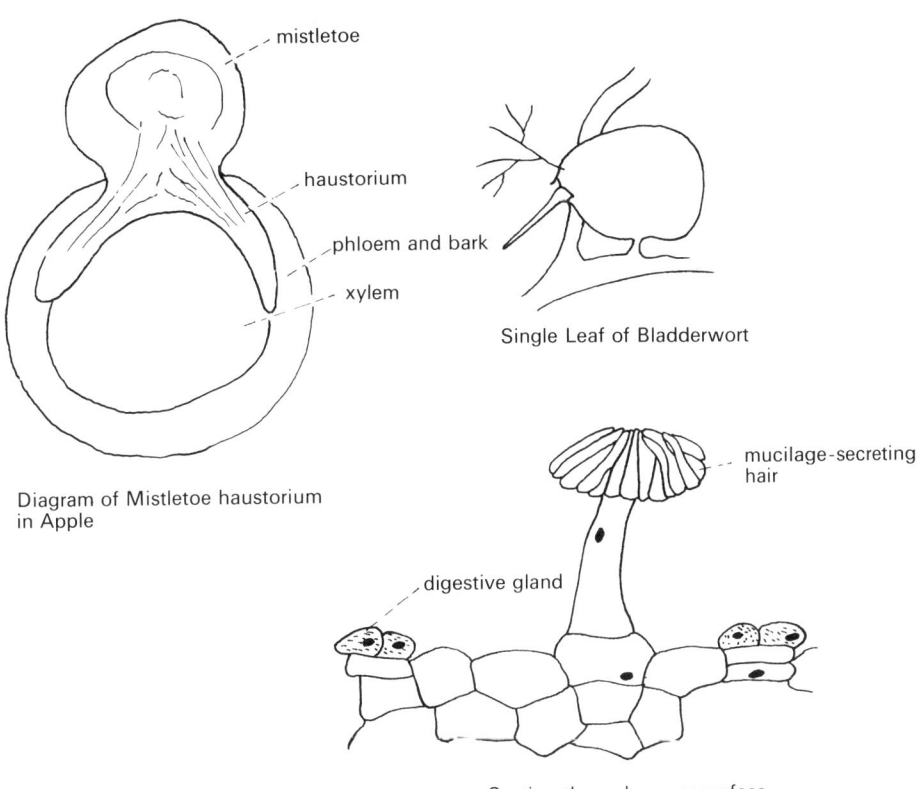

mistletoe

haustorium

phloem and bark

xylem

Diagram of Mistletoe haustorium in Apple

Single Leaf of Bladderwort

mucilage-secreting hair

digestive gland

Section through upper surface of Bladderwort Leaf

Fig. 206.

transverse sections through the leaves, and draw the glandular hairs.

Examine the water plant Bladderwort *(Utricularia)*. Draw the complete plant and work out the morphology of the bladder. Look at a single bladder under the microscope and draw it.

Secretory substances and by-products

Calcium oxalate is frequently found as insoluble crystals in cells.

Cut sections or examine prepared slides of the following and see the various forms of calcium oxalate crystals.
Dog's Mercury *(Mercurialis)*—stem and leaf—large cluster crystals.
Enchanter's Nightshade *(Circaea)*—leaf—long, pointed crystals *(raphides)* Rhubarb *(Rheum)*—petiole.

Tannins are fairly widespread in plants. Carry out the following tests on various materials, especially Oak *(Quercus)* bark, and fruit (acorns), and unripe pears *(Pyrus)*.

Grind the material in a pestle and mortar, under water. Using the extract add a solution of ferric chloride which will give a blue-black or green colour with tannins. To some more of the extract add iodine solution and a 10% solution of ammonia which gives a red colour with tannins.

Many plants produce scented oils, many of which are of commercial value.

Examine the skin of an orange or lemon *(Citrus* spp.) with a hand-lens, and observe the translucent glands on the surface. Cut sections of the skin and look at the glands under the microscope.

The colours of petals are frequently due to the presence of *anthocyanins* which vary in colour according to the pH of the cell sap.

Extract any flower petals in water and see the colour changes caused by adding a few drops of sodium hydroxide solution of dilute hydrochloric acid. This effect can be seen well if you use beetroot *(Beta vulgaris)* juice.

Examine sections of the following material to see the secretory canals and vessels:
Rue *(Ruta graveolens)*—leaf—lysigenous cavities.
Ivy *(Hedera helix)*—leaf petiole—schizogenous cavities.
Pine *(Pinus sylvestris)*—leaf—schizogenous cavities.
Spurge *(Euphorbia* spp.)—stem—lactiferous cells.
Dandelion *(Taraxicum)*—root—lactiferous vessels.

Translocation of elaborated materials

The following two experiments are open to many criticisms, but they do give some idea of the translocation in plants. Can you think of any criticisms?

Cut some pieces of willow *(Salix* spp.), (which roots easily when left in water), about 18 in. to 2 ft. long. Ring half of them just below the leaves, and place the cut ends of each group in water. The exposed tissue should be protected from drying out by smearing with 'Vaseline'. Leave them for 2–3 weeks, and observe what happens. Does the development of roots give any evidence of translocation?

Repeat the experiment, but remove the 'bark' from about three-quarters the way around the stem, with a similar strip removed about ½ in. above it, so that the intact bark is on opposite sides of the stem as shown in Figure 207. Explain the observations you make.

Test the leaves of a woody shoot for starch. When the result is positive, pick the twig and ring the bark between the leaves. Place the cut end in water, and leave the shoot in the dark so that it can be de-starched. After 24 hours test the leaves above and below the ring for starch. Why are the results different? What control is needed in this experiment?

Transpiration and water economy
Functions of the root

Experiment: To demonstrate root pressure.
Apparatus: Potted *Pelargonium* or *Fuchsia;* glass tubing and rubber tubing of suitable

diameter to set up the experiment as shown in Figure 208; cotton.

Method: Place the potted plant in a sink of water, and leave it until the soil is thoroughly saturated. Cut the top off the plant, and as

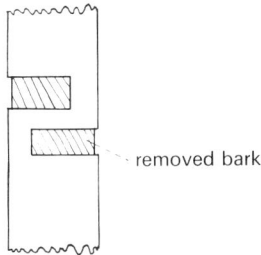

Fig. 207. Method of removing bark.

quickly as possible attach the tubing as shown in the diagram. Remove the plant from the sink. Observe the level of the water in the tube at various times. How do you account for the changes in the water level?

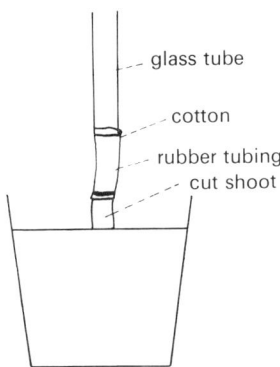

Fig. 208.

N.B. This experiment works best when the plants are growing actively. It is useless to try it in the winter.

The experiment can be repeated using a young sapling growing in the ground. If you do this, pour some water into the tube to keep the cut end of the tree moist, and avoid **air-locks at the joints. This is rather difficult to do.**

Channels of transport

Experiment: To show the tissue through which water is transported in vascular plants, and to show that insoluble substances are not absorbed.

Apparatus: Groundsel *(Senecio vulgaris)* or White Deadnettle *(Lamium album)* plants, beakers, eosin (red ink), carmine powder, razor, cover-slips and slides, microscope.

Method: Dig (not pull) up the plants and wash the roots free of earth. Place one plant in the eosin solution and the other in a suspension of carmine in water. (Carmine is insoluble in water). Leave them for a few hours or overnight. Which parts of the plant are stained? Cut transverse sections of the roots, stems and leaves. In which tissue is the red colour seen when examined microscopically? Account for the difference between the two plants.

Repeat the experiment with a Brussel's Sprout, and cut it longitudinally. The coloration shows up exceptionally well.

Functions of the leaves

Experiment: To show that the leaves are responsible for the water-loss from a plant.

Apparatus: Two leafy twigs, beakers and oil (or two potted plants which have been well-watered and the soil surface covered with a water-proof cover); two bell-jars.

Method: Remove the leaves from one twig (or plant), and place each in separate beakers of water, the surface of which is covered with oil to prevent evaporation. Place under separate bell-jars. Test any liquid appearing on the leaves or bell-jar with cobalt chloride paper, identify it and account for its presence or absence.

Experiment: To show which surface of a dicotyledonous leaf is mainly responsible for the loss of water.

Apparatus: Four evergreen leaves (e.g. Laurel), 'Vaseline', balance and weights.

Method: Remove the leaves from the plant and 'Vaseline' the cut ends of the petioles of all of them to prevent water loss. Leave one leaf untreated as a control, 'Vaseline' both surfaces of another, and the upper and lower surface respectively of the other two. Weigh all four separately, and hang them up to allow a free circulation of air around them. Re-weigh them after 24 hours. Calculate any loss in weight as a percentage of the original weight and draw your conclusions as to the surface responsible for water loss.

Experiment: To show the surface of the leaf responsible for water loss.
Apparatus: Well-watered potted plant, cobalt chloride paper two glass microscope slides, and two clips strong enough to hold the slides together.
Method: Make sure that the slides and the surface of the leaf are dry. Clip a piece of cobalt chloride paper on to the upper and lower surfaces of the leaf as shown in the diagram (Figure 209). Which piece of paper changes colour first, and why?

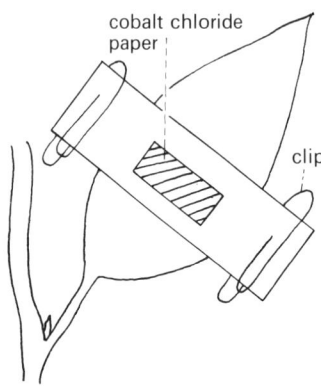

cobalt chloride paper

clip

Fig. 209.

Experiment: To show that the atmosphere is continuous from the exterior through the lamina and petiole of a leaf.
Apparatus: Evergreen leaf (e.g. Laurel), hard glass tube, and 2 one-holed bungs to fit it, glass tubing, water pump.
Method: Seal the petiole of the leaf in one of the bungs as shown in (Figure 210), and the glass tube through the other. Place the bung

containing the leaf in one end of the tube and half fill the tube with water. Attach the other end to the water pump. Turn on the pump and see the bubbles of air coming from the cut end of the petiole. The flow of air, due to the reduced pressure above the surface of the water, will continue indefinitely, showing that the air must be being drawn in from the outside.

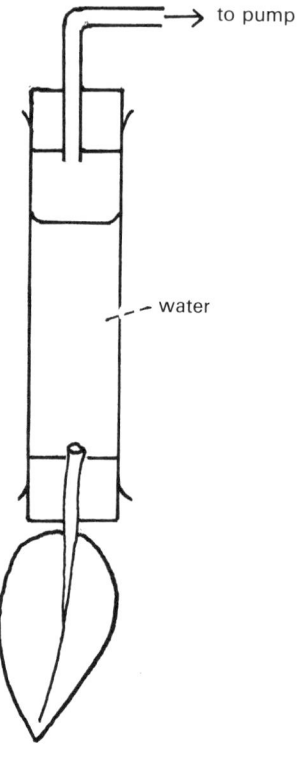

to pump

water

Fig. 210.

Experiment: To show the force of the transpiration pull.
Apparatus: Woody shoot, wide glass tube (an old burette is ideal) about 2 ft. long, one-holed bung to fit the top of the tube, dish of mercury, clamp.
Method: Cut the wood shoot, and push it through the hole in the bung, making sure that it is a good fit. Fill the tube with water, and push the bung containing the shoot into one end. If the joint is sound, you should now

be able to take your finger off the bottom of the tube without the water pouring out. Clamp the tube vertically with the lower end in the dish of mercury. Leave the apparatus for 24 hours and measure the height of the column. Calculate its height in terms of water. (The density of mercury is 13.6 g/ml.). Why is mercury and not some other liquid used in this experiment?

Stomata

Experiment: To examine the stomata on leaves.
Apparatus: Laurel (or some other firm dicotyledonous leaf), monocotyledonous leaf e.g. *Iris*, Daffodil *(Narcissus)*, concentrated calcium chloride solution, razor, forceps, coverslips and slides, microscope.
Method: Strip the epidermis from both surfaces of both leaves and mount them in distilled water on separate slides. Examine and draw them. Look particularly for the presence or absence of chlorophyll. Irrigate the slides with the concentrated calcium chloride solution and observe what happens to the stomata. Account for your results. What conclusions do you draw about the arrangement of stomata on leaves?

The epidermis is easily stripped from the leaf by bending it over the index finger of the left hand, and nicking it with the razor. Hold the cut edge with the forceps and pull gently to remove the epidermis.

Experiment: To demonstrate the distribution of stomata on the surface of a leaf.
Apparatus: Beaker of warm water, various leaves.
Method: Dip the leaves in the warm water singly. The heat will cause the air inside the leaf to expand, forcing it out as bubbles through the stomata. The position of the bubbles on the leaf surface indicate the general position of the stomata.

The experiment to show the continuity of the atmosphere from the exterior through the petiole and lamina of a leaf also demonstrates the presence of stomata. Can you design your

own experiment, using this principle, to compare the rate of air flow through leaves of different types of plants?

The presence of stomata is necessary for gaseous exchange in general, not only for water-loss.

Experiment: To show that open stomata are necessary for photosynthesis.
Apparatus: Potted plant, 'Vaseline', apparatus to test for starch.
Method: De-starch the plant by leaving in the dark for 24 hours. 'Vaseline' both surfaces of some leaves, the upper surface of others and the lower surface of others, and leave the fourth group as a control. Leave the plant in the light for 24 hours then wipe off the 'Vaseline' and test each group of plants for starch. Account for your results.

Water economy

Experiment: To measure the water-loss from a cut shoot.
Apparatus: Two boiling tubes, two twigs, chemical balance and weights, oil.
Method: Remove the leaves from one of the twigs and place a twig in each tube. Partly fill the tubes with approximately the same amount of water, and pour oil on the water to prevent evaporation. Hang the tubes on opposite arms of the balance and add weights until they balance. Leave for 24 hours and adjust the weights again until the two tubes balance. Any additional weights are equal to the weight of water lost by the shoot. If the weighings are made daily, the results can be graphed.

A similar experiment can be carried out as follows:
Apparatus: Potted plant, Buchner (or similar) balance, waterproof (polythene) sheet.
Method: Water the plant well and cover the pot and soil with the waterproof sheet to prevent evaporation. Stand the plant on the balance pan and take the initial reading. Take successive readings at 24-hourly intervals. Graph your results. Leave the experiment for several days, and interpret

your graph to explain the water-loss from the plant.

The rate of water uptake by a shoot is measured by a *potometer*. (*N.B.* The rate of water uptake is not necessarily the same as water loss.) There are many types of potometer, but they all work on the same principle (see Figure 211). A cut shoot is attached directly or indirectly to a graduated capillary tube, along which the water is drawn as it moves into the shoot. The rate of movement along the capillary is measured so that the rates of water absorption can be compared.

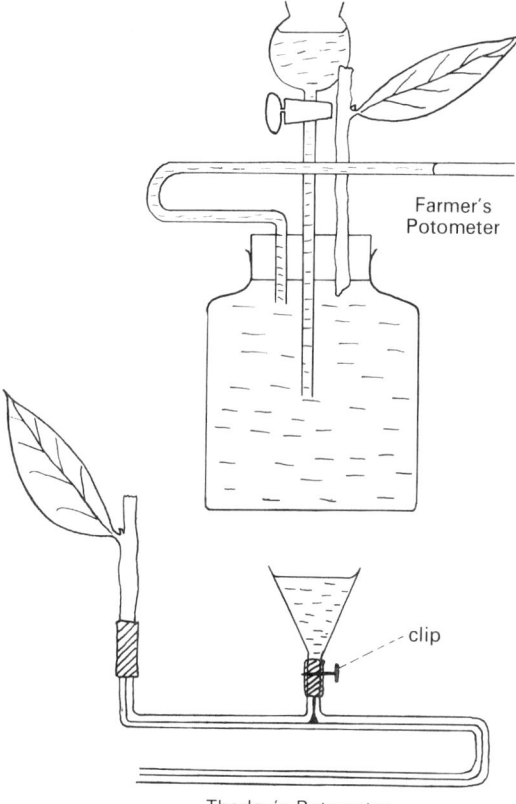

Farmer's
Potometer

clip

Fig. 211.   Thoday's Potometer

It is essential that the shoot used in the potometer is cut and fitted into the apparatus under water, to avoid air locks. Air bubbles are the main difficulty in setting up the apparatus.

The readings are usually taken by timing

the movement of the meniscus between two fixed points on the capillary. Several readings are taken in each experiment, and averaged. The apparatus is zeroed after each reading by releasing water into it from the reservoir.

The apparatus can be used for measuring the water uptake of the same shoot in different situations, or different plants in the same situation. But for a valid comparison in the latter case, it is necessary to calculate the leaf areas of the shoots used, and express the result as uptake per unit leaf area.

Mineral requirements

Experiment: To demonstrate that various minerals are necessary for growth.
Apparatus: Eight culture jars as shown in the diagram (Figure 212), barley *(Hordeum)* or Maize *(Zea)* seed, or rooted cuttings *(Tradescantia* works well), aerating pump, dark paper, and the chemicals needed for the culture solution.
Method: Germinate the seeds in damp sawdust or sand for about 10 days. Wash all the glassware with concentrated nitric acid, then tap water, then distilled water to make sure that it is clean. Sterilize the corks and cotton wool by autoclaving or boiling, then oven-dry them before use. This is to reduce the chances of the plants becoming attacked by fungi. Cover each jar with black or brown paper to stop the growth of algae in the solution. Fill each jar with its appropriate solution and fix one seedling through the cork of each jar so that the roots only are in the culture solution.

Aerate the solutions daily and top them up as required. Completely change the solutions once a fortnight. The complete solution should contain:

Calcium sulphate	0.25 g.
Calcium phosphate	0.25 g.
Magnesium sulphate	0.25 g.
Sodium chloride	0.08 g.
Potassium nitrate	0.70 g.
Ferric chloride	0.005 g.

Dissolved in 1 litre of *distilled* water (after Sachs). These weights refer to *crystals* of the salts, i.e. include the water of crystallization where appropriate. Only pure, e.g. 'Analar' chemicals should be used.

Make up sufficient of each solution to last for the complete experiment, and store them in darkened airtight bottles. The following substitutions will give comparable solutions omitting the element mentioned:

Omit nitrogen, replace potassium nitrate with 0.52 g. potassium chloride.

Omit phosphorus, replace calcium phosphate with 0.16 g. calcium nitrate.

Omit potassium, replace potassium nitrate with 0.59 g. sodium nitrate.

Omit calcium, replace calcium sulphate with 0.2 g. potassium sulphate, and calcium phosphate with 0.71 g. sodium phosphate.

Omit magnesium, replace magnesium sulphate with 0.17 g. potassium sulphate.

Omit sulphur, replace calcium sulphate with 0.16 g. calcium chloride, and magnesium sulphate with 0.21 g. magnesium chloride.

Omit iron, leave out ferric chloride.

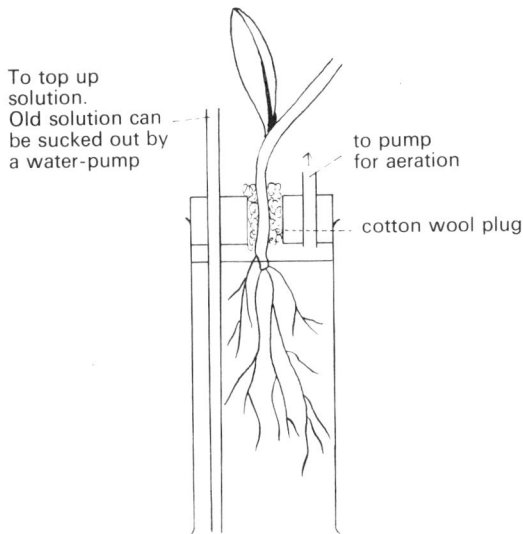

Fig. 212.

Observe the growth rate of both leaves and roots, and the leaf colour.

These solutions can also be made up in 2% agar and plated in Petri dishes as described

for bacteria and fungi, and used to compare the growth of these organisms. Before you do this you will have to find out the particular carbohydrate requirements of the fungus.

Experiment: To show the need of iron to produce chlorophyll.
Apparatus: Cress seed, culture solution deficient in iron as described in the previous experiment, small test-tubes, cotton wool, filter-paper, ferric chloride.
Method: Germinate some cress seed on filter-paper and transplant them singly into small test-tubes of iron-deficient solution, but gently wedging them into the top of the tube with cotton-wool. Allow them to grow and they should produce chlorotic leaves. When this occurs add a drop of ferric chloride solution (0.5 g./l.) to each tube and observe the effect. Can you suggest a control to this experiment to ensure that it is the iron and not the chloride ions that are having the effect? This is a short term experiment, so the tubes will not need aerating.

Growth

The necessity for certain elements during growth has been demonstrated in the previous section.

Increase in length

Experiment: To measure the increase in length of a potted plant.

There are various forms of lever which can be used to exaggerate the growth of a plant, and consequently make its measurement easier. Most of them involve tying a light beam or weight to the growing tip, which will cause it to stretch. Also, no allowance is made for the stretching of the cotton used. In consequence most of these methods are very inaccurate.
Apparatus: Auxonometer, actively growing potted plant, cotton (or thin nylon) thread.
Method: Set the apparatus up as shown in Figure 213, making sure that the cotton is tied

to the growing part of the shoot, and that it is vertical in two planes in relation to the needle of the auxonometer. Take daily readings on the scale, and graph your results.

If the absolute increase in length is required, it can be calculated from the formula $a/b = d/c$ (similar triangles), a being the distance required, and using the notation shown on the diagram.

Experiment: To examine the growth of a plant radicle.
Apparatus: Broad bean seed, suitable stoppered jar, pins, cotton-wool, paper-clip, cotton.

about the zone of maximum elongation of the root.

Increase in weight
Experiment: To show the change in weight during germination and growth.
Apparatus: Seed box, damp sawdust, pea seeds, steam oven, balance.
Method: Fill the seed box with sawdust which must be kept damp throughout the experiment. Plant about 200 pea seeds at equal distances from each other. Take 20 dry seeds, cut them into small pieces and dry them in the oven, weigh them in a weighed

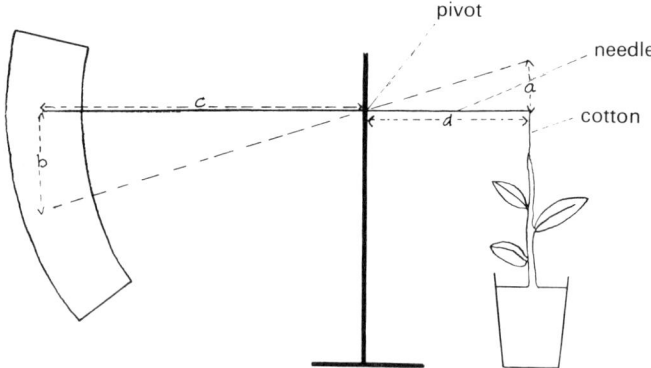

Fig. 213. Auxonometer.

Method: Soak the bean seed and allow it to germinate. When the radicle appears, pin the seed to the cork so that the radicle is pointing downwards, wet the cotton-wool and put it in the jar. Stopper the jar with the cork. This will ensure that the radicle grows straight. When the radicle is about 1.5 cm. long, mark it at millimetre intervals in the following way. Make a miniature bow, by opening out the paper-clip and tying the cotton across the two ends. Moisten the cotton with Indian ink and draw it gently across the radicle in the appropriate place.

Replace the bean in the damp chamber. Measure the distance between the first and last marks each day, and graph the result. Notice any change in the distance between neighbouring markings. Make any deductions

dish. Repeat the drying and weighing until a constant weight is obtained. Gently remove 20 peas from the box 24 hours after planting and find their dry weight as above. After this remove seedlings at weekly intervals and find their dry weight. Graph your results, weigh against time, accounting for any irregularities in the curve.

N.B. To obtain a valid result, it is imperative that the seeds and seedlings be selected *at random* from the box. A convenient way to do this is to number the seeds on planting, and select those which are to be used for a particular experiment by drawing them as numbered pieces of paper out of a hat. A primitive, but valid method of selection! Discard the numbers of the plants after the plant has been used in the experiment.

Experiment: To find the daily increase in the weight of leaves.

Apparatus: Cork borer, balance, steam oven, weighed evaporating basin.

Method: Select and mark fifty developing leaves on a shrub or bush (*Rhododendron* or Laurel are useful). Remove a disk from each leaf each day, using a cork-borer. This will mean that you are using the same leaf-area each day. Avoid the inclusion of the mid-rib in your sample, and take the sample at the same time each day as there are diurnal fluctuations due to photosynthesis. Oven-dry each daily sample of fifty disks together as soon as possible after collection, weighing and re-weighing until the weight is constant. Graph your results, weight against time.

Change in shape during growth

Experiment: To show the change of shape of a leaf during growth, and to show the relationship between the growth of the lamina and petiole.

Apparatus: Ruler, graph paper.

Method: Select an actively growing plant, (Nasturtium *(Tropaeolum)*, Runner Bean *(Phaseolus)* or Ivy *(Hedera helix)* are suitable) with developing leaves. Select and mark a leaf. Draw an outline of the leaf on the graph paper at weekly intervals, and at the same time measure the length of the petiole, and some convenient measurement of the leaf, e.g. the length of the main vein, or the leaf diameter.

When the measurements are complete, re-draw the leaf outlines on graph-paper, but convert them to a common length. This will show the change in shape.

Graph the petiole lengths against your other measurement. What shape curve does this give? Draw another graph of log petiole length against log of the second measurement. What sort of curve does this give? What do these two curves show about the growth relations between these two parts of the leaf?

The effect of the environment

Experiment: To investigate the effect of light and temperature on the growth of plants.

Apparatus: Four 6-in. flower pots, broad bean *(Vicia faba)* seeds, two opaque covers to go over the plant pots and allow the plants to grow inside them.

Method: Fill each pot with soil, and sow about ten seeds in each. Soak the soil, and keep the pots well watered throughout the experiment. Place one pot in a cool place (if this has to be in the shade, slight shading will not affect the experiment), and the other beside it covered. Place another in a warm place with a covered one beside it. When the seeds are germinated, and you are sure that you have five plants per pot, remove the others.

Measure the height of each plant, and the number of leaves each week, and average the figures for each pot. At the end of the experiment, calculate the average internode length for each treatment, and the average leaf area for each treatment.

N.B. This experiment does not separate the effects of a single factor, e.g. light or temperature. If you wish to investigate only one of these, use only two pots, e.g. one in the dark and one in the light, at the same temperature.

The effect of light on the growth of plants (Phototropism)

Experiment: To show the effect of light on the growth of shoots.

Apparatus: Five small seed boxes, or flower pots, Oat *(Avena)* seed, covers for seed boxes, tin foil.

Method: Place some damp soil in the seed boxes, and sow the oat seed fairly thinly. Make sure that the seeds are sown at the same depth in each box, so that they will emerge at the same time. Keep the boxes well-watered. When the coleoptiles emerge, cover one box to exclude all the light, and place another where it can receive light all around it. Place the other three so that they receive light from one side (near a window will do). Of these three, leave one, make caps of tin foil to cover the tips of the seedlings only in the second box, and cut the top millimetre of the tops of the coleoptiles in the third box. Leave the seedlings to continue growing. From your experiment deduce the effect of light on the directional growth of the young shoot, and the point of the shoot that is affected by the light.

Experiment: To find the wave-length of the light which causes the response of the shoot.
Apparatus: Oat *(Avena)* (or other) seed, light filters (green, blue, red, yellow), five boxes of suitable size, small seedboxes or pots.
Method: Cut one side out of the boxes (old shoe-boxes work perfectly well, although one can buy special wooden boxes to do this experiment) and replace it with the light filter suitably stuck on, leaving one without the filter as a control. Sow the seed in damp soil, and cover each seed box with a shoe box, arranging them so that the light is unilateral, and from the same side for all treatments.

Which coloured light has the greatest effect on the growth of the shoot?

Experiment: To show the effect of light on the growth of roots.
Apparatus: Cress *(Lepidium)* seed, beaker, piece of cardboard, filter-paper, Petri dish.
Method: Germinate the cress seed on moist filter-paper in the Petri dish. When the radicles are long enough, place them through small holes in the cardboard. Put the cardboard over a beaker of water, so that the cress roots are in the water. Cover the side of the beaker with brown paper, leaving a slit about 1 in. wide down one side. Place the beaker so that the light shines through the slit. Observe the roots after 24 hours.

The effect of gravity on the growth of plants (Geotropism)

Experiment: To show the effect of gravity on the growth of shoots, and the effect of removing the leaves or apical bud.
Apparatus: Three stands, nine clamps, nine large test-tubes and one-holed rubber bungs to fit them, nine young shoots (Mint

(Mentha), or *(Coleus)* are useful).
Method: Cut the nine young shoots under water, and fit them through the holes in the bungs. Fill the test-tubes with water and place the bungs in them. Clamp three tubes to each stand so that on each stand, one shoot points upwards, one downwards, and one horizontally. From the shoots on one stand remove all the leaves, and from those on another remove the apical buds. Leave the third group complete. Place all three in similar light conditions. Account for the position of the shoots after a few days.

Experiment: To show the effect of gravity on the growth of the young radicle.
Apparatus: Broad Bean *(Vicia)* or Pea *(Pisum)* seed, three wide-necked jars, corks to fit the jars, paper-clip, cotton, Indian ink, cotton wool, pins.
Method: Grow the beans in moist saw-dust until the radicles are about 1 in. long. Straighten the paper-clip, and make it into a minature bow, by tying the cotton across the two ends. Soak the cotton in the Indian ink, and use it to mark the radicles at 2 mm. intervals along its length.

Place moist cotton wool in the bottom of each jar. Pin one bean to the cork of one jar, so that the radicle is horizontal. Place the cork in the jar. Cut about 3 mm. from the tip of the radicle of the second bean, and pin it similarly in a jar. Lay the third bean so that the radicle is horizontal. Leave it for about 2 hours. Cut 3 mm. from the tip of the radicle and pin the bean with the radicle vertical in the third jar.

Observe the experiment after a day, and account for the results. Note the regions of growth as indicated by the spacing of the black lines.

Experiment: To show the effect of equalizing the effect of gravity around a young radicle on its growth.
Apparatus: Clinostat, germinated bean or pea seeds with radicles about 2 cm. long.
Method: Set the clinostat so that the axis

is horizontal. Pin the germinated seed to the cork base (Figure 214). Note the angles of the radicles to the base. Line the plastic cover with damp filter-paper and replace it. Run the clinostat for 24 hours. Note any change in the direction of growth of the radicles. Stop the clinostat and leave the seeds attached to it for a further 24 hours. Account for any results.

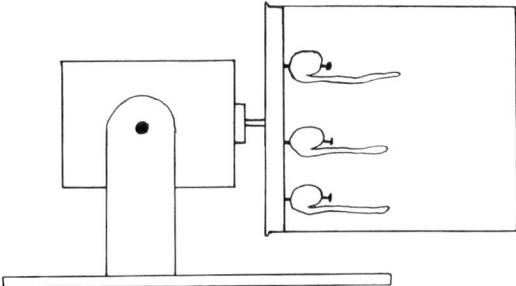

Fig. 214. Clinostat.

The stage of the clinostat revolves slowly (usually 4 r.p.h.) so that the gravitational pull affects all the sides of the root for an equal length of time in any hour.

The effect of water on the growth of the plant (Hydrotropism)

Experiment: To show the hydrotropic response of roots.
Apparatus: Gauze, saw-dust, cress *(Lepidium)* seed, cotton.
Method: Bend up the sides of the gauze so as to make a tray. Fill it with the saw-dust, and soak it thoroughly. Sow the cress seed in the saw-dust, and, using the cotton, suspend the gauze so that it slopes (see Figure 215). Keep the saw-dust uniformly moist throughout the experiment. Observe the growth of the young roots. Why is the gauze hung at an angle?

The growth in response to chemical stimulants (Chemotropism)

Experiment: To show the growth of pollen tubes in response to various stimulants.
Apparatus: Gelatine, slides, microscope, flower, e.g. Lily, Hyacinth *(Scilla nutans),*

Petri dish, short pieces of glass rod to go inside the Petri dish.
Method: Make a 2–3% solution of the gelatine, and place a drop on a slide. Place a few grains of pollen from the flower in the centre of the drop just before it sets, and, at the same time, place some ovules and the stigma around one side of the drop. Place the slide on the glass rods in the Petri dish in which there are a few drops of water. Place the covered dish in the dark, and observe the direction of growth of the pollen tubes after several hours.

The experiment can be repeated using crystals of sugar, various salts, or growth-regulating substances, e.g. indoleacetic acid or gibberellic acid, instead of the ovules and stigma.

A similar experiment can be performed by replacing the pollen grains with fungus spores.

The effect of growth-regulating substances on growth

Experiment: To examine the effect of indole-acetic acid on the growth of shoots and roots.
Apparatus: Paste of 1/3000 parts of indole-acetic acid in one of lanoline, maize *(Zea)*

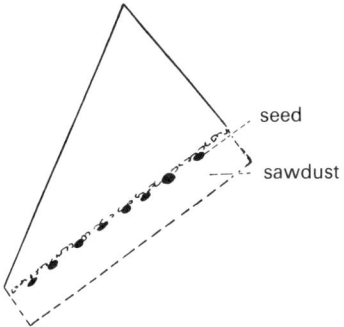

Fig. 215.

and broad bean *(Vicia)* or pea *(Pisum)* seeds, plant pots and soil, wide-mouthed jar with cork, pin, cotton wool, agar.
Method: Germinate the seeds in saw-dust, planting the beans so that the radicles will

grow vertically downwards. Remove the beans seeds and pin them on the corks of the jars which contain damp cotton wool, so that the radicles are horizontal in half of them and vertical in the other half.

Make a 2% solution of agar, and pour it to a depth of 1 mm. in a dish. Allow it to set, and cut the resulting jelly into squares of about 5 mm. side.

Apply the following treatments to the maize seedlings: (1) Leave one as a control, (2) apply the paste of indole-acetic acid to one side of the shoots, (3) remove the tips, about 1 mm., (4) remove the tips and replace them with a piece of agar, (5) remove the tips and replace them with a piece of agar, and on the agar place a little of the indole-acetic paste, (6) remove the tips and replace with a little indole-acetic acid paste.

Apply the same treatments to the vertical and horizontal bean radicles.

What conclusions can you draw concerning the effect of indole-acetic acid on the growth of plants? Can you suggest any other experiments you could carry out? Read about other growth-regulating substances, e.g. gibberellins, and carry out other similar experiments using them.

11 Variety in Plants

ALGAE

Depending on the method of classification adopted, the algae are included either with other, more complex forms, in the Thallophyta, or in a separate Kingdom—the Protista, with the fungi and the unicellular animals. Whichever classification you adopt, compare the structure of the unicellular algae with that of the unicellular animals. Note the similarities in structure. Examine electron microscope photographs of cell organelles, especially flagella, and note that they are similar in both groups of organisms.

Pleurococcus *(Protococcus)*

This occurs commonly as a green dust or powder on damp wood, e.g. tree bark or fencing. Mount some of this powder in a drop of water on a slide and examine it under the low-power of the microscope. Notice that some of the cells occur in small clusters, due to their rapid division. Examine a single cell under the high power of the microscope (see Figure 216).

The chloroplast is cup-shaped, so that if it is seen from the bottom, the whole of the cell contents will appear to be green. The cellulose cell-wall is fairly thick. A nucleus is present in the centre of the cytoplasm, but will probably be obscured by the chloroplast.

Chlamydomonas

Chlamydomonas occurs in stagnant fresh water. Examine a drop of such water and see if you can find any (Figure 216). You will certainly find various other small algae, some of which, although they look relatively simple, are not even closely related to these simple green algae. If you wish to pursue the question of their classification further, refer to a standard text on the algae.

Chlamydomonas is very small (about 20 μ long) so that you will get only a rough idea of its structure from a low-power examination. Place a few strands of cotton-wool on your slide before adding the drop of pond water, to slow down the movement of the algae. Irrigating the slide with iodine solution may also help.

Examine this, and prepared slides, under the high power of the microscope (preferably $\frac{1}{6}$ in. objective). See the flagella, the cup-shaped chloroplast and pyrenoid. The chloroplast may appear to fill the cell,

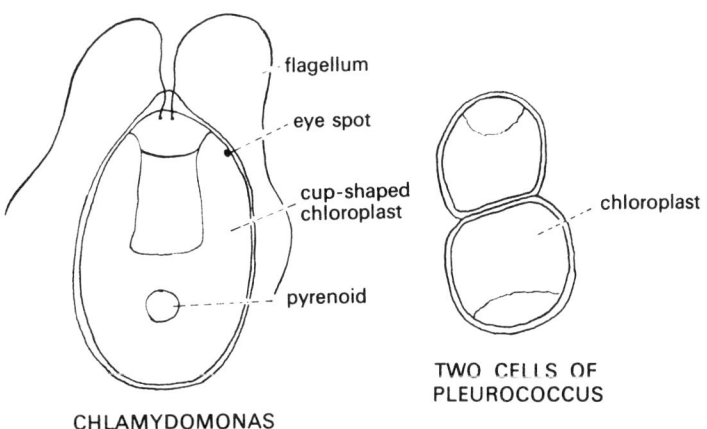

CHLAMYDOMONAS

TWO CELLS OF
PLEUROCOCCUS

Fig. 216.

depending on the angle from which it is viewed, and it may obscure the nucleus which lies near the centre of the cell. You may see the other structures labelled in the diagram (Figure 216) when you examine several specimens.

Re-examine a slide of *Euglena*.

Volvox

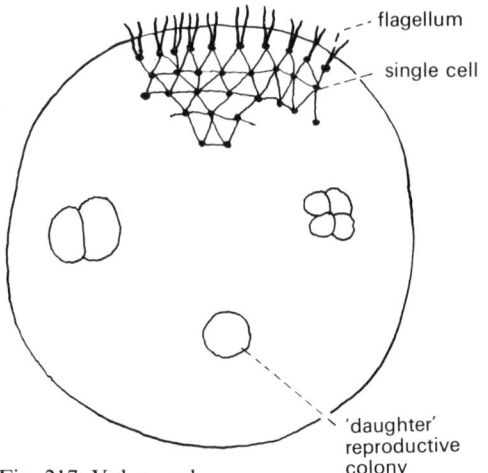

Fig. 217. Volvox colony.

This is a colonial green alga found in stagnant water. The colony, called a *coenobium*, is made-up of a large number (usually several thousand) *Chlamydomonas*-like units, with their flagella outside the colony. This is an interesting organism in that it shows division of labour. Each colony has specialized sex-cells, and may be dioecious or monoecious. The various parts are shown in Figure 217. Compare this with prepared slides.

Spirogyra

Spirogyra is found near the surface of ponds and slow-flowing streams. Collect some and examine it under the microscope. Under low power, notice the lack of branching. Examine under high power and look for the features shown in the diagram (Figure 218). The following methods may help to show up the structures more clearly: (1) Irrigate the slide with iodine solution to stain the starch on the pyrenoids. (2) Irrigate the slide with aceto-carmine and gently warm it. The nuclei will show up red. (3) Irrigate the slide with ethanol to remove the chlorophyll, then irrigate with Schultz's solution which will stain the cell-walls and starch grains blue, and the pyrenoids brown. (4) Remove the chlorophyll with ethanol as above, and similarly stain the slides with safranin *or* haemotoxylin. This will stain the nucleus and protoplasmic strands purple-red.

Examine prepared slides of conjugating material, and material containing zygospores.

Spirogyra may form zygospores if filaments are kept in a small amount of 2% sucrose solution, which is then allowed to dry out slowly.

Other filamentous algae

If you collect filamentous green algae from streams or ponds you are likely to find *Ulothrix* or *Oedogonium*. These are not in the same class as *Spirogyra* as both produce motile spores at some time during their life-cycle. Examine slides of the material you

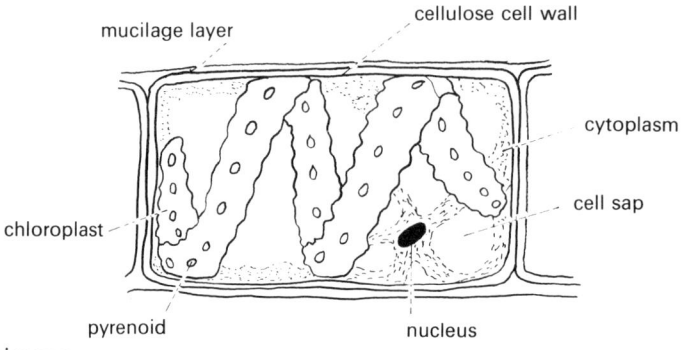

Fig. 218. A cell of Spirogyra.

collect in the same way as you did *Spirogyra*. In *Ulothrix* the chloroplast forms a ring around the circumference of the cell, so that it possibly will appear as a bar along the long axis of the cells, which are shorter than they are wide (Figure 219). The cells of *Oedogonium* are longer than they are wide, and the chloroplast forms a network around the inside of the cell.

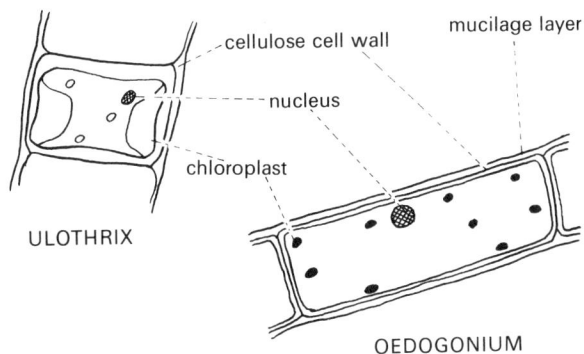

Fig. 219.

Ulva (The Sea Lettuce (Figure 220))

This is a marine green alga, usually found in sea-water which has been polluted, and therefore containing a high proportion of nitrogenous material.

Examine the flat thallus. Look at it under the microscope, and notice that it is only two cells thick, each cell having a thick wall and a single chloroplast.

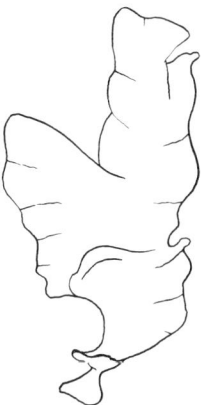

Fig. 220. Ulva.

Fucus

There are several species of *Fucus* and they are all marine. They are very common, especially in the inter-tidal zone, or the shallow water of rock coasts. The flattened, branched thallus shows some differentiation of the tissues into a flattened mid-rib. The swollen tips—*receptacles*—bear the reproductive cavities—*conceptacles*—which contain the female *oögonia* or the male *antheridia*. There are no asexual reproductive structures (see Figure 221).

Draw a portion of the complete plant. Cut a transverse section of the vegetative part of the thallus, and draw it under low power, noting the different regions. Cut sections of male and female conceptacles, and draw them under low and high power, noting the structures shown in the diagram. If you happen to use *F. spiralis*, you will find that the conceptacles are hermaphrodite.

FUNGI

The following are a selection of *Phycomycetes* which you will obtain easily. The vegetative hyphae of the Phycomycetes is recognizable because it is a coenocyte (there are no cross-walls).

Leave some damp bread covered by a belljar for three or four days, and you will certainly find a fine grey mass of mycelium developed, with black *sporangia*. This will be *Mucor* (Figure 222), or *Rhizopus*. *Rhizopus*

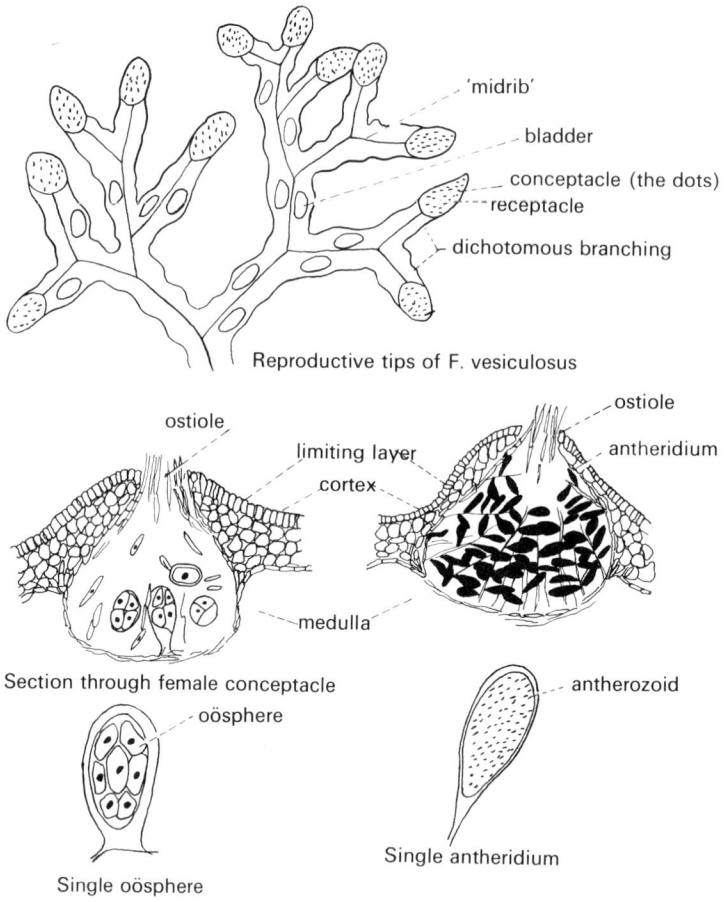

'midrib'

bladder

conceptacle (the dots)

receptacle

dichotomous branching

Reproductive tips of F. vesiculosus

ostiole

limiting layer

cortex

ostiole

antheridium

medulla

Section through female conceptacle

oösphere

antherozoid

Single oösphere

Single antheridium

Fig. 221. Fucus.

can be distinguished from *Mucor* by the stolon-like growth of the hyphae of the former. Mount some of the material in water on a slide and examine under high-power. Draw some of the hyphae and the sporangia.

Draw prepared slides of the sexual reproduction of *Mucor*.

Grow some cress *(Lepidium)* seedlings very close together, and keep them very damp. They will 'damp-off'—the hypocotyl becomes rotted. This is caused by several fungi, one of which is *Pythium*.

Take a piece of the diseased material and crush it on a slide, using the cover-slip. Stain by irrigating the mount with lactophenol-blue. Notice the coenocytic hyphae and the terminal *zoosporangia*.

The 'damping-off' may be caused by species of *Phytophthora*, related to the Potato Blight fungus, which is distinguished from *Pythium* by a branched sporangiophore, with the tip of each branch bearing lemon-shaped sporangium.

Cystopus is a parasite on the Cruciferae, and is fairly common on Shepherd's Purse *(Capsella bursa-pastoris)*. Examine some of these plants and look for smooth blisters and distortion.

Cut a fairly thick section of one of these zones and crush the section under a cover-slip on a slide. Stain with lactophenol-blue. Look for the hyphae running between the host cells (intercellular) giving off small haustoria into them; the enlarged host cells (hypertrophy),

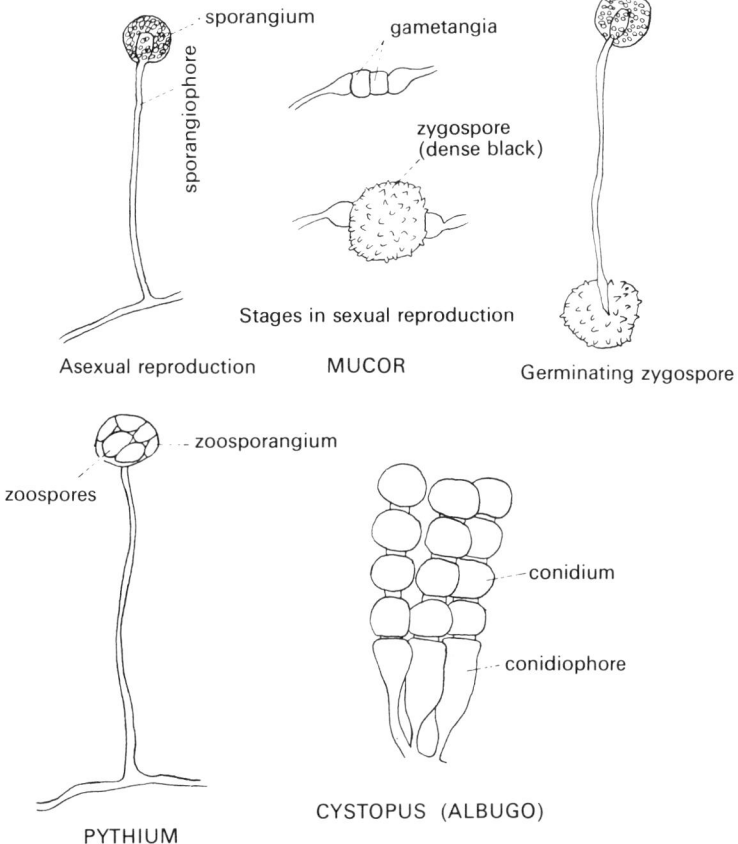

Fig. 222.

and the chains of conidia produced below the host epidermis. Deeper in the host tissue, you may find *oögonia* and *oöspores*.

The *Ascomycetes* are a class of fungi characterized by the sexually produced spores *(ascospores)* being produced in a closed sac or *ascus*. The mycelium is septate, and *conidia* asexual spores, are produced by a large number of genera, especially in the simpler forms.

Aspergillus (green) and *Penicillium* (blue-green) are often found growing with *Mucor* on damp bread, and they are often found on citrus fruits.

Mount some of these fungi on a slide, and examine them under low and high power (see Figure 223).

You are more likely to see the shape and

distribution of the conidia if you put only a little material on the slide, and tease it out carefully with a pair of needles.

If you allow the culture to dry out, you may be lucky enough to find some *perithecia*, in which the asci develop.

Many of the Ascomycetes cause diseases of other plants. These parasites include the mildews.

Examine mildewed grass with *Erysiphe* on it, roses with *Sphaerotheca,* and Lilac *(Syringa)* with *Microsphaera.* Look at the diseased material. Scrape off some of the mycelium from the surface of the leaf and examine it under the microscope. See the septate mycelium, and the simple unbranched conidiophores.

Examine prepared slides of the perithecia,

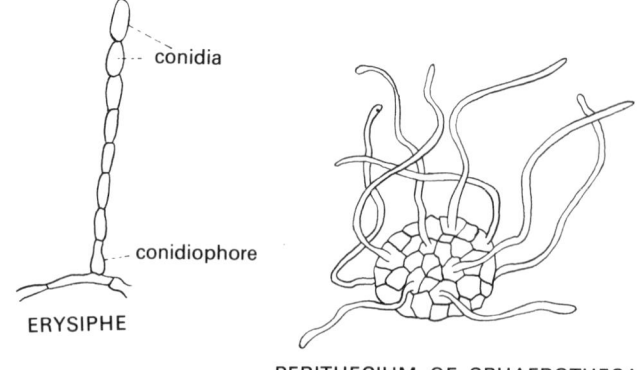

ERYSIPHE

PERITHECIUM OF SPHAEROTHECA

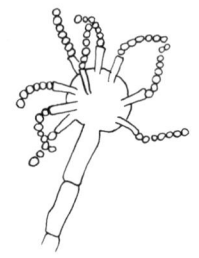

PENICILLIUM (Simplified)

ASPERGILLUS (Simplified)

Fig. 223.

noting the thick wall and the appendages.

Yeasts (Saccharomycetes) are also Ascomycetes, although they reproduce largely by budding, complete cells may take place in sexual fusion, resulting in the formation of an ascus (Figure 224). You are not likely to see the sexual phase, but yeasts are easy to grow in the laboratory.

Make up some medium for growing yeasts (see appendix) and pour about 200 ml. into a large sterilized conical flask. Place a little bakers' yeast into it, and bung it with a cotton wool plug. Keep it for a few days at 25–30°C, and the yeast will grow prolifically. Take a drop of the culture and examine it under the microscope.

Bakers' yeast will grow successfully in a 5% sucrose solution, but if you wish to try the following experiment, the medium suggested will give a better result.

Make up the medium, and inoculate it by adding a grape, raisin or plum. Examine the culture after a few days and see the yeast. Yeasts occur naturally on the surface of several fruits.

Many Phycomycetes and Ascomycetes can be grown in culture, using similar techniques

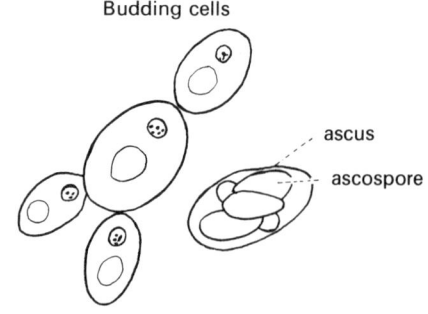

Fig. 224. Yeast.

to those described later for the bacteria, and the media mentioned in the appendix. See if you can grow any fungi you find and identify them. Culturing Basidiomycetes is more difficult, and should only be undertaken if you wish to do it as a special project.

The *Basidiomycetes* include the fungi which bear their sexually produced spores externally on *basidia*.

The *Mushroom (Psalliota)* is a typical Basidiomycete. Draw some fresh specimens (see Figure 225), and examine prepared slides through the gills.

more using a bright light. If you have difficulty in seeing these organisms, cut down the light.

You will not see the fine structure of bacteria under the normal student microscope, so you will have to rely on using micro-photographs, and electron microscope photographs.

Make a hay infusion by boiling some dry hay in a beaker of water. Leave this to stand for a few days at room temperature. It will become turbid due to the presence of *Bacillus subtilis*. Examine a drop of the scum,

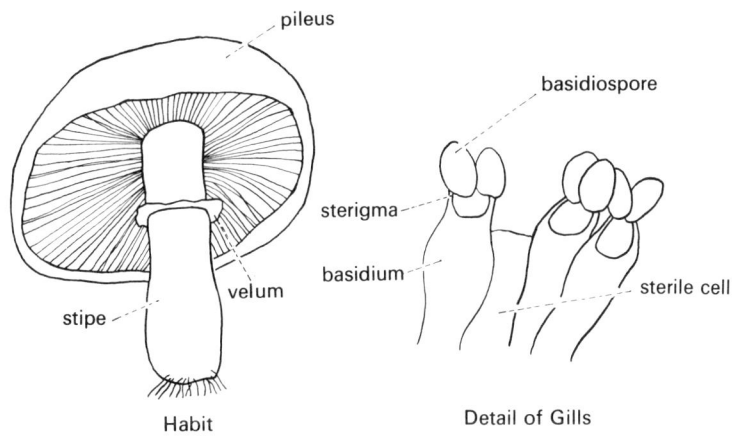

Fig. 225. A Mushroom.

Some fungi form special associations with the roots of other plants. These associations are called *mycorrhiza* (fungus-roots). Examine prepared slides of the mycorrhiza found on Pine *(Pinus)* or Heather *(Erica)*.

BACTERIA

The techniques required for the various experiments with bacteria are described in the appendix. They are similar to those used in growing fungi and other micro-organisms.

Examine some prepared slides of bacteria, so as to get an idea of the size of the organisms you are handling. All observations on bacteria should be carried out using the high power of the microscope. Do not make the mistake of thinking that you will see

and you should see the bacteria moving about in it.

You can achieve a similar result by leaving some meat or peas to rot in water for several days. By these means you will get a mixture of bacteria.

Pour some milk into a flask, and measure its pH by using indicator paper, or by removing a drop and testing it with an indicator solution. Leave the milk exposed to the air for a few days at about 25–30°C. Test the pH daily. Notice the smell, which is caused by lactic acid. The bacteria entering the milk from the air ferment the milk sugar (lactose) to lactic acid.

Sterilize two plugged flasks, containing milk (use 250 ml. flasks with 50 ml. milk).

Leave one as a control, and into the other place a few grains of garden soil. Leave both for a few days at 25–30°C. Test the pH of both flasks. Take a drop of milk from each flask and dilute it on a slide with an equal volume of water. Examine under the microscope for the presence of bacteria.

The bacteria present on flies can be shown by pouring Petri dishes (plates) of nutrient agar, placing the legs, head and wings of the fly separately on the cooled agar and allowing it to incubate for a few days. Bacteria and fungi will develop.

Pour another agar plate. Take a test-tube of sterile water (distilled water, sterilized by autoclaving) and into it put one of the bacterial cultures from the fly (or a drop of milk containing bacteria from the previous experiment). Shake thoroughly. This bacterial suspension can be used for making a streak culture on the plate. This is done by sterilizing a platinum wire by heating it in a flame; allow it to cool, and dip it into the bacterial suspension; then open the plate just sufficiently to place the needle under the lid, and draw a zig-zag line on the surface with the needle. Leave the plate for a few days at room temperature, and the bacteria will grow along the line drawn by the needle.

To investigate the growth of bacteria, pour an agar plate as before and prepare a suspension in sterile water from a few bacteria from the streak culture. Using a sterilized needle, place a small drop of this suspension in the centre of the plate. Measure the diameter of the culture at daily intervals, and graph your results.

You can make cultures of different bacteria by putting a suspension of soil in water and using this to make a streak culture which you can use in the same way.

The antagonism between different organisms can be demonstrated by drops of suspensions of the organisms placed about 2 cm. apart on the same plate. If antagonism exists, they will grow as shown in the diagram (Figure 226). If there is no antagonism, the cultures will grow together, and even intermingle. Try

this experiment, using two different bacteria, or a bacterium and a fungus, e.g. *Penicillium*. Different bacteria frequently produce colonies of different colours, so that they can be separated from a mixed culture without difficulty.

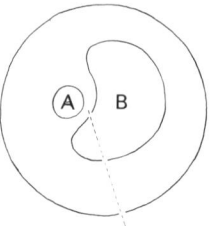

zone of inhibition

Fig. 226. Example of inhibition.

Some bacteria will feed on substances which are used as disinfectants, if the concentrations are suitable. This is shown in the following experiment.

Dissolve 10 g. of phenol (carbolic acid) in 100 ml. of water. Use this solution to make the following dilution series: 10%, 5% 2.5%, 1.25%, 0.625%, 0.312%, 0.156%, 0.078%, 0.039%, 0.019%. Place 10 ml. of these solutions into separate large test-tubes which are to be plugged with cotton wool. Shake about 50 g. of soil in 500 ml. of water to which has been added 2 g. ammonium sulphate and 1 g. potassium dihydrogen phosphate (to provide the mineral salts for the bacteria). Allow the large particles to settle and add 1 ml. of the suspension to each of the test-tubes. Leave the tubes to incubate for a few days. Notice any change in the colour or smell of the cultures. Examine each for the presence of bacteria. A brown colour in the culture indicates that the phenol has been used by the bacteria.

The bacteria in root nodules of legumes can be examined in the following way: Wash the roots of the legume free of soil. Crush one of the nodules in a few drops of water in a watch-glass. Smear a drop of the liquid on a slide and allow it to dry. Stain the smear with crystal violet, and examine under the high power of the microscope. You should see the irregularly shaped bacterial cells.

BRYOPHYTA

Class: **Hepaticae** (Liverworts)

The Liverworts are quite common on damp soil and near water. Collect some and examine the thallus. The type described here is *Pellia* which is found on damp soil in woodland and under hedges.

Examine the whole thallus using a hand-lens. The branching is dichotomous by the division of an apical cell at the base of the dent at the top of each branch. Look for the stages in the development of the sporangium, which is diploid. Turn the specimen over and examine the *rhizoids*. Notice their position. Pick off a few rhizoids and examine them under the microscope. They are unicellular.

Cut a transverse section of the thallus, and see the distribution of the chloroplasts and the absence of conducting tissue.

Examine prepared slides of longitudinal sections of the thallus to show the *archegonia* at the top, and the *antheridia* further back (see Figure 227).

Examine a prepared slide of a *sporangium,* formed by the fertilization of an *oösphere.* Some of the sporangial tissue has remained

sterile, and become modified as *elaters* which alter shape hygroscopically, possibly aiding the dispersal of the spores.

Class: **Musci** (The Mosses)

There are many genera of mosses, occurring in damp places. They are recognizable by the upright gametophyte being differentiated into a thickened central axis bearing flattened green appendages. These are *not* stem and leaves in the true sense of the word, but will be referred to as such as a matter of convenience. The moss described here is *Funaria* (Figure 228), which is found on damp soil.

Examine the complete plant, noticing the arrangement of the 'leaves', the rhizoids, and the 'stem'.

Look at the 'leaves' under the microscope and notice their relatively simple structure, and the 'midrib'. Compare this structure with the structure of the leaves of higher plants and see the parallel development associated with the similar function.

Cut a transverse section of the 'stem', using pith, and see the distribution of the differentiated thickened tissue, which is for support, and *not* conduction.

Pick off a few rhizoids and examine them

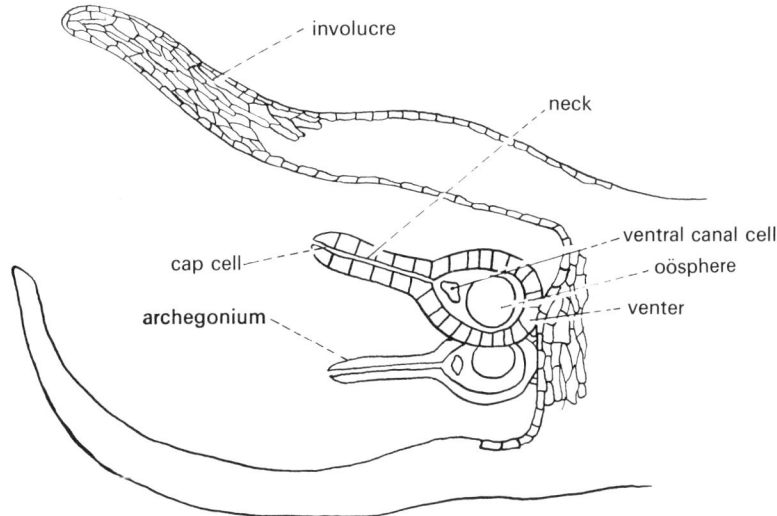

Fig. 227. L.S. Thallus tip of Pellia to show archegonia.

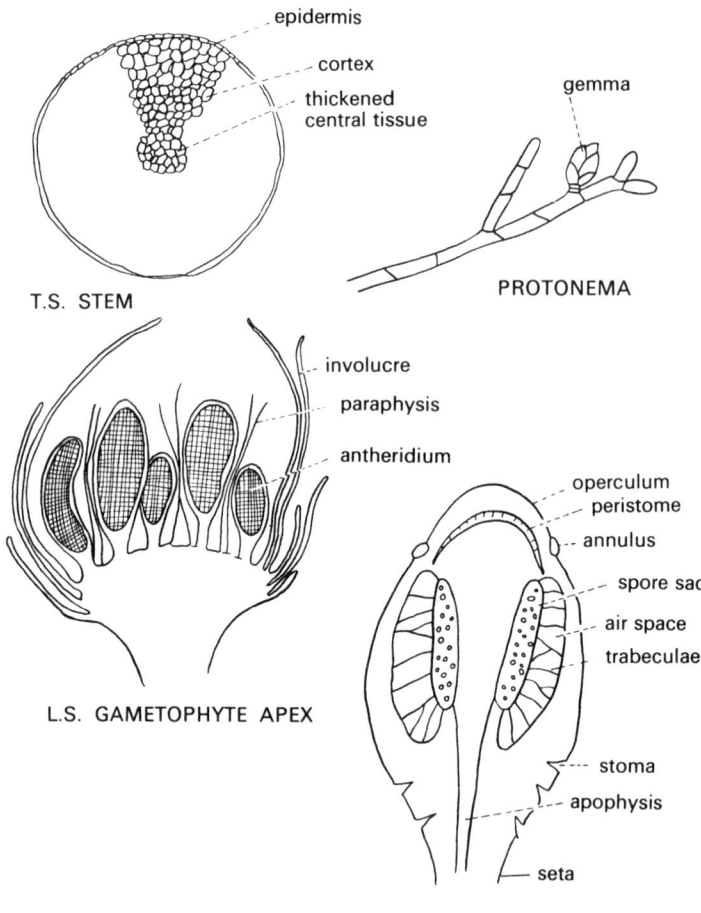

epidermis
cortex
thickened
central tissue
gemma

T.S. STEM

PROTONEMA

involucre
paraphysis
antheridium

operculum
peristome
annulus
spore sac
air space
trabeculae
stoma
apophysis
seta

L.S. GAMETOPHYTE APEX

L.S. SPOROGONIUM

Fig. 228. A moss.

under the microscope. Compare these structures with those of the liverwort rhizoids.

Look at prepared slides of longitudinal sections through the apices of *Funaria* showing *antheridia* and *archegonia*. What is the function of the *involucre* and sterile *paraphyses?*

Examine a prepared slide of a longitudinal section through a sporangium, and compare it with that of *Pellia*. Elaters are absent, but the sporangium is more complex than that of the liverwort, becoming so at the expense of the sporangial tissue.

Take some soil from around the moss plants, and place a little in a watch-glass of water. Tease it out and you should find some *protonema*. Look at them, they are easily

distinguished from the rhizoids and other debris by their having diagonal cross-walls.

Remove some spores from a dehiscing sporangium, (this will ensure that they are ripe), and place them on some damp filter-paper in a covered dish. The spores will germinate to produce two branches, one at each end. This is the protonema. It will be differentiated into a branch containing chloroplasts, and one or more rhizoids without chlorophyll. Buds will form on the chlorophyll-containing branch, and these will develop into new moss plants.

Water passes up the outside of moss plants. This can be demonstrated by cutting an upright 'branch' bearing leaves, and placing the cut end into an eosin (red ink) solution.

After a very short time, the red ink has travelled to the top of the plant, and will colour a small piece of paper which is just touched on the top cluster of 'leaves'. The water travels from the base of one 'leaf' to that of the next by capillarity.

PTERIDOPHYTA

Class: **Lycopodinae**

Selaginella

Most species of *Selaginella* are tropical so that you will have to use preserved material. Examine the complete plant which is the sporophyte. The leaves are simple with a small membranous *ligule* on the upper surface. Specialized branches of the stem *(rhizophores)* bear adventitious roots, while the sporangia are borne in loose cones *(strobili)* at the tips of the shoots.

Examine a prepared slide of a longitudinal section through a strobilus (see Figure 229). The sporangia are borne in the axils of its leaves. You will see that there are two kinds of sporangia—the *microsporangia* containing numerous small spores and the *megasporangia* with four large spores. The microspores develop into prothalli bearing antheridia, and tne megaspores develop into prothalli bearing archegonia.

Class: **Filicinae**

Dryopteris felix-mas (The Male Fern)

Examine and draw a complete plant. This is the sporophyte generation. The underground part is a *rhizome* bearing adventitious roots, and covered with brown scales *(ramenta)* and closely packed leaf-bases. There is no apical bud, but the young leaves are coiled in a *circinate vernation.*

The leaves (fronds) are compound, with a long *rachis* bearing the *pinnae* which are themselves divided into *pinnules*.
On the underside of the older pinnules there are the kidney-shaped, brown *indusia* each of which covers a *sorus* of sporangia. Lift an indusium with a needle, and you will see the sporangia.

Cut a transverse section (or use a prepared slide) of a pinnule bearing sori and examine it under the microscope. Notice that the vegetative structure of the leaf is essentially the same as that of the leaf of a flowering plant. Each sporangium has a band of specialized cells running from the top of the stalk, over it. Some of the cells are thickened apart from their outer sides, forming the *annulus*, leaving a strip of unthickened cells— the *stomium*. During drying the annulus straightens, rupturing the stomium and

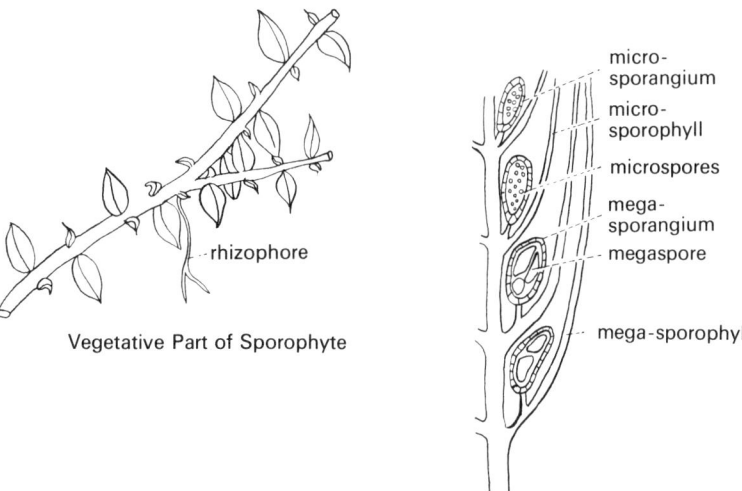

rhizophore

Vegetative Part of Sporophyte

micro-sporangium
micro-sporophyll
microspores
mega-sporangium
megaspore
mega-sporophyll

L.S. through Strobilus

Fig. 229. Selaginella.

releasing the spores. There is a *water gland* on the stalk of the sporangium.

Examine a prepared mount of a *prothallium* —gametophyte using a hand-lens and the low power of the microscope. See the rhizoids on the central cushion, the growing point, and the position of the antheridia and archegonia. Compare the structure of the prothallium and its organs with that of *Pellia*.

Look at transverse sections of the prothallium in the regions of the antheridia and archegonia, and examine these in detail.

If you mount a mature prothallium in water on a slide, you may see the antherozoids swimming in the water.

SPERMATOPHYTA

Class: **Gymnospermae**
Pinus sylvestris (The Scots Pine)
Examine and draw a piece of the stem. The

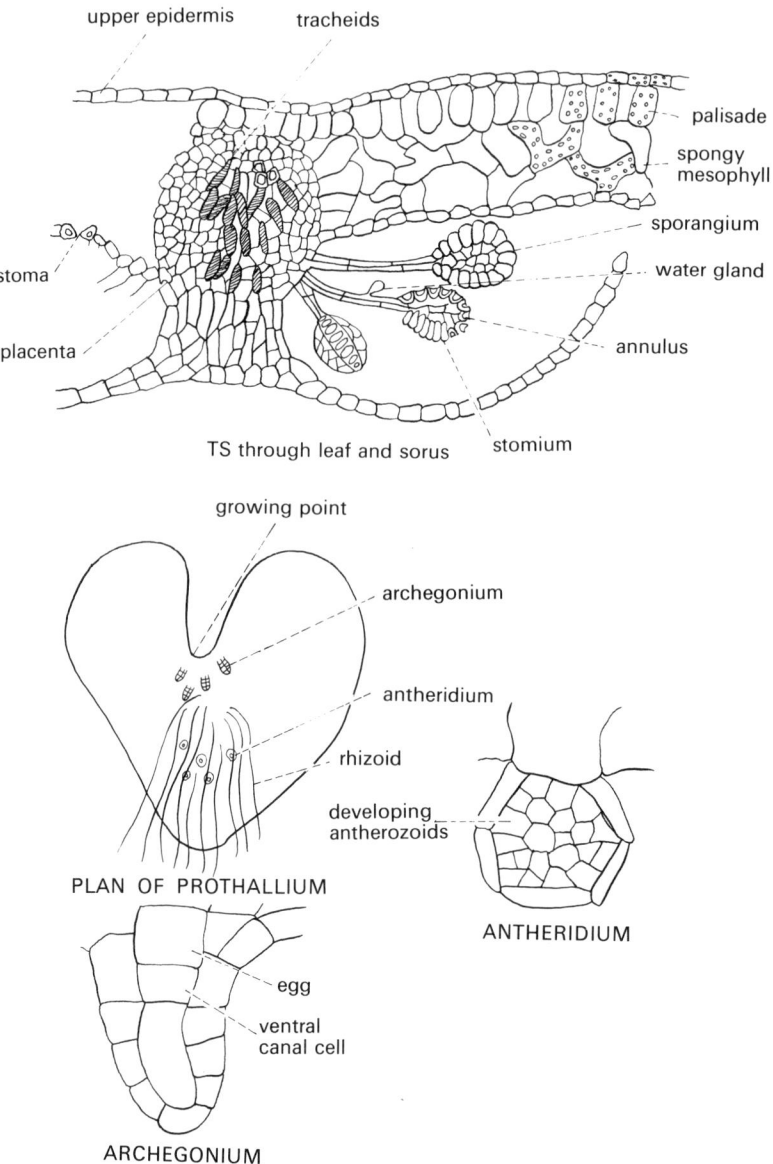

Fig. 230. Male Fern.

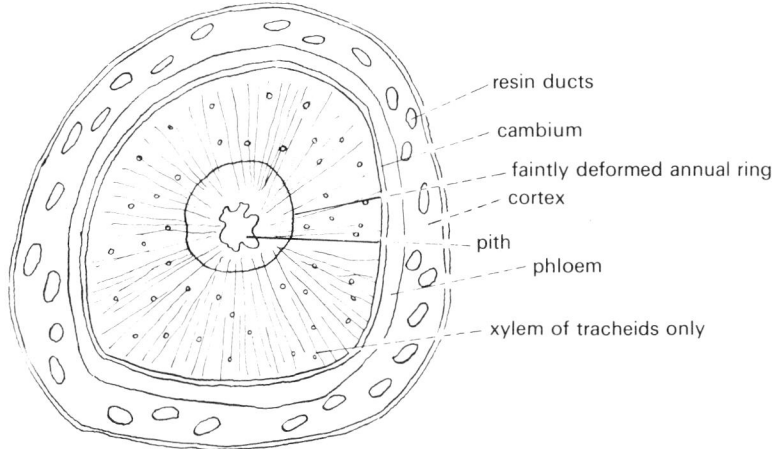

DIAGRAM OF T.S THROUGH A SHOOT

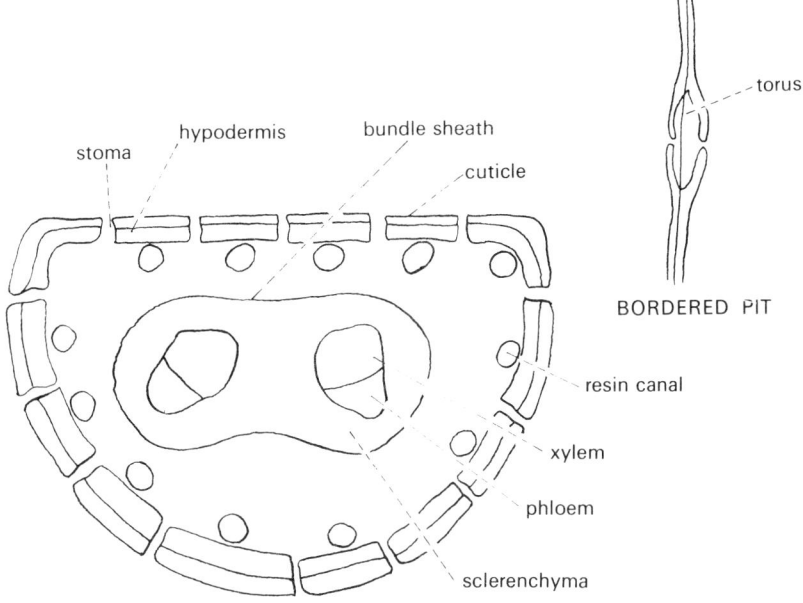

DIAGRAM OF T.S. THROUGH A LEAF

Fig. 231. Pinus.

leaves are borne in clusters on the tips of short shoots of limited growth. Cut transverse and longitudinal sections of the stem, mount them on slides and stain them with aniline hydrochloride or phloroglucinol. The structure of the wood is very even. This is because there are no vessels in it, and because the annual rings are far less obvious than in a deciduous angiosperm. Notice the *resin canals* in the cortex. In the longitudinal section, look for *bordered pits* between the tracheids.

Cut and examine a transverse section through the leaf. The distribution of the

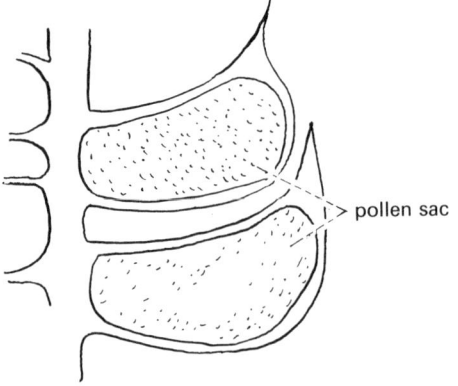

MALE CONE IN SECTION

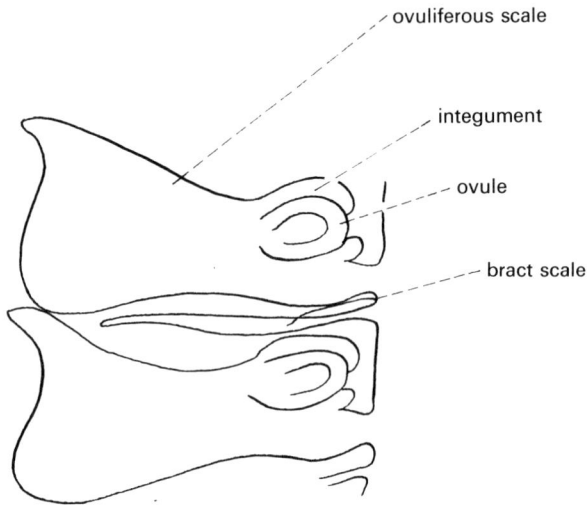

FEMALE CONE IN SECTION

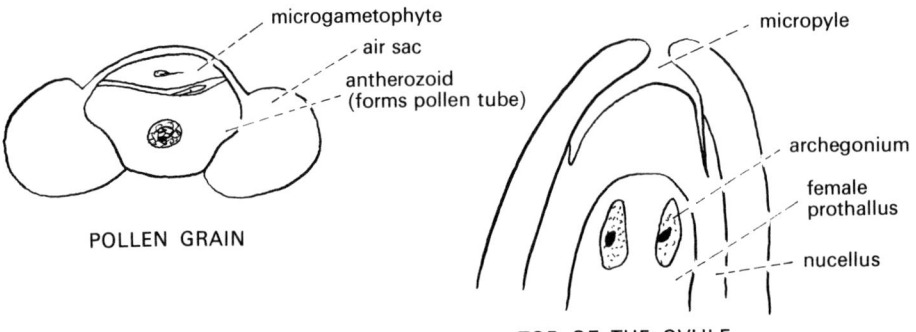

POLLEN GRAIN   TOP OF THE OVULE

Fig. 232. Pinus.

tissues is shown in Figure 231. You will have done this already when you were considering the modifications of the leaf.

Look at the staminate and pistillate cones (strobili) (see Figure 232). See the spiral arrangement of the *sporophylls*.

Dissect out and draw a *microsporophyll* to see the *pollen sacs*. Place a little pollen on a slide and examine it under the microscope.

Dissect out a *megasporophyll* from a pistillate cone and see the *ovules* on fine scales on its upper surface. Examine prepared slides of longitudinal sections through the ovule.

Draw some of the winged seeds, and dissect them to see if you can find the embryo.

12 Ecology

THE SOIL

The Soil profile

To examine the layers in a soil (profile) you need to dig a pit. The depth of the pit will vary considerably with the type of soil, but you should dig down until you reach a stony layer, which probably will be the C horizon. Dig the pit large enough to be able to work in it, observing the following points: (1) make sure that the face that you are going to examine is vertical, so that your measurement will be accurate, (2) always examine the shaded face of the pit, for direct sunlight can give a false idea of the colour.

Having cut the profile, look for the various zones and measure their depths. Look at the colours and notice the relative amounts of the different constituents, e.g. there is a large amount of humus in the upper zones, and a preponderance of stones in the lower ones. Look for evidence of leaching and water-logging. Usually the dark streaks are humic material, a blue-green colour indicates water-logging, and a yellow-brown colour shows good aeration. The blue-green colour is due to the presence of unoxidized iron compounds, and the yellow-brown indicates that there is sufficient oxygen to oxidise the iron to ferric compounds.

If there is time available carry out the appropriate experiment in the next section on samples from each zone. This will give a clear indication of the differences in their constituents.

Soil constituents

Experiment: To estimate the relative amounts of the particles of different sizes in the soil (see Figure 233).

Apparatus: Large measuring cylinder (500 or 1000 ml.), soil, sodium carbonate, large beaker.

Method: Fill the cylinder with soil to a depth of about 10 cm. *N.B.* This is not the volume of the soil, as it will contain air-spaces.) Dissolve about 10 gm. sodium carbonate in a litre of water, and pour the solution into the measuring cylinder, until it is about three-quarters full. Shake vigorously, and allow the sample to settle. The large particles will settle quickly, but the total sedimentation may take a week. Measure the heights of the different layers. This will give a rough idea of the amounts of the different constituent particles.

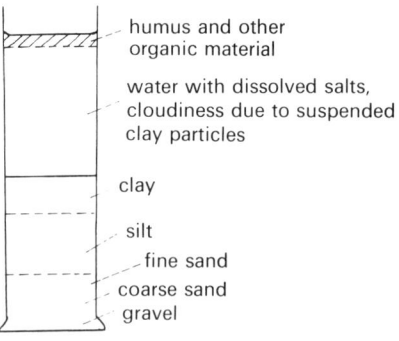

Fig. 233.

Take a little of each layer of the soil, and examine it microscopically, noting the sizes and shapes of the particles. Similarly examine some of the water which will still contain clay particles. They will show *Brownian movement.*

Experiment: To weigh the various particles of different size.

Apparatus: Graded soil sieves, balance, air-dried soil (prepared by spreading out the sample thinly and allowing it to dry for 3–4 days at room temperature).

Method: Arrange the sieves so that the one with the largest mesh is at the top, and that with the smallest mesh is at the bottom. Place the soil sample on the top sieve, and shake them. Weigh the contents of each sieve. You may have to break down some of the clay aggregates on the top sieve.

Experiment: To calculate the weight of water and humus in a soil sample.

Apparatus: Evaporating basin, balance and weights, oven, burner, tripod and gauze.

Method: Weigh the evaporating basin (a), empty, then about one-quarter full of soil (b), dry the soil in the oven, weighing repeatedly until the weight is constant (c). You may well have to break-down the larger lumps to make sure that they are completely dry. Now heat the soil with the burner, gently at first, finally playing the flame directly on the soil. This will burn or vapourize the humic material. Cool and weigh (d). Repeat the weighings until a constant weight is obtained. Calculate the percentage water, and the percentage humus in your sample. % water = density $(b-c) \times 100/(b-a)$; % humus = $(c-d) \times 100/(b-a)$.

Experiment: To calculate the volume of air in a soil sample.

Apparatus: Round tobacco (or similar) tin, large basin, beaker of diameter larger than than of the tin, measuring cylinder.

Method: Punch a large number of holes in the bottom of the tin. Calculate the volume of the tin. Push the tin into a bare patch of soil, so that it cuts out a cylinder of soil inside it. Remove the soil from around the tin, and lift it and its contained soil. You now have a known volume of soil. Gently and slowly lower the tin of soil into the basin of water, so that the water percolates slowly through the soil. Hold the tin with the soil surface level with the water surface until all the air has stopped bubbling out. Have the beaker as near to the basin as possible; lift the tin out of the water, and hold it over the beaker until the water has stopped draining out of it. Measure the volume of water drained out. This represents the volume of air which is held in the original volume of soil when it is saturated.

Experiment: To investigate the chemical constituents of soil.

Apparatus: Soil indicator, bench hydrochloric acid, reagents and apparatus for testing for elements as shown in the appendix, watch-glass.

Method: Place a little of the soil sample on a watch-glass, and add a few drops of hydrochloric acid. The amount of effervescence gives some idea of the amount of calcium carbonate in the soil. The gas evolved is carbon-dioxide, which can be tested with lime water.

Test the pH of your soil with the indicator according to the maker's instructions. The reading is only approximate, as the pH varies considerably with the condition of the soil, e.g. the plants growing on it, or the amount of water it contains.

Shake 100 g. of soil in 100 ml. of distilled water. Filter the resulting solution until clear. Evaporate this to a small bulk and test the solution for the various ions, as in the appendix.

Experiment: To find the larger animals that inhabit the soil.

Method: Select a fairly large area of ground, e.g. a playing field, and choose ten places on it at random. At these places remove a square foot (40 cm.^{2}) of turf or top soil, to a depth of about 5 cm. Break down this soil on a sheet of paper, and count the various types of organisms you find, e.g. millipedes, worms. Take further 5 cm. samples at successive depths and again count the animals. By comparing the different sites and the different depths, draw conclusions as to the type of soil and the degree of saturation which are most suitable for these organisms.

If you wish to make a more detailed and accurate collection of the animals use the following piece of apparatus. Make a cone of tin or stout cardboard of 20 cm. side, having a hole of 5 cm. diameter at the bottom. Place a piece of coarse mesh gauze over the hole. Fill the cone with the soil sample, and hang it over a jar. Suspend an electric light bulb near the surface of the soil and leave it on. The bulb will dry out the soil, and the organisms will migrate deeper into it. Many of them will fall through into the jar, while the others will become concentrated in the lower part of the cone.

You have already demonstrated the presence of micro-organisms in the soil in your study of bacteria, but the following experiment shows their presence rather neatly.

Experiment: To show the presence of micro-organisms in the soil.
Apparatus: Two 250 ml. conical flasks, rubber bungs to fit the flasks, cotton, two pieces of gauze about 10 cm. square, evaporating basin, burner, tripod and gauze, lime water.
Method: Take two samples of soil, of approximately the same weight. Heat one strongly in the evaporating basin and allow it to cool. Tie both samples separately in the two pieces of gauze. Pour a little lime water in each of the flasks, and suspend the gauze bags inside the flasks over the lime water. Cork securely. Note any change in the lime water, and account for it.

Physical properties

Experiment: To compare the water-retaining properties of soils and their permeabilities.
Apparatus: Clay soil, loamy soil, sandy soil, 3 filter funnels, measuring cylinder, 3 beakers, cotton wool.
Method: Place a piece of cotton wool at the base of each funnel and fill them separately with the same volume of the different soils, but do not press the soil too firmly into the funnel. Pour equal volumes of water on to each sample, and collect it as it permeates through. Record the time at which each funnel starts to drip. This will give an indication of the relative permeabilities of the soils. When the funnels have stopped dripping, measure the volumes of water which have passed through the soils. This

volume taken away from the volume poured on to the sample will show the amount retained by the soil.

Experiment: To demonstrate the capillary action of soils.
Apparatus: Clay soil, loamy soil, sandy soil, 3 glass tubes of a convenient bore, gauze, trough or large beaker.
Method: Tie a piece of gauze over the ends of each of the tubes. Fill each to the same level with the different soils, but do not leave any large air-spaces. Place the covered ends of the tubes in the trough of water and observe the water rising through the soil. Measure the final height of the water in each tube, and the time taken to reach this maximum. What conclusions do you draw?

Experiment: To compare the rates at which air penetrates through soils.
Apparatus: Three burettes, 3 beakers, 3 filter funnels, cotton wool, clay soil, loamy soil, and sandy soil, rubber tubing.
Method: Fill each of the burettes up to the

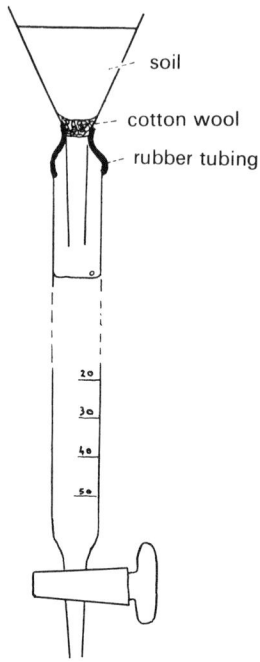

Fig. 234.

same reading with water. Place a piece of cotton wool in each funnel and fill them with the soil samples. Set up the apparatus as shown in Figure 234. Open the burettes and compare the rates at which the water falls in them. This is the rate at which the air penetrates the soil. If the burettes are closed at any time, the volume of water that has passed out is the same as the volume of air that has passed through the soil.

Experiment: To compare the rates of heating and cooling of soils.
Apparatus: Two similar calorimeters, water-bath, two thermometers equal weights (to approximately fill the calorimeters) of clay soil and sandy soil, burner, stand.
Method: Place the soil in the calorimeters, and put a thermometer in each to the same depth. Put the calorimeters in the water-bath and heat it. Record the temperature of each soil at two minutes intervals, until they each reach about 60°C. Remove the calorimeters from the water and allow them to cool, recording the temperatures as before. Graph your results, and draw conclusions about the heating and cooling of the soils.

Experiment: To show the **absorptive** properties of clay soil.

Apparatus: Clay soil, eosin (red ink), filter-funnel, filter-paper, test-tube.
Method: Pour some water into the test-tube and colour it with the eosin. Add some of the clay soil. Shake. Filter. Note the colour of the filtrate and account for it.

Adaptation to the environment

You will no doubt undertake some form of ecological survey as a long term project during your course in biology. Such surveys are beyond the scope of this book, but the following simple exercises may be instructive.

Select any common herbaceous plant, and collect specimens of it from different habitats, e.g. shady, sunny, wet, etc. Calculate the average leaf-area for each type of habitat, and compare the anatomy of the stems and leaves. What conclusions do you draw?

Study specimens or drawings of parasitic members of the plant and animal phyla with which you are familiar. How do they differ from the typical members?

Re-examine the Mollusca in the light of this being a phylum which contains aquatic and terrestrial members. What are the basic similarities, and what differences can be associated with adaptation to the land habitat?

Similarly examine the larvae of the Diptera.

13 Genetics and Heredity

Meiosis

Examine prepared slides of Lily *(Lilium)* anthers in various stages of development. Look for and identify the stages in meiosis, looking especially for any indication of crossing-over. You may find it difficult to distinguish the corresponding stages in the first and second meiotic divisions, but this difficulty is decreased if you remember that the chromatids that separate during the second division, are thinner than the chromosomes that separate during the first; and that the cells in which the second division is occurring will initially be in pairs, and in the late stage of the second telophase form tetrads.

Examine a series of prepared slides of the uterus of *Ascaris*. The advantage of using *Ascaris* is that the diploid number of chromosomes is only four. High in the oviduct are the *primary oöcytes* which are diploid. Meiosis does not begin until they are penetrated by the sperm cell, the nucleus of which remains passive within the developing egg until meiosis of the female nucleus is complete. After penetration by the sperm cell a thick wall is formed around the egg, and the first meiotic division takes place. At the anaphase the pairs of homologues separate, and one pair is extruded as the *first polar body*. The second meiotic division then takes place, and the *second polar body* is extruded, leaving the mature egg nucleus, with which the male nucleus then fuses to form the zygote, which then divides by mitosis. Look for these various stages in your slides.

You may like to examine other material, e.g. other anthers, worm *(Lumbricus)* or insect testes, to see the dividing chromosomes. This can be done in the following way: Place the material on a slide and squash it firmly but gently. Add a few drops of aceto-carmine, and remove any excess tissues with needles. Put a cover-slip on the slide, making sure that it is flat on the slide. Warm gently. Examine the slide under the microscope.

Variation

No two individuals are exactly alike, with the possible exception of identical twins. This difference between individuals is called *variation*. As such variation occurs in groups of individuals it is nearly meaningless to use an average to indicate any attribute unless the range of the population is also used. You will see this done by the use of a $\pm$ sign. E.g. the average height of boys in a class may be 1.65 ± 0.15 m. i.e. the average is 1.65 m. but there are individuals of 1.5 m. and 1.8 m. This variation is worked out by a statistical process which is beyond the scope of this book, but you may like to pursue it if you are mathematically-minded. If you do, you will find that the statement given in the example is not absolutely true, but it will serve to make the point, to the non-mathematician.

Unless there is some other factor affecting the population, which you have not taken into account, most of the population should be about the average, while there should be very few at the two extremes. So that if you graph the measurement, against the number of individuals having that measurement, the graph should look like the one in Figure 235. This is the graph of a *normal distribution* of a factor within a population.

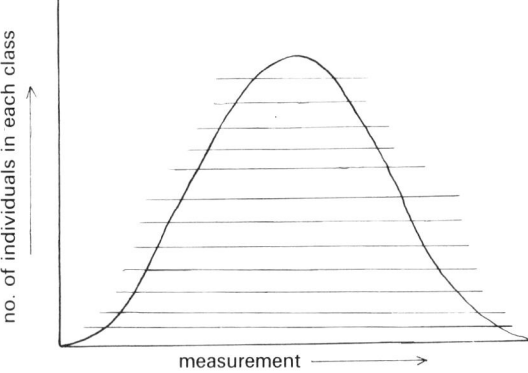

Fig. 235.

If you are dealing with small populations, which you will be, you may get only one or two individuals of any particular measurement, so to make the figure meaningful it is better to group your individuals, e.g. with an example of population height, place all individuals say between 1.5 m. and 1.56 m. together, with the next group 1.56–1.62 m. Graph your figure as shown in Figure 236. This is called a *histogram*.

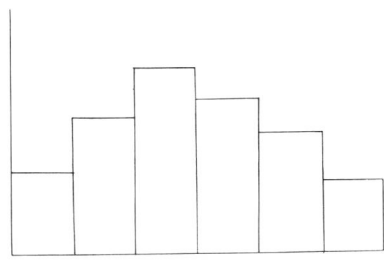

Fig. 236. A histogram.

This raises the problem of border-line cases. In the previous example, where would you place an individual of 1.56 m.? Does he go in the 1.5–1.56 m. group or in the 1.56–1.66 m. group? The answer is that your classes should include the upper limit, so that the 1.56 m. individual, should go in the 1.5–1.56 m. group, i.e. the group contains individuals over the lower limit, but not over the upper limit.

With these ideas in mind try the following experiments, which may well be undertaken as a class project if a number of measurements are involved.

Measure the length of broad bean pods, and graph the results. Open the pods and count the number of seeds in each pod. Graph these figures. Measure the length of each seed using a pair of dividers, and graph these figures.

Weigh about 100 small fruit, e.g. hazel nuts or rose hips, and construct a distribution curve.

Use a single 'seed' head of a large flower, e.g. *Dahlia* and measure the lengths of the seeds. Plot a distribution curve.

Count the number of florets in the head of a Composite 'flower'. Graph the shoe-sizes of the individuals in the class.

These are only a few suggestions, you can probably think of a lot more.

Heredity

A simple experiment to illustrate the random assortment of gametes is as follows:

Have two containers each containing equal numbers of black and white marbles, or pieces of paper, etc., so that there are about 100 to each container. One represents the female gametes, and the other the male gametes. Now select one marble from each box simultaneously, and score them as BB when both are black, Bb when one is black and the other white, and bb when both are white. Replace the marbles after each selection, and mix them thoroughly. Make at least 100 selections, and score the totals of the different selections. These should fall approximately into the ratio of 1 : 2 : 1.

Many plants and animals can be used for breeding experiments (e.g. Sweet Pea, or Rat) but the results take a long time to obtain, e.g. until the next season with flowering plants. Because of this the experiment described below uses the Fruit Fly

(Drosophila sp.) which completes its life-cycle in about 14 days at room temperature.

Different pure-bred types of the fly are obtainable from any biological supplier.

The flies are cultured easily in milk bottles using the following technique:

Sterilize the milk bottles. Make a thick paste of cooked maize meal and treacle, and add to it a little yeast. Cover the bottoms of the bottle to a depth of about an inch with the paste, and fit a gauze or stiff paper above it for the flies to rest on. Sterilize some cotton wool plugs for the bottles.

You will need a fine camel-hair brush, a dissecting microscope or good hand lens, and a white tile for examining and handling the insects.

The flies have to be anaesthetized with ether before handling, but be careful not to allow them to be exposed to the vapour for too long, or to come in contact with the liquid ether. The flies can be etherised in another milk bottle by inverting the culture bottle mouth-to-mouth over the empty milk bottle and tapping it so that all the flies fall into the lower bottle. Plug this with a cotton wool plug on to which has been poured a few drops of ether. As soon as the last fly has stopped moving, remove the plug and tip the flies on to the tile for examination.

The sexes can be distinguished by the following features:

1. The females are longer than the males.

2. The female abdomen is longer and more pointed than the male.

3. The male has a thick, dark band at the posterior of the abdomen; the female has not.

4. The male has sex-combs on the first tarsal joint of the fore-leg; the female has not.

5. External genitalia are visible on the male, but not on the female.

In carrying out the actual experiment, employ the following technique:

Start with two pure strains of the flies (obtained from the suppliers). Remove all the adults into another container and keep them as stock. Use any flies which emerge after this for the experiment, collecting them not later than 8 hours after emergence. This is to ensure that the females have not already mated. Place equal numbers of virgin females (about 10) and males of another strain into another culture bottle and leave them. After about a week, larvae and pupae will be present. Remove the original parents. As soon as the new flies begin to emerge (about 10 days after mating) record their phenotypes and tabulate the results, thus discovering the Mendelian principle involved.

Do not delay your examination of the culture for longer than 10 days as confusion may be caused by the presence of the F_2 generation.

14 Embryology

The flowering plant

Collect some young fruits of Shepherd's
Purse *(Capsella bursa-pastoris)*. Remove the
seeds. Place one on a microscope slide and
squeeze it gently with a pair of forceps. This
will squeeze out the embryo (Figure 237).
Place a drop of water on the embryo, and
cover it with a cover-slip. Examine it under
the low power of the microscope.

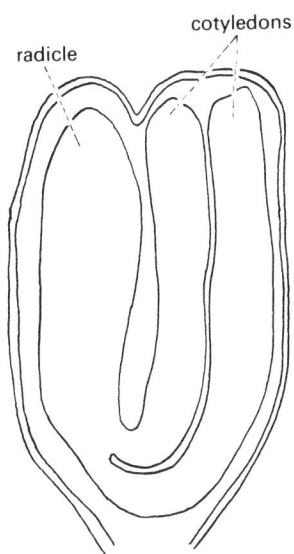

Fig. 237. Embryo of Shepherd's Purse.

The frog

Living frog spermatozoa can be obtained in
the following way: Using a freshly-killed male
frog, quickly dissect it to remove the testes.
Place them in water and macerate them.
Allow the suspension to stand for about 10
minutes, after which time the sperm will
become active. Place a drop of the suspension
on a microscope slide and look at it under
the microscope.

If you wish to fertilize some frog's eggs
artificially, this suspension can be used, but
the ability to penetrate the eggs is lost
rapidly after about half-an-hour.

Female frogs can be induced to ovulate
artificially by injecting them with pituitary
extract. This injection is made into the
abdominal cavity, and after 24–48 hours at
room temperature the female will produce
eggs. The eggs can be withdrawn from the
female by holding the animal in the right
hand with the legs between the thumb and
fore-finger. Bend the legs back up over the
ventral side of the body, with the left hand,
and press gently with the right hand. The
eggs will then be extruded.

If the eggs are needed for fertilization,
extrude them into a sperm suspension. Leave
the eggs in the suspension for about 10
minutes, then pour off the water, replacing
it with clean water. This should be removed
again after about 20 minutes to remove the
tissue debris thoroughly.

The eggs should be examined hourly for
the first 8 hours, then at approximately
8-hourly intervals (see Figure 238).

These stages can also be examined in
prepared slides.

The chick

Use some freshly fertilized eggs. Mark the
shell on one side and incubate them at 39°C.
This can be done in an electric oven if the
atmosphere inside it is kept humid by means
of a beaker of water. Turn the eggs through
180° twice each day. This prevents the
embryo from becoming stuck to the shell.
After 36 hours take one of the eggs and
examine the embryo as described below,
using a good hand-lens. Repeat this procedure
with another egg, after 72 hours. Compare
your embryos with prepared slides.

Using a fine scissors and a forceps, remove
a cap of shell from the upper side of the egg.
This will reveal the embryo on the top of the
yolk. Using a pipette, remove the albumen

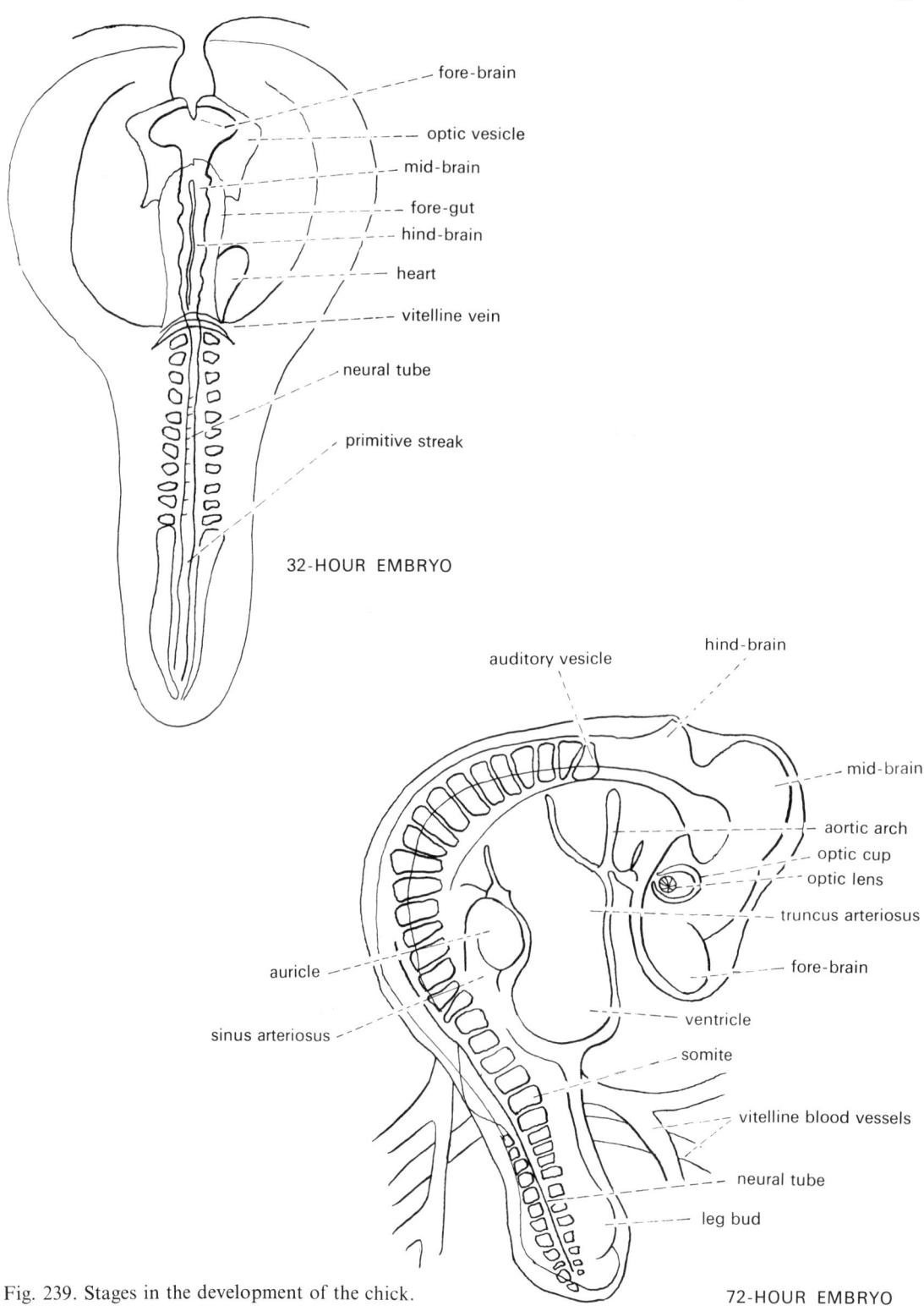

Fig. 239. Stages in the development of the chick.

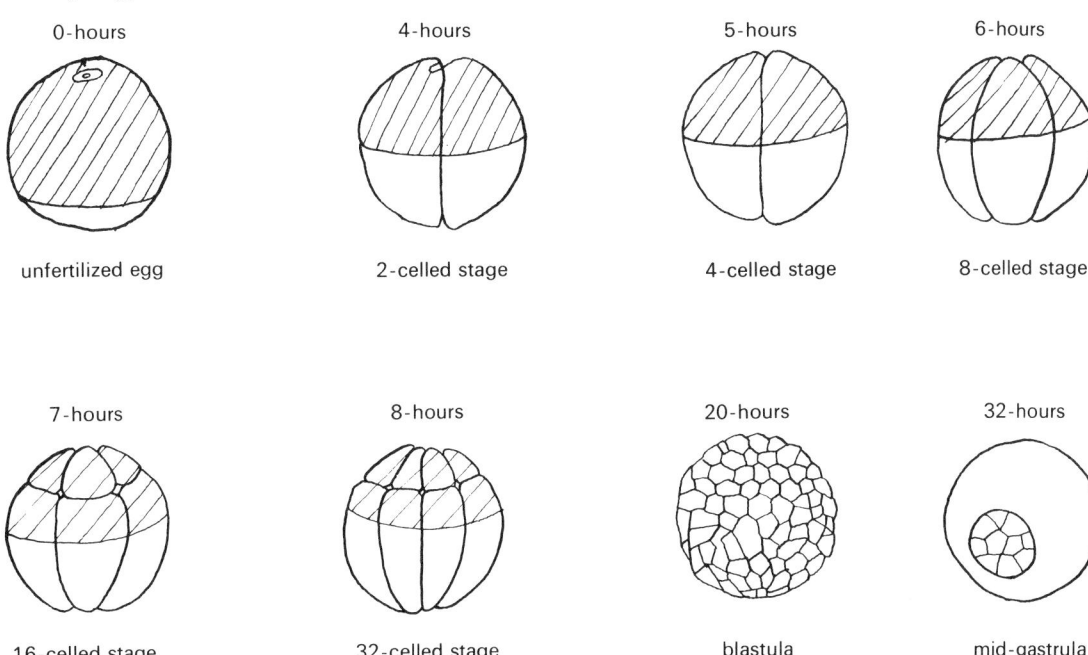

Fig. 238. Stages in the development of the Frog's egg.

from around the yolk, so that the yolk is not covered. Place a ring of filter-paper over the embryo so that it, and the marginal blood sinus are enclosed within the ring. This gives a firm edge for the manipulation of the embryo. Holding the filter-paper ring, and the membrane underneath it, with a forceps, cut the membrane around the outside of the filter-paper. Lift the ring and the enclosed embryo into a Petri dish of physiological saline, and examine the embryo (see Figure 239).

Incubate further eggs for 10–21 days. Carefully crack the shell, and remove it, taking care not to damage the egg membrane. The later stages in development can be seen clearly.

Appendix 1.
Reagents, Media, etc.

Iodine in Potassium Iodode solution

Dissolve 2 g. of potassium iodide in as little water as possible; to this add 1 g. of iodine and shake it until the iodine is dissolved. Make up to 100 ml. with distilled water.

This solution gives a yellow coloration with cellulose, and a blue-black colour with starch.

Test for Chlorine ions

Dissolve the solid in water, or dilute nitric acid. Add a little silver nitrate solution, which will form a white precipitate, which is soluble in ammonium hydroxide.

Test for Sulphate

Dissolve the solid in water or dilute nitric acid. Add a little barium chloride solution which will form a white precipitate which is insoluble in dilute hydrochloric acid.

Congo Red solution

Dissolve 0.5 g. Congo Red in 100 ml. distilled water.

Methyl Violet

Dissolve 1 g. methyl violet in 100 ml. water by boiling. Filter before use.

Benedict's solution

Dissolve 170 g. of sodium citrate in 600 ml. of water. Add 90 g. of sodium carbonate and dissolve, warming if necessary.

Dissolve 17 g. of copper sulphate in 140 ml. of water. Mix these two solutions and make up to 1 litre.

If this solution is mixed with half the volume of test solution, and warmed, an orange precipitate is formed in the presence of reducing sugars.

Fehling's solution

Solution A—Dissolve 73 g. of copper sulphate in 1 litre of water. Solution B— Dissolve 100 g. of sodium hydroxide, and 350 g. of sodium potassium tartrate in 1 litre of water.

Equal quantities of these solutions should be mixed immediately before use, and 1 part added to 3 parts of the test solution. Warm. An orange-red precipitate is formed with a reducing sugar.

Chlor-zinc iodide solution

Solution A—Dissolve 20 g. of zinc chloride in 9 ml. of water, warming if necessary. Solution B—Dissolve 1 g. potassium iodide in 20 ml. water, and add 0.5 g. iodine. Add solution B to solution A slowly until a precipitate of iodine is formed.

Lime water

Shake some quick lime (calcium oxide) in water and filter.

Sudan III

Make up a saturated solution in 70% ethanol.

Osmic acid solution (Osmium tetroxide)

This is best bought as a 2% solution.

Millon's reagent

Dissolve 20 g. mercury in 100 ml. concentrated nitric acid, preferably in a fume-cupboard. When cool add 100 ml. of water.

Buffer solutions:

These are best purchased as ready-made tablets.

Ninhydrin solution

Dissolve 0.2 g. ninhydrin in 100 ml. water.

Warm a few drops with the test solution. A blue colour indicates the presence of α-amino acids. This solution will keep only a few days.

Methylene blue solution

Dissolve 0.5 g. methylene blue in 100 ml. of warm water. Add 30 ml. of ethanol, and 0.01 g. of potassium hydroxide. Filter.

Aniline hydrochloride (or sulphate) solution

Make a saturated solution of the aniline hydrochloride (or sulphate) in water, and add concentrated hydrochloric (or sulphuric) acid, until the solution is acid.

Phloroglucinol solution

Dissolve 10 g. phloroglucinol in 100 ml. ethanol. Flood the preparation with this for 2–3 minutes. Drain off the liquid and add a few drops of concentrated hydrochloric acid. This stains lignin red.

Aceto-carmine solution

Mix 45 ml. glacial acetic acid with 55 ml. water. Saturate this with carmine. Heat under a condenser for 4 hours. Filter.

Ringer's Solution

Dissolve 0.8 g. sodium chloride, 0.02 g. calcium chloride, 0.02 g. potassium chloride, 0.02 g. sodium bicarbonate, all in 100 ml. of water.

Congo Red solution

Dissolve 0.5 g. Congo red in 100 ml. of water.

Haematoxylin (Delafield's)

This is best purchased ready made, as it has to be kept for at least two months after making, before it can be used.

Physiological saline

For mammalian tissue (except blood), dissolve 0.9 g. of sodium chloride in 100 ml. of water. For blood and invertebrate tissue, use 0.6 g. sodium chloride, and for amphibian tissue 0.75 g. of sodium chloride.

Lactophenol blue

Make a 0.1% solution of cotton blue in lactophenol, which can be purchased ready prepared.

Neutral Red

This is best purchased ready made.

Alkaline pyrogallol (Potassium pyrogallate)

Dissolve 5.0 g. of pyrogallol in 95 ml. of water. Dissolve 25 g. of potassium hydroxide in 15 ml. of water. Mix these two solution; stir well and use immediately.

Schultz's solution

Saturate a small amount of concentrated nitric acid with potassium chlorate.

Safranin solution

Dissolve 1 g. of safranin in 100 ml. of $70\frac{1}{2}$ ethanol.

Yeast culture solution (Pasteur's solution, modified)

Dissolve the following in water, finally making the solution up to 1 litre; 1.0 g. potassium phosphate (KH_2PO_4), 0.1 g. calcium phosphate, 0.1 g. magnesium sulphate, 5.0 g. ammonium tartrate, 75.0 g. glucose.

Bacteria culture solution

Dissolve the following in water, and make up to 1 litre; 10.0 g. bacteriological peptone, 10.0 g. meat extract, 5.0 g. sodium chloride. If a solid medium is required, 20.0 g. of agar, or 150 g. of gelatine can be added.

Fungus culture medium

Most fungi will grow satisfactorily on this medium. Dissolve 20 g. of malt extract in 500 ml. of water. Dissolve 20 g. agar in 500 ml. of water. Mix these two solutions, stir well and autoclave.

Crystal violet solution

Dissolve 1 g. crystal violet in 100 ml. of water.

Tests for elements in soil and plant ash

Flame test—Place a little of the material on a platinum wire and heat it in a Bunsen flame. A yellow flame indicates the presence of sodium. Look at the flame through a cobalt glass; a lilac flame indicates potassium and a brick red one calcium.

Dissolve the material in nitric acid, and to separate portions add the following solutions: ammonium molybdate—yellow colour or precipitate indicates phosphate; potassium ferrocyanide—blue colour or precipitate indicates iron; silver nitrate—a white precipitate indicates chloride.

Dissolve some of the material in concentrated hydrochloric acid. Dilute and add barium chloride solution—a white precipitate indicates sulphate.

Dissolve some of the material in water. Make a 0.5% solution of diphenylamine in concentrated sulphuric acid. Add a few drops of this to the salt solution. A blue colour indicates nitrate.

Appendix 2
Apparatus

Accumulators
Ammeter
Aquarium
Aspirator
Auxonometer

Balance
Beakers,   100 ml.
          250 ml.
          500 ml.
          1000 ml.
Bell-jars
Bottles,   500 ml.
          1000 ml.
Bowl
Buchner funnel
Buchner flask
Bunsen burner
Burettes

Cage
Calorimeter
Cellophane
Cellotape
Clamps
Clinostat
Clips, for rubber tubing
Coloured light filters
Corks
Cork borer
Cotton
Cotton wool
Cover slips
Crucible
Culture jars

Dissecting instruments
Dropper

Evaporating basin

Filter-funnel
Filter-paper
Filter-pump
Flasks, conical, 250 ml.
Flasks R.B. 1 l.

Gas-jars
Gauze
Glass beads
Glass rod
Glass wool
Glass plates

Key 9 (electrical)

Lens

Measuring cylinders
Microscope
Muslin

Oven

Paper clips
Parchment
Petri dishes
Pins
Pipette,   1 ml.
          10 ml.
          25 ml.

Polythene sheet
Potometer

Revolving drum
Rheostat
Rubber bungs
Ruler

Sand
Soil sieves
Slides
Spectrometer
Stands
Stop watch

Test tubes, $\frac{1}{2}$ in.
            1 in.
Thermometers
'Thermos' flasks
Thistle funnel
Tin foil
Tripod
Trough
Tubing, glass of bores up
    to $\frac{1}{2}$ in.
Tubing, rubber, to fit glass
    tubing

Watch glasses
Water bath
Water pump
Weights, to 500 g.
         1–5 Kg.

Wire, copper

Y-piece

Appendix 3
Chemicals

Acetic acid
Acetone
Acetylchlorine
Adrenaline chloride
Agar
Ammonium hydroxide
Ammonium molybdate
Ammonium sulphate
D-amphetamine sulphate
Aniline hydrochloride
Aniline sulphate

Barium chloride
Butanol

Calcium chloride
Carbon disulphide
Carmine
Casein
Chalk
Charcoal
Chloropromazine
Clove oil
Cobalt chloride
Congo red
Copper sulphate
Crystal violet

Diphenylamine

Ethanol

Formaldehyde
Fructose
Fusion mixture

Gelatine
Gibberellic acid

Glucose
Glue
Glycerine
Guaiacum gum

Haematoxylin
Histamine acid phosphate
Hydrochloric acid
Hydrogen peroxide

Indian ink
Indole acetic acid
Iodine
Lactic acid
Lactophenol blue
Lanoline
Lead acetate
Lime water
Litmus paper

Magnesium oxide
Maltose
Methylene blue
Methyl violet
Methylated spirit

Neutral red
Ninhydrin
Nickel sulphate
Nitric acid

Olive oil
Osmic acid

Pancreatin
Pepsin

Petroleum ether
Phenol
Phenolphthalene
Phenol Red
Phloroglucinol
Potassium dihydrogen
    phosphate
Potassium ferryocyanide
Potassium hydroxide
Potassium oxalate
Potassium permanganate
Pyrogallol

Quinine sulphate

Red ink
Rooting powder

Safranin
Sand
Silver nitrate
Soda-lime
Sodium bicarbonate
Sodium chloride
Sodium citrate
Sodium hydroxide
Sodium nitrate
Sodium sulphate
Sucrose
Sudan III
Sulphuric acid

Urea

'Vaseline'
Vinegar

Appendix 4
Biological Material

Acacia
African violet
Aphid
Apple
Arrow head
Ascaris
Ash
Aspergillus

Bacterium culture
Barley
Bee (all stages)
Beech nut
Beetroot
Begonia
Berberis
Bile salts
Blackberry
Bladderwort
Blowfly (all stages)
Bones
Broad bean
Broomrape
Brussels sprout
Buttercup (and *R. scleratus* seed)
Butterwort

Carrot (and fruit)
Castor oil seed
Celery
Centipede
Chlamydomonas
Cinquefoil
Cockroach
Coleus
Columbine
Couch grass

Crayfish
Cress seed
Cystopus

Daffodil
Dahlia
Dandelion
Daphnia
Dead nettle
Delphinium
Dogs' mercury
Dock
Dodder
Dogfish
Dogfish skeleton
Drosophila

Earthworm
Eggs
Elder
Elodea
Enchanter's Nightshade
Erysiphe
Euglena
Eyes

Flea
Fresh-water mussel
Frog
Frog skeleton
Frog spawn
Fuchsia
Fucus
Funaria

Geranium fruit
Gladiolus

Gooseberry
Gorse
Groundsel

Hakea
Hawthorn
Hazel nut
Heather
Heterodera
Hop
Honesty
Horse chestnut
Hyacinth
Hydra

Iris
Ivy

Jellyfish
Jerusalem Artichoke

Lard
Lesser Celandine
Lily
Lily of the valley
Liverfluke
Locust
Lousewort
Maize
Male fern
Mare's tail
Marram grass
Marrow
Mice
Microsphaera
Minnow
Mint

Mistletoe
Monk's hood fruit
Mosquito (all stages)
Mucor
Mushroom

Nereis
Nettle

Oak
Obelia
Oedogonium
Onion and seed
Orange

Pansy
Paramecium
Parsnip
Passion flower
Peas
Pear
Pelargonium
Pellia
Penicillium
Phytophthora
Pieris (all stages)
Pigeon
Pine and seed

Pleurococcus
Planaria
Potamogeton
Potato
Privet (and variegated)

Radish (and fruit)
Raspberry
Rat (and skeleton)
Rhizopus
Rhubarb
Rice
Rose
Rue
Runner Bean

Selaginella
Skulls (dog, sheep, rabbit)
Sloe
Snail
Sphaerotheca
Spirogyra
Spurge
Starfish
Strawberry (and fruit)
Sundew
Sunflower
Sycamore fruit

Tapeworm
Thyroid extract
Tobacco seed
Tomato
Toothwort
Tradescantia
Turbatrix
Tulip

Ulva

Vine
Virginia creeper
Volvox
Wallflower fruit
Water crow's foot
Wheat
White dead nettle fruit
Willow
Willow-herb seed
Woodlice

Yeast
Yellow archangel
Yellow rattle